Gels for Oil and Gas Industry Applications (2nd Edition)

Gels for Oil and Gas Industry Applications (2nd Edition)

Guest Editors

Qing You
Guang Zhao
Xindi Sun

MDPI

Basel • Beijing • Wuhan • Barcelona • Belgrade • Novi Sad • Cluj • Manchester

Guest Editors
Qing You
School of Energy Resources
China University of Geosciences (Beijing)
Beijing
China

Guang Zhao
School of Petroleum Engineering
China University of Petroleum (East China)
Beijing
China

Xindi Sun
Physics and Engineering Department
Slippery Rock University of Pennsylvania
Slippery Rock
United States

Editorial Office
MDPI AG
Grosspeteranlage 5
4052 Basel, Switzerland

This is a reprint of the Special Issue, published open access by the journal *Gels* (ISSN 2310-2861), freely accessible at: www.mdpi.com/journal/gels/special_issues/oil_gas_2.

For citation purposes, cite each article independently as indicated on the article page online and using the guide below:

Lastname, A.A.; Lastname, B.B. Article Title. *Journal Name* **Year**, *Volume Number*, Page Range.

ISBN 978-3-7258-2884-5 (Hbk)
ISBN 978-3-7258-2883-8 (PDF)
https://doi.org/10.3390/books978-3-7258-2883-8

Contents

Preface

This Special Issue, titled "Gels for Oil and Gas Industry Applications (2nd Edition)", features a collection of high-quality papers highlighting advancements in gel applications within the oil and gas industry. The included studies showcase recent developments in gels designed for use as hydraulic fracturing agents, drilling fluid additives, and conformance control solutions. The guest editor of this Special Issue wishes to thank all the contributing authors as well as editors of the journal Gels, who have addressed the myriad of publication issues involved in the production of this Special Issue.

Qing You, Guang Zhao, and Xindi Sun
Guest Editors

gels

MDPI

Article

A Supramolecular Reinforced Gel Fracturing Fluid with Low Permeability Damage Applied in Deep Reservoir Hydraulic Fracturing

Yongping Huang [1,2], Xinlong Yao [1,2], Caili Dai [1,2,*], Yining Wu [1,2], Lin Li [2] and Bin Yuan [2,*]

[1] Shandong Key Laboratory of Oilfield Chemistry, China University of Petroleum (East China), Qingdao 266580, China; hypupc@163.com (Y.H.); yaoxl_upc@163.com (X.Y.); wuyining@126.com (Y.W.)

[2] School of Petroleum Engineering, China University of Petroleum (East China), Qingdao 266580, China; lilin@upc.edu.cn

* Correspondence: daicl@upc.edu.cn (C.D.); yuanbin@upc.edu.cn (B.Y.)

Abstract: Gel fracturing fluid is the optimum fracturing fluid for proppant suspension, which is commonly applied in deep reservoir hydraulic fracturing. The content of polymers and crosslinkers in gel fracturing fluid is usually high to meet the needs of high-temperature resistance, leading to high costs and reservoir permeability damage caused by incomplete gel-breaking. In this paper, a supramolecular reinforced gel (SRG) fracturing fluid was constructed by strengthening the supramolecular force between polymers. Compared with single network gel (SNG) fracturing fluid, SRG fracturing fluid could possess high elasticity modulus ($G' = 12.20$ Pa) at lower polymer (0.4 wt%) and crosslinker (0.1 wt%) concentrations. The final viscosity of SRG fracturing fluid was 72.35 mPa·s, meeting the temperature resistance requirement of gel fracturing fluid at 200 °C. The gel-breaking time could be extended to 90–120 min using an encapsulated gel breaker. Gel particles are formed after the gel fracturing fluid is broken. The median particle size of gel particles in the SRG-breaking solution was 126 nm, which was much smaller than that in the industrial gel (IDG) breaking fluid (587 nm). The damage of the SRG-breaking solution to the core permeability was much less than the IDG-breaking solution. The permeability damage of cores caused by the SRG-breaking solutions was only about half that of IDG-breaking solutions at 1 mD.

Keywords: supramolecular reinforced gel; gel fracturing fluid; deep reservoir; gel-breaking solution; low permeability damage

Citation: Huang, Y.; Yao, X.; Dai, C.; Wu, Y.; Li, L.; Yuan, B. A Supramolecular Reinforced Gel Fracturing Fluid with Low Permeability Damage Applied in Deep Reservoir Hydraulic Fracturing. *Gels* **2024**, *10*, 2. https://doi.org/10.3390/gels10010002

Academic Editor: Mario Grassi

Received: 2 December 2023
Revised: 14 December 2023
Accepted: 16 December 2023
Published: 20 December 2023

1. Introduction

The development of conventional oil and gas reservoirs has gradually entered the middle or late period [1–4]. How to efficiently develop deep reservoirs with abundant reserves attracts the interest of researchers [5–7]. The buried depth of the deep reservoir determines that most of its permeability is low, and the process of oil or gas gathering to the bottom of the well will be subject to great flow resistance [8,9]. As a result, the production of the well is low, which makes it difficult to reach the level of economic development. Although some EOR methods, such as asphaltene control and nano flooding, can effectively improve the recovery of low permeability reservoirs, hydraulic fracturing is an important means to realize the efficient development of deep reservoirs [10–12]. An instantaneous pressure higher than the formation fracture pressure is generated at the bottom of the well by pumping fracturing fluid into the formation with a high flow rate during hydraulic fracturing, forming many fracturing fractures [13,14]. The fracturing fluids then carry proppant throughout the fracture, maintaining the high permeability of fractures after the pressure is released. The flow resistance of oil or gas can be greatly reduced, and the effective flow area of the oil or gas can be expanded after hydraulic fracturing [15].

Fracturing fluids play an important role in hydraulic fracturing. Their main role is to transfer pressure and carry proppant [16]. Compared with slick water fracturing fluid [17], viscous water fracturing fluid [18], and viscoelastic surfactant fracturing fluid [19,20], gel fracturing fluid is considered to be the strongest water-based fracturing fluid for proppant suspension under high temperatures [21]. Guar gum fracturing fluid is the most widely used among gel fracturing fluids, which is mainly crosslinked using boron crosslinker [22]. High temperatures, some of which can be higher than 180 °C, are common in deep reservoirs. Neither the glucoside bond of the guar gum nor the boron crosslinker can be stabilized at such high temperatures [23,24]. Currently, polyacrylamide-based gels are mainly used to replace guar gum to solve the problem of high-temperature resistance (>180 °C) of gel fracturing fluid. Polyacrylamide-based gels are usually formed by crosslinking a polyacrylamide-based polymer with a transition metal crosslinker (Zr-crosslinker or Ti-crosslinker) [25,26]. Generally, the monomer containing the sulfonic acid group is added to the polyacrylamide to improve the temperature and salt resistance of the polymer, and the cyclic monomer is added to improve the thermal stability of the polymer at high temperatures [27].

Although polyacrylamide-based gels have been able to meet the high-temperature resistance requirements of most fracturing fluids applied in deep reservoirs, the content of polymers and crosslinkers in the gels is generally high in this condition. The high content of polymers and crosslinkers will lead to high costs and incomplete gel-breaking [28]. The permeability damage to the reservoir caused by incomplete gel-breaking will be a problem, which will significantly reduce the effectiveness of deep reservoir fracturing [29,30]. Therefore, it is very important to develop a fracturing fluid with high-temperature resistance and low permeability damage for deep reservoir hydraulic fracturing [31].

Incomplete gel-breaking is due to the high density of crosslinking bonds and tight network structure of gels containing high polymer and crosslinker. Therefore, the key to solving the above problems is to reduce the density of crosslinking bonds without reducing the gel strength. The bond energy of supramolecular forces such as hydrogen bonding, electrostatic interaction, and hydrophobic interaction is much higher than the van der Waals force, which has been widely used in the research and development of hydrogels in recent years [32–34]. Therefore, it is feasible to reduce the content of crosslinker by strengthening the supramolecular interaction between polymers to replace some crosslinking bonds [35]. In this paper, a supramolecular reinforced gel fracturing fluid is constructed by introducing functional groups with supramolecular forces into polymers, which possess high-temperature resistance and low reservoir permeability damage after gel-breaking. The research in this paper will provide new insights and theoretical guidance for the development and application of gel fracturing fluids in deep reservoir hydraulic fracturing.

2. Results and Discussion

2.1. Rheological Properties of Gel Fracturing Fluid

The main function of gel fracturing fluid is to suspend and transport proppant during the hydraulic fracturing process. However, the high-strength gel fracturing fluid is needed for gel fracturing fluid. The modulus is an important property in judging the strength of the gel fracturing fluid [36,37]. Therefore, the strength of gel fracturing fluid was determined by comparing the modulus of the single network gel (SNG) fracturing fluid and supramolecular reinforced gel (SRG) fracturing fluid. The viscoelastic modulus of SNG (0.4 wt% polymers) fracturing fluid with different concentrations of crosslinkers is shown in Figure 1. The elastic modulus (G') and viscous modulus (G'') of the SNG fracturing fluid increased with the acceleration of oscillation frequency (Figure 1a,b) [38]. The G' increased significantly due to the crosslinking network structure of the SNG fracturing fluid becoming tighter with the increase in Zr-crosslinker concentration, but the G'' was unchanged. The G'' was the highest, achieving 5.15 Pa when the concentration of Zr-crosslinker was 1.0 wt% (Figure 1c). Compared to 0.4 wt%, the G' and G'' were greatly improved when the polymer concentration was increased to 0.6 wt%. Similarly, the G' and the G'' also increased with the acceleration of the oscillation frequency (Figure 2a,b).

Although the G′ of the SNG fracturing fluid increased with the increase in the crosslinker concentration, the G″ decreased. When the concentration of the Zr-crosslinker was 1.0 wt%, the G′ reached 6.91 Pa (Figure 2c).

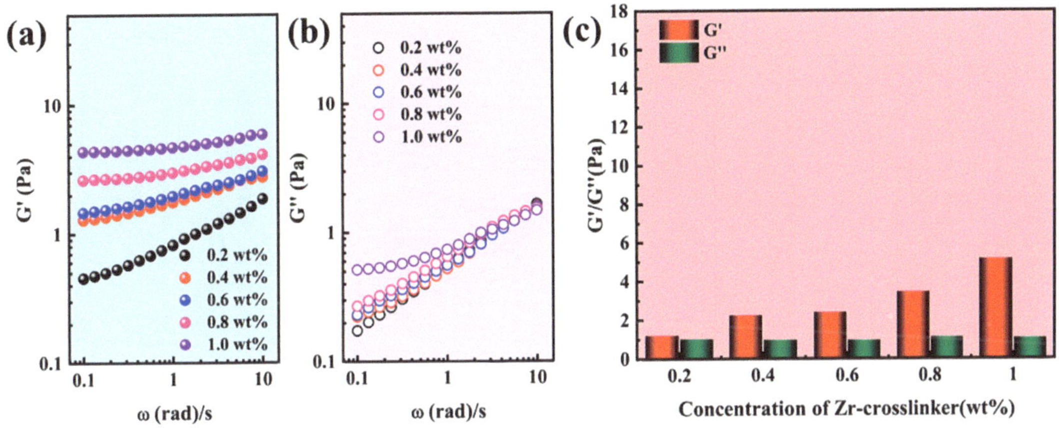

Figure 1. Viscoelasticity of the SNG fracturing fluid (0.4 wt% polymer) at different concentrations of Zr-crosslinker. (**a**) The G′ with frequency. (**b**) The G″ with frequency. (**c**) The G′ and G″ of SNG fracturing fluid at 0.5 Hz.

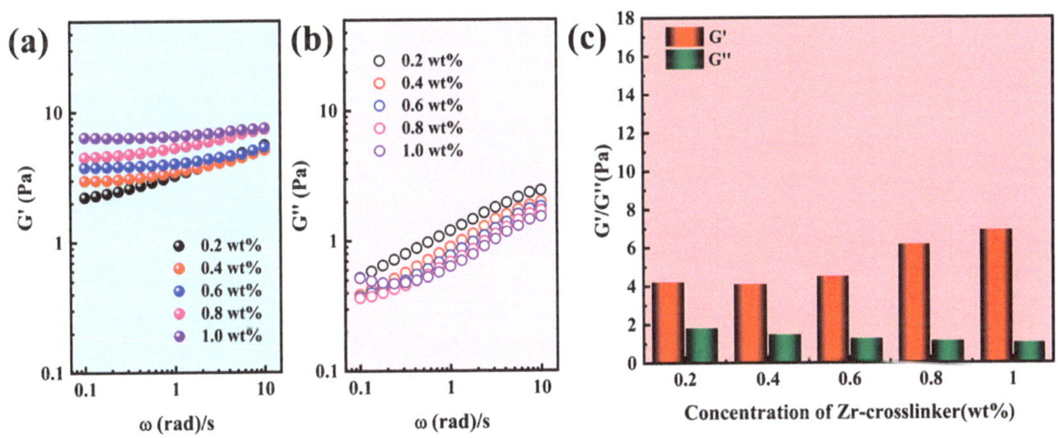

Figure 2. Viscoelasticity of the SNG fracturing fluid (0.6 wt% polymer) at different concentrations of Zr-crosslinker. (**a**) The G′ with frequency. (**b**) The G″ with frequency. (**c**) The G′ and G″ of SNG fracturing fluid at 0.5 Hz.

The G′ and G″ of SRG fracturing fluid are shown in Figure 3. Both the G′ (Figure 3a) and G″ (Figure 3b) increased with the increase in the oscillation frequency, except that the G″ increased at low frequencies when the concentration of Zr-crosslinker was 0.2 wt% and 0.3 wt% [24,25]. Unlike SNG fracturing fluid, SRG fracturing fluid could form high-strength gels at lower Zr-crosslinker concentrations. When the concentration of the crosslinker was 0.04 wt%, the G′ was 6.20 Pa. The G′ reached the maximum (16.80 Pa) at 0.2 wt% Zr-crosslinker and then decreased to 12.57 Pa when the Zr-crosslinker increased to 0.3 wt% [23]. The high G′ of SRG fracturing fluid at a lower Zr-crosslinker concentration was due to the network structure formed by the supramolecular interaction between the polymers;

however, whether the shear viscosity could be maintained at a high temperature needed further demonstration.

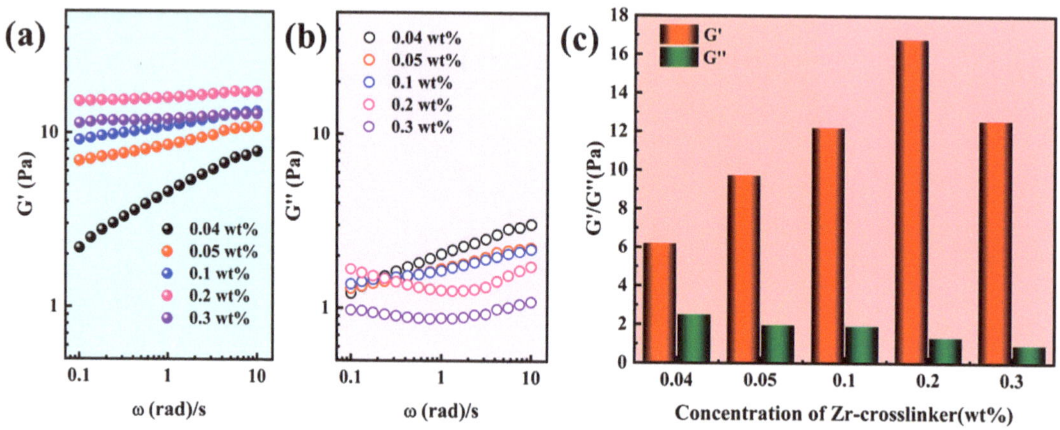

Figure 3. Viscoelasticity of the SRG fracturing fluid (0.4 wt% polymer) at different concentrations of Zr-crosslinker. (**a**) The G′ with frequency. (**b**) The G″ with frequency. (**c**) The G′ and G″ of SRG fracturing fluid at 0.5 Hz.

2.2. *Shear Resistance of Gel Fracturing Fluid at High Temperatures*

As could be seen from Figure 3c, when the Zr-crosslinker was 0.1 wt%, the G′ of SRG fracturing fluid reached 12.20 Pa (>10 Pa). Therefore, the concentration of Zr-crosslinker was fixed at 0.1 wt% to conduct the shear experiment at high temperatures, and then it was judged whether the SRG fracturing fluid could meet the high-temperature requirements of deep hydraulic fracturing. The experimental results of shear viscosity at high temperatures are shown in Figure 4. As can be seen from Figure 4, the viscosity stabilized at about 600 mPa·s after the initial shear. During this period, the viscosity did not change with the increase in temperature. When the temperature exceeded 100 °C, the viscosity began to decrease, and the viscosity dropped to 197.8 mPa·s as the temperature reached 200 °C. The viscosity eventually stabilized at 72.35 mPa·s as the shear time continued to 2 h. The criterion of gel fracturing fluid is that the viscosity is greater than 50 mPa·s [25,27]. The final viscosity was sufficient to meet the proppant suspension performance of the gel fracturing fluid, so SRG fracturing fluid fully meets the need for temperature resistance of the gel fracturing fluid applied in deep reservoir hydraulic fracturing.

2.3. *Gel-Breaking Properties of Gel Fracturing Fluid*

Compared with oxidative-breaking agents such as ammonium persulfate (APS), the encapsulated gel-breaker was beneficial in delaying the rapid degradation of the gel fracturing fluid at high temperatures [39]. The gel-breaker was selected, and the properties of the gel-breaking solution were explored in this section. The conductivity of different gel-breaker solutions is shown in Figure 5. The encapsulated gel-breaker (Capsule-A and Capsule-B) was mainly composed of a gel-breaker wrapped in the capsule. The release of the gel-breaker from the capsule could be judged by the conductivity [40]. It could be seen from Figure 5a that the conductivity of the three gel-breakers (APS, Capsule-A, and Capsule-B) did not change with the increase in soaking time at 25 °C. The conductivities of APS, Capsule-A, and Capsule-B were 1131 μS/cm, 67.5 μS/cm, and 20.85 μS/cm at 600 min, respectively (Figure 5b), indicating that Capsule-A and Capsule-B did not release the gel-breaker. The conductivity of Capsule-A solution at different temperatures is shown in Figure 5c. There was a small increase in conductivity at the initial moment when the temperature was 60 °C or 90 °C. The conductivity increased with the extension of soaking time. A rapid increase in conductivity occurred at 420 min when the temperature was 60 °C, meaning that Capsule-A began to release a large amount of gel-breaker at this moment. The release time was reduced to 240 min when the temperature increased to 90 °C, indicating that the release rate of Capsule-A accelerated with the increase in temperature. The conductivity trend of Capsule-B was similar to that of Capsule-A, as shown in Figure 5d. The difference was that the conductivity of Capsule-B had not a rapid increase in conductivity at 60 °C, and the increased amplitude was much smaller than that of Capsule-A. However, the rapid increase in conductivity also occurred at 240 min when the temperature was increased to 90 °C. From the above experiments, it could be concluded that the conductivity of the encapsulated gel-breaker solution would increase with the increase in time and temperature, which meant that the release rate of the gel-breaker in the capsule was affected by temperature and soaking time. The higher the temperature, the faster the dissolution rate of the surface of the capsule. Therefore, the magnitude of the conductivity is controlled by temperature [41]. Capsule-B has a longer delay time than Capsule-A under the same amount of gel-breaker release. Therefore, Capsule-B was selected as the gel-breaker to conduct the following gel-breaking experiment.

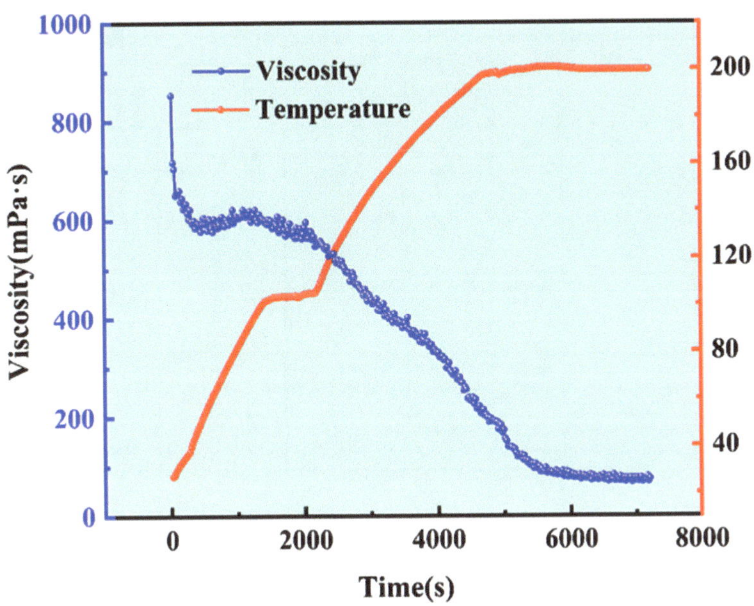

Figure 4. Temperature and shear resistance of SRG fracturing fluid (0.4 wt% polymer + 0.1 wt% Zr-crosslinker). Shear rate: 170 s^{-1}.

It could be seen from the conductivity experiment that Capsule-A and Capsule-B both had the property of delaying the release of the gel-breaker, but the mode of the delaying release needed further research. The micro-morphology of Capsule-A is shown in Figure 6. The Capsule-A with untreated and soaked in water at 25 °C are shown in Figure 6a,b. No cracks were observed on the surface of the capsules, but the difference was that the surface of the untreated Capsule-A was relatively rough. The reason was that the surface of Capsule-A might be attached to a small amount of gel-breaker, which could also explain why Capsule-A would have a very small conductivity after immersion in water. Some tiny cracks appeared on the surface of Capsule-A when the temperature was 60 °C, as shown in Figure 6c, indicating that Capsule-A began to release the gel-breaker at 60 °C. When the temperature increased to 90 °C, there were more cracks on the surface of Capsule-A, as could be seen in Figure 6d, and the width of the cracks was wider, indicating that the release rate of the gel-breaker should be faster at 90 °C. The above conclusions were consistent with those obtained in Figure 5c. The experiment showed that the surface of Capsule-A would generate cracks after being soaked in high-temperature water and then releasing the gel-breaker [41].

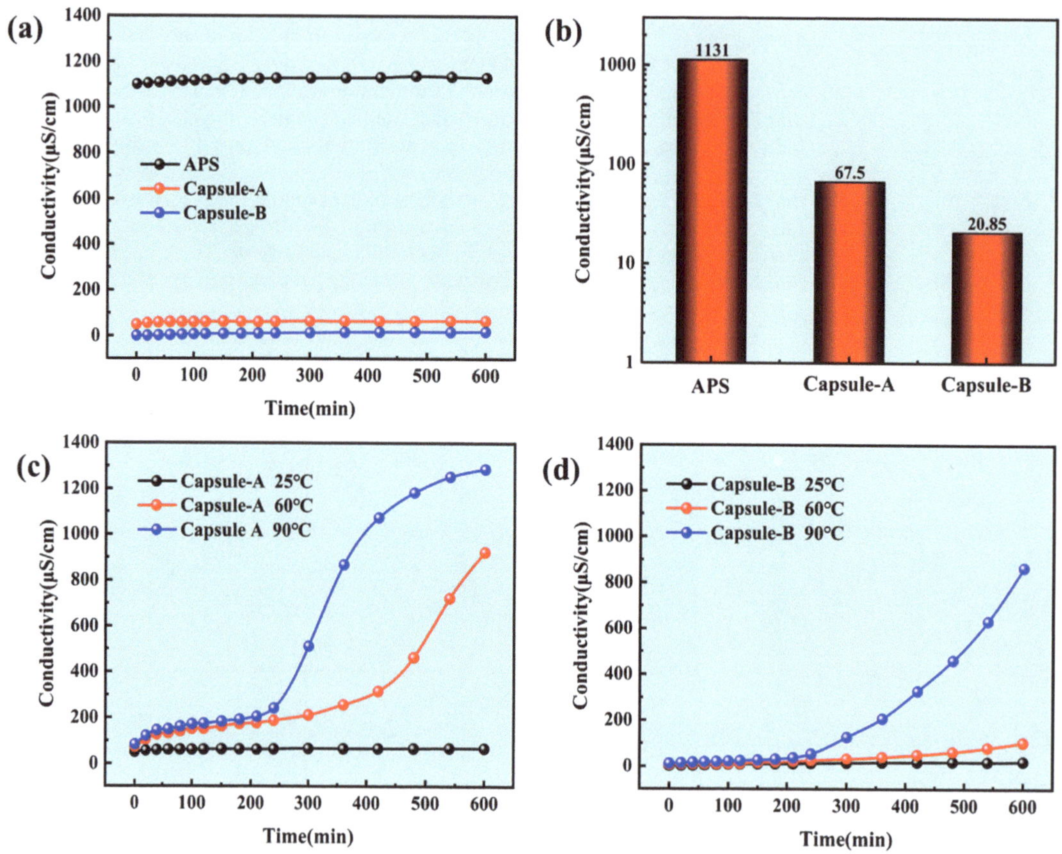

Figure 5. The conductivity of different gel-breaker solutions. (**a**) The conductivity varies with time at 25 °C. (**b**) The conductivity at 600 min (25 °C). (**c**) The conductivity of Capsule-A. (**d**) The conductivity of Capsule-B.

The micro-morphology of Capsule-B is shown in Figure 7. There were no obvious cracks or holes on the surface of Capsule-B that was untreated or soaked in 25 °C water, as shown in Figure 7a,b. However, tiny holes were observed when the temperature was 60 °C (Figure 7c). It was found that some Capsule-B even had complete rupture, as shown in Figure 7d, when the temperature increased to 90 °C. It could be concluded from the experiment that the surface of Capsule-B gradually formed holes and might eventually rupture to release the gel-breaker after being soaked in high-temperature water [42].

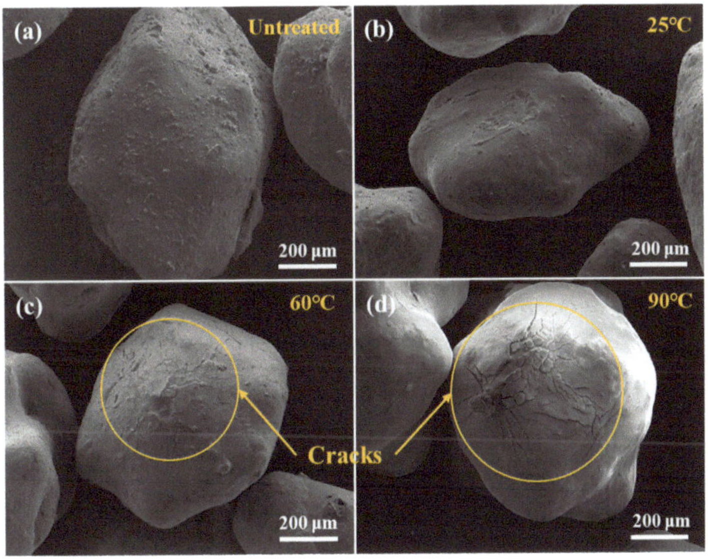

Figure 6. Surface morphology of Capsule-A after being soaked in the water at different temperatures. (**a**) Untreated. (**b**) 25 °C. (**c**) 60 °C. (**d**) 90 °C.

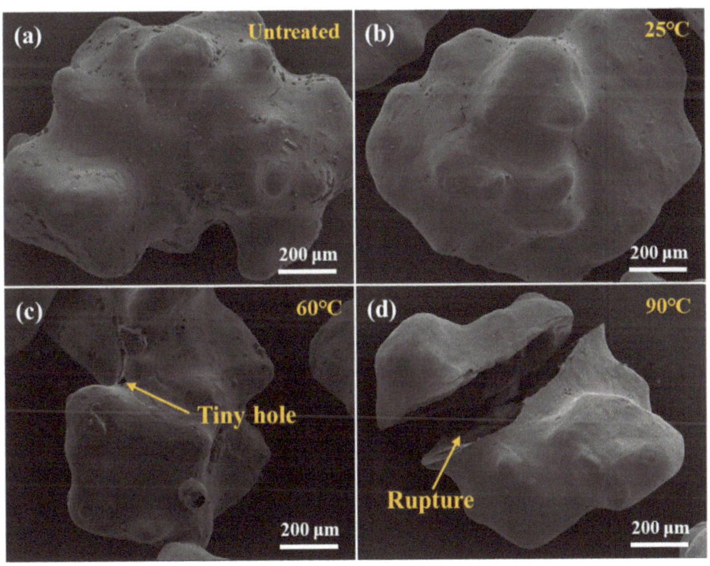

Figure 7. Surface morphology of Capsule-B after being soaked in water at different temperatures. (**a**) Untreated. (**b**) 25 °C. (**c**) 60 °C. (**d**) 90 °C.

The breaking time of the SRG fracturing fluid and IDG fracturing fluid under the action of the APS or Capsule-B and the viscosity of the gel-breaking solution are seen in Table 1. It could be seen from the results that the gel-breaking time of the two gel fracturing fluids was about 30–60 min when the gel-breaker was APS. However, when the gel-breaker was Capsule-B, the gel-breaking time of the two gel fracturing fluids was extended to 90–120 min. The gel-breaking time was related to the type of gel-breaker. Compared with

APS, Capsule-B had a longer gel-breaking time and could avoid the too-quick degradation of the gel fracturing fluid at 200 °C [43]. The viscosity of all the gel-breaking solutions was less than 2 mPa·s, which could be observed from Table 1.

Table 1. Gel-breaking time of gel fracturing fluid under different gel-breakers.

Gel Fracturing Fluid	Gel-Breaker	Gel-Breaking Time/min	The Viscosity of Gel-Breaking Solution/mPa·s
SRG	APS	30–60	1.28
SRG	Capsule-B	90–120	1.57
IDG	APS	30–60	1.71
IDG	Capsule-B	90–120	1.96

The gel-breaking solutions of the SRG fracturing fluid and industrial gel (IDG) fracturing fluid under the action of Capsule-B were investigated. The viscosity of both gel-breaking solutions was low, as can be seen in Figure 8a, which proves that the gels are all basically broken. However, the particle size of the SRG-breaking solution was much smaller than that of the IDG-breaking solution, as shown in Figure 8b. The distribution size of the SRG-breaking solution and the IDG-breaking solution was 25~658 nm and 186~1814 nm, respectively. The median particle size of the SRG-breaking solution was 126 nm, and the median particle size of the IDG-breaking solution was 587 nm. The difference between the two gel-breaking solutions was 4.66 times. The morphology of the two gel-breaking solutions could be obtained by an atomic force microscope, as shown in Figure 8c,d. The gel particles appeared in the solution after gel-breaking. Compared with the IDG-breaking solution, the particle size of the SRG-breaking solution was smaller, which was consistent with the result in Figure 8b. The above experiments showed that the SRG-breaking solution containing smaller gel particles would be cleaner, which should be conducive to reducing the permeability damage of the reservoir in hydraulic fracturing [44].

2.4. Reservoir Permeability Damage under Gel Fracturing Fluid

The damage of different gel-breaking solutions to the permeability of cores was studied according to the diagram of the experimental device shown in 4.2.7, and the results are shown in Table 2. It could be seen from the results that the permeability damage of the cores increased with the decrease in the initial permeability of the two gel-breaking solutions [45]. The permeability damage of 0.1-1, 1-1, and 10-1 was 13.22%, 9.7%, and 7.22%, respectively. The permeability damage of 0.1-2, 1-2, and 10-2 was 26.17%, 16.96%, and 11.04%, respectively. The average permeability damage was 10.05% for the SRG-breaking solution and 18.06% for the IDG-breaking solution. The lower the core permeability, the smaller the pore throat size. The gel particles contained in the gel-breaking solutions would gather in the pore throat and lead to permeability damage. However, the permeability damage of the cores under the SRG-breaking solutions was less than that of the IDG-breaking solutions. The permeability damage of the cores caused by the SRG-breaking solutions was only about half that of the IDG-breaking solutions at 1 mD. The gel particle size in the SRG-breaking solutions was much smaller than that of the IDG-breaking solutions. The larger the gel particle size was, the more likely it was to damage the core permeability [46].

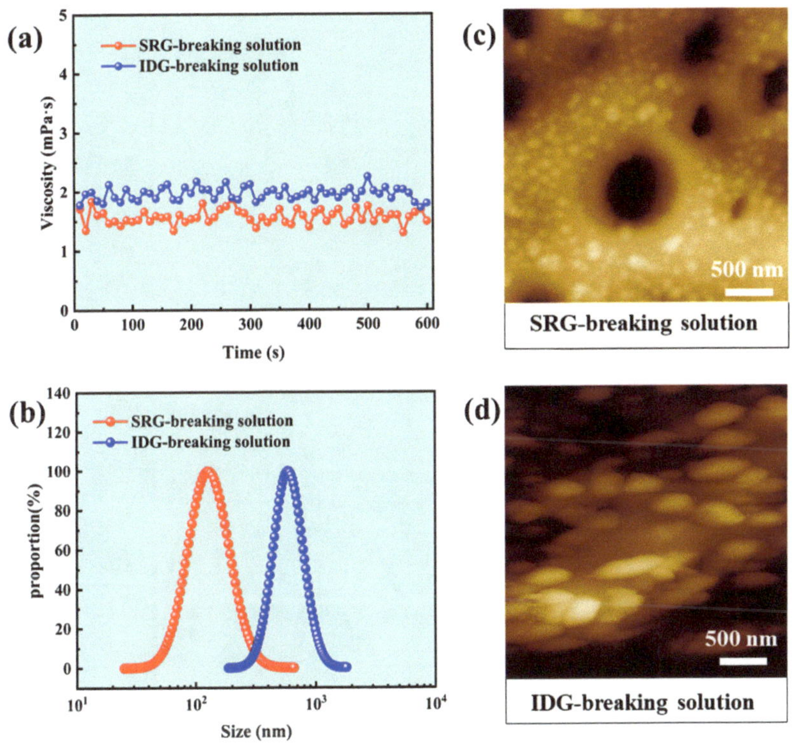

Figure 8. Properties of the gel-breaking solutions. (**a**) Viscosity; (**b**) particle size distribution; (**c**) microtopography of the SRG-breaking solution; (**d**) microtopography of the IDG-breaking solution.

Table 2. Permeability damage under different gel-breaking solutions.

Core Label	Gel-Breaking Solution	Permeability K_1/mD	Permeability K_2/mD	Permeability Damage, η/%
0.1-1	SRG	0.121	0.105	13.22
1-1	SRG	1.691	1.527	9.70
10-1	SRG	10.983	10.190	7.22
0.1-2	IDG	0.107	0.079	26.17
1-2	IDG	1.834	1.523	16.96
10-0	IDG	9.511	8.461	11.04

The surface morphology inside the cores after the permeability damage test in Table 2 was observed using SEM, and the results are shown in Figure 9. The surfaces of core 10-1 and core 1-1 were clean, and a small amount of debris could be seen in core 0.1-1, respectively, indicating that the SRG-breaking solution only had obvious damage to core 0.1-1. A certain amount of debris accumulated at the surface of core 0.1-2, core 1-2, and core 10-2, respectively, and more debris was observed with the decrease in core permeability, indicating that the IDG-breaking solution had obvious damage to the permeability of the three cores. It could also be concluded that the damage of the SRG-breaking solution to the core permeability was much lower than that of the IDG-breaking solution. The concentration of the polymer and crosslinker contained in the SRG was lower than that of IDG, and the gel particles after gel-breaking were smaller in size, so there was less damage to the core permeability. Low permeability damage would contribute to the conductivity of the fractured fractures [28].

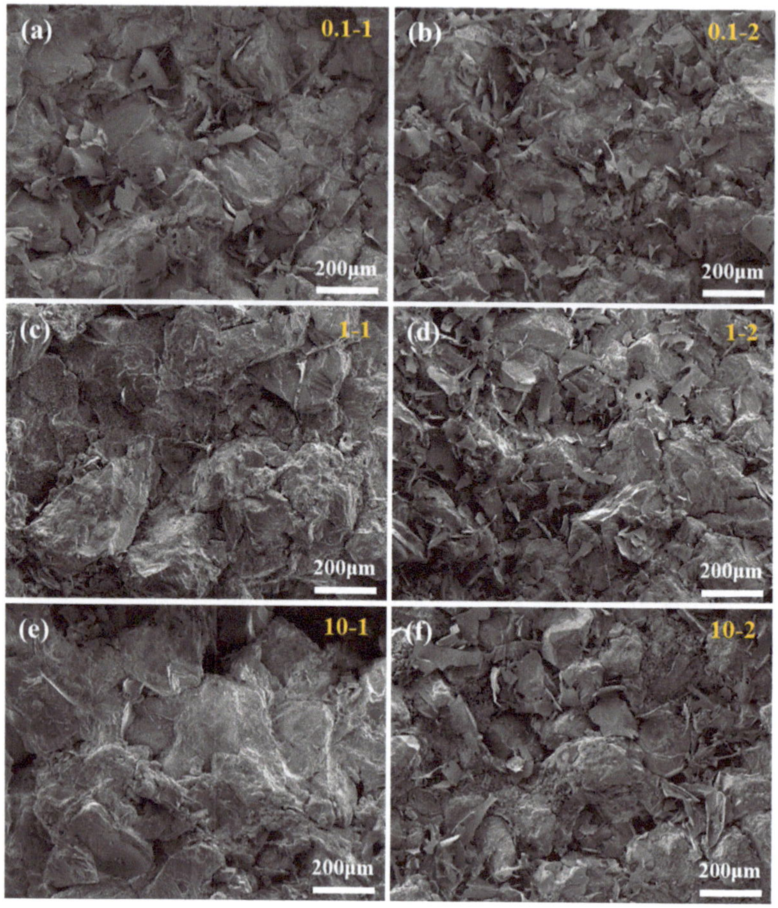

Figure 9. Micromorphology of the different cores after the permeability damage experiment. (**a**) 0.1-1; (**b**) 0.1-2; (**c**) 1-1; (**d**) 1-2; (**e**) 10-1; (**f**) 10-2.

3. Conclusions

In this study, an SRG fracturing fluid was constructed for deep hydraulic fracturing. Compared to the SNG fracturing fluid, the SRG fracturing fluid could obtain higher G' under lower concentrations of polymer and crosslinker. The network structure formed by the interaction of supramolecular forces between polymers in the SRG replaced the part network structure formed by the crosslinkers. The viscosity of the SRG fracturing fluid after 2 h shearing at 200 °C was 72.35 mPa·s, indicating that it could effectively suspend proppant at 200 °C. Capsule-B did not release the gel-breaker at room temperature. Small holes were gradually formed on the surface of Capsule-B, and some Capsule-B completely ruptured to release the gel-breaker with the increase in temperature. The gel-breaking time of the gel fracturing fluid could be delayed to 90–120 min at 200 °C, avoiding too-quick degradation of gel fracturing fluid. The distribution size of the SRG-breaking solution and the IDG-breaking solution was 25~658 nm and 186~1814 nm, respectively. The median size of the gel particles in the SRG-breaking solution was 126 nm, which was 0.21 times that of the gel particles in the IDG-breaking solution. The smaller particle size of the gel particles in the gel-breaking solution might cause less damage to the reservoir permeability. The permeability damage of cores under the SRG-breaking solution only occurred signally at very low core permeability, and the damage to permeability was much less than that of the IDG-breaking solution. In summary, the SRG-fracturing fluid could maintain the temperature and shear resistance under low polymer and crosslinker concentrations, which also contributed to lower reservoir permeability damage. The novelty of this paper is obtaining a clean solution after the high-strength gel is broken. The gap is that the gel-breaking time is still short at high temperatures, even with the encapsulated gel-breaker.

4. Materials and Methods

4.1. Materials

The polymers used in this paper are all polyacrylamide-based. Polymer A and polymer B are provided by our laboratory. Chemical enterprises provided industrial polymers and Zr-crosslinkers. Ammonium persulfate (APS) (>98%) was purchased from Sinopharm Group Chemical Reagent Co., LTD, Shanghai, China. Deionized water was provided in our laboratory. Encapsulated gel-breaker A (Capsule-A) was purchased from Dongying Baiyang Petroleum Technology Co., LTD, Dongying, China. Encapsulated gel-breaker B (Capsule-B) was provided by Beijing Shida Ode Technology Co., LTD, Beijing, China. Cores with different permeability were purchased from Beijing Tiandi Kaiyuan Geological Technology Co., LTD, Beijing, China.

4.2. Methods

4.2.1. Preparation of the Gel Fracturing Fluids

The composition and shortened form of gel fracturing fluids used in the paper are shown in Table 3. The single network gel (SNG) and industrial gel (IDG) fracturing fluids were formed by adding a Zr-crosslinker to a pre-prepared polymer solution and stirring. To obtain a supramolecular reinforced gel (SRG) fracturing fluid, polymer A and polymer B should be evenly mixed before adding the Zr-crosslinker. The polymer solution was prepared with deionized water.

Table 3. The composition and shortened form of the gel fracturing fluids.

Gel Type	Polymer	Crosslinker	Shortened Form
Supramolecular reinforced gel	polymer A: polymer B = 2:1	Zr	SRG
Single network gel	Polymer A	Zr	SNG
Industrial gel	Industrial polymer	Zr	IDG

4.2.2. Rheological Performance Test

All rheological performance tests were conducted using the HAKKE RS600B rheometer (Thermo Hakke Corporation, Waltham, MA, America). A frosted plate module was used for viscoelastic testing at 25 °C to obtain a stable value of the viscoelastic modulus. The viscoelastic modulus of the gel fracturing fluids was obtained using frequency scanning. The shear resistance of the gel fracturing fluids was tested using a drum module. The temperature was gradually raised from 25 °C to 200 °C and then continued to shear for 2 h at 170 s^{-1}. Then, 2.75 MPa was pressurized to prevent evaporation before the experiment.

4.2.3. Conductivity Experiment

The three different kinds of gel-breakers were prepared into a 0.1 wt% solution. The solution was placed in ovens at 25 °C, 60 °C, and 90 °C, respectively, and removed at intervals for conductivity measurements. The conductivity of the solution was measured using a conductivity meter (DDS-307, Shanghai Lei Magnetic Instrument Factory, Shanghai, China). The probe of the conductivity meter was thoroughly cleaned with deionized water before each measurement.

4.2.4. Micro-Morphology Experiment

Capsule-A and Capsule-B were immersed in water at 25 °C, 60 °C, and 90 °C for 24 h, respectively. After most of the water was removed, it was frozen with liquid nitrogen for

5 min and then placed in a freeze dryer (Scientz-10N, Ningbo Xinzhi Biotechnology Co., LTD, Ningbo, China) for freeze-drying. The cores of the permeability damage experiment were freeze-dried according to the above steps. The surface structures of the capsules and cores were characterized using scanning electron microscopy (SEM, instrument model: JSM-7610F, magnification: 25–106 times, produced by Nippon Electronics, Showima City, Tokyo, Japan).

4.2.5. Gel-Breaking Test

We added 0.001 wt% APS or 0.001 wt% Capsule-B into the prepared SRG fracturing fluid and IDG fracturing fluid and evenly mixed them in a pressure bottle. The pressure bottle was subsequently placed in the oven at 200 °C, and the time when the gel completely broke was recorded as the breaking time.

4.2.6. Particle Size Measurement of Gel-Breaking Solution

After being set for 24 h, the particle size of the gel-breaking solution was measured using the Nano Brook Omni (Brookhaven Instruments Corporation, New York, NY, America). The experimental temperature was 25 °C.

4.2.7. Permeability Damage Experiment

Six cores with different permeability (Table 2) were used in the reservoir damage experiment, and the experimental diagram is shown in Figure 10. The green tank contained the gel-breaking solution, and the yellow tank contained the brine solution (2.0 wt% KCl + 5.5 wt% NaCl + 0.45 wt% $MgCl_2$ + 0.55 wt% $CaCl_2$). We loaded the core into the core holder and adjusted the confining pressure to 2.5 MPa. The brine solution was reversely injected into the core at 0.1 mL/min flow rate to obtain a stable injection pressure P_1. Then, the gel-breaking solution was injected, and the pump was stopped to stabilize for 2 h. Finally, the brine solution was reversely injected into the core at 0.1 mL/min flow rate to obtain a stable injection pressure P_2. The core permeability K_1 and K_2 corresponding to injection pressure P_1 and P_2 were calculated using Darcy's law, respectively. The formula for calculating the reservoir permeability damage was as follows:

$$\eta = \frac{K_1 - K_2}{K_1} \times 100\% \quad (1)$$

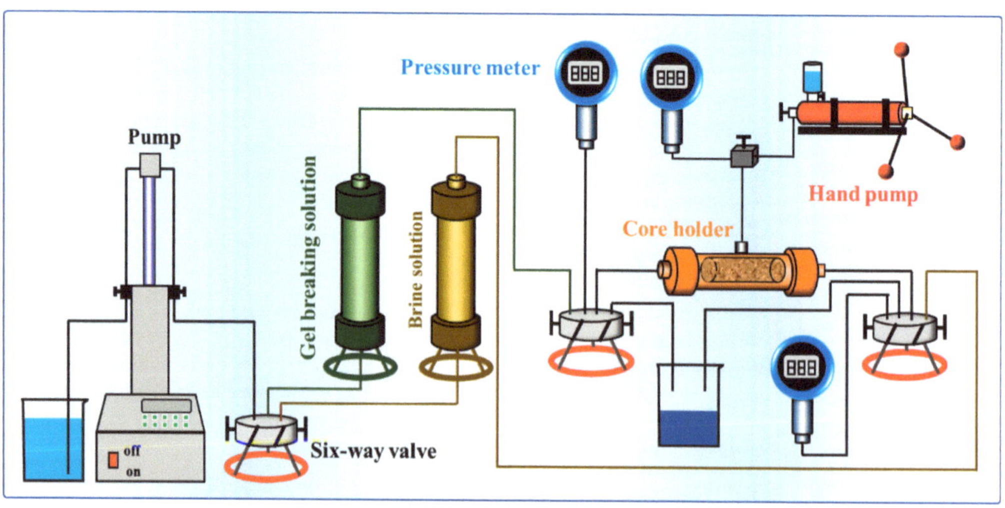

Figure 10. Schematic diagram of permeability damage experiment.

Author Contributions: Conceptualization, Y.H., C.D. and B.Y.; Data curation, Y.H. and X.Y.; Formal analysis, Y.W.; Funding acquisition, C.D., Y.W. and B.Y.; Investigation, Y.H. and X.Y.; Methodology, X.Y., Y.W. and L.L.; Software, Y.H., X.Y. and Y.W.; Supervision, Y.W., C.D. and L.L.; Validation, L.L., X.Y. and B.Y.; Writing—original draft, Y.H.; Writing—review and editing, C.D. and B.Y. All authors have read and agreed to the published version of the manuscript.

Funding: This study is financially supported by the National Natural Science Foundation of China (NO. 52120105007), the National Key R&D Program of China (NO. 2022YFE0129900), the National Natural Science Foundation of China (NO. 52374062, NO: 52288101) and the Innovation Fund Project for Graduate Student of China University of Petroleum (East China) supported by "the Fundamental Research Funds for the Central Universities" (NO. 23CX04047A).

Institutional Review Board Statement: Not applicable.

Informed Consent Statement: Not applicable.

Data Availability Statement: The data presented in this study are openly available in article.

Acknowledgments: The authors would like to thank the National Natural Science Foundation of China and the National Key R&D Program of China.

Conflicts of Interest: The authors declare no conflict of interest.

References

1. Olajire, A.A. Review of ASP EOR (alkaline surfactant polymer enhanced oil recovery) technology in the petroleum industry: Prospects and challenges. *Energy* **2014**, *77*, 963–982. [CrossRef]
2. Tavakkoli, O.; Kamyab, H.; Shariati, M.; Mohamed, A.M.; Junin, R. Effect of nanoparticles on the performance of polymer/surfactant flooding for enhanced oil recovery: A review. *Fuel* **2022**, *312*, 122867. [CrossRef]
3. Wu, Y.; Huang, Y.; He, Z.; Hu, Y.; Zhao, X.; Sun, L.; Ren, H.; Dai, C. Cation-Dependent Oil-Rock Interactions in Nanopores: Insights from Core Flooding Experiments and Molecular Dynamics Simulations. *Energy Fuels* **2023**, *37*, 16564–16572.
4. Malozyomov, B.V.; Martyushev, N.V.; Kukartsev, V.V.; Tynchenko, V.S.; Bukhtoyarov, V.V.; Wu, X.; Tyncheko, Y.A.; Kukartsev, V.A. Overview of Methods for Enhanced Oil Recovery from Conventional and Unconventional Reservoirs. *Energies* **2023**, *16*, 4907. [CrossRef]
5. Hao, F. Enrichment Mechanism and Prospects of Deep Oil and Gas. *Acta Geol. Sin.-Engl. Ed.* **2022**, *96*, 742–756. [CrossRef]
6. Han, S.B.; Xiang, C.H.; Du, X.; Xie, L.F.; Bai, S.T.; Wang, C.S. Logging evaluation of deep multi-type unconventional gas reservoirs in the Songliao basin, northeast China: Implications from continental scientific drilling. *Geosci. Front.* **2022**, *13*, 101451. [CrossRef]
7. Shen, W.; Zuo, L.; Ma, T.; Chen, C.; Qin, C.; Yang, L.; Xie, K. Quantitative Studies on the Characterization and Evaluation of Adsorbed Gas and Free Gas in Deep Shale Reservoirs. *Energy Fuels* **2023**, *37*, 3752–3759. [CrossRef]
8. Cao, P.; Liu, J.S.; Leong, Y.K. General Gas Permeability Model for Porous Media: Bridging the Gaps Between Conventional and Unconventional Natural Gas Reservoirs. *Energy Fuels* **2016**, *30*, 5492–5505. [CrossRef]
9. Ali, A.; Potter, D.K. Thermomagnetic Analyses of the Permeability-Controlling Minerals in Red and White Sandstones in Deep Tight Gas Reservoirs: Implications for Downhole Measurements. *Spe Reserv. Eval. Eng.* **2011**, *14*, 557–565. [CrossRef]
10. Khormali, A. Effect of water cut on the performance of an asphaltene inhibitor package: Experimental and modeling analysis. *Pet. Sci. Technol.* **2022**, *40*, 2890–2906. [CrossRef]
11. Lei, Q.; Xu, Y.; Yang, Z.; Cai, B.; Wang, X.; Zhou, L.; Liu, H.; Xu, M.; Wang, L.; Li, S. Progress and development directions of stimulation techniques for ultra-deep oil and gas reservoirs. *Pet. Explor. Dev.* **2021**, *48*, 221–231. [CrossRef]
12. Wang, F.J.; Xu, H.; Liu, Y.K.; Meng, X.H.; Liu, L.C.F. Mechanism of low chemical agent adsorption by high pressure for hydraulic fracturing-assisted oil displacement technology: A study of molecular dynamics combined with laboratory experiments. *Langmuir* **2023**, *39*, 16628–16636. [CrossRef] [PubMed]
13. He, J.M.; Lin, C.; Li, X.; Zhang, Y.X.; Chen, Y. Initiation, propagation, closure and morphology of hydraulic fractures in sandstone cores. *Fuel* **2017**, *208*, 65–70. [CrossRef]
14. Zhang, Y.J.; Zou, Y.; Zhang, Y.; Wang, L.Y.; Liu, D.Q.; Sun, J.; Ge, H.K.; Zhou, D.S. Experimental study on characteristics and mechanisms of matrix pressure transmission near the fracture surface during post-fracturing shut-in in tight oil reservoirs. *J. Pet. Sci. Eng.* **2022**, *219*, 111133. [CrossRef]
15. Wu, Y.T.; Pan, Z.J.; Zhang, D.Y.; Down, D.I.; Lu, Z.H.; Connell, L.D. Experimental study of permeability behaviour for proppant supported coal fracture. *J. Nat. Gas Sci. Eng.* **2018**, *51*, 18–26. [CrossRef]
16. Barati, R.; Liang, J.T. A Review of Fracturing Fluid Systems Used For Hydraulic Fracturing of Oil and Gas Wells. *J. Appl. Polym. Sci.* **2014**, *131*, 40735. [CrossRef]
17. Yang, B.; Zhao, J.Z.; Mao, J.C.; Tan, H.Z.; Zhang, Y.; Song, Z.F. Review of friction reducers used in slickwater fracturing fluids for shale gas reservoirs. *J. Nat. Gas Sci. Eng.* **2019**, *62*, 302–313. [CrossRef]
18. Shi, S.; Sun, J.; Lv, K.; Liu, J.; Bai, Y.; Wang, J.; Huang, X.; Jin, J.; Li, J. Comparative Studies on Thickeners as Hydraulic Fracturing Fluids: Suspension versus Powder. *Gels* **2022**, *8*, 722. [CrossRef]

19. Wu, Y.M.; Zhang, J.; Dong, S.B.; Li, Y.F.; Slany, M.; Chen, G. Use of Betaine-Based Gel and Its Potential Application in Enhanced Oil Recovery. *Gels* **2022**, *8*, 351. [CrossRef]
20. Li, Z.; Kang, W.; Zhao, Y.; Yang, H.; Li, M.; Kang, X.; Zhu, T.; Zhou, B.; Sarsenbekuly, B.; Aidarova, S. Organic Acid-Enhanced Viscoelastic Surfactant and Its Application in Fracturing Fluids. *Energy Fuels* **2021**, *35*, 3130–3139. [CrossRef]
21. Liu, P.; Dai, C.; Gao, M.; Wang, X.; Liu, S.; Jin, X.; Li, T.; Zhao, M. Development of the Gemini Gel-Forming Surfactant with Ultra-High Temperature Resistance to 200 °C. *Gels* **2022**, *8*, 600. [CrossRef] [PubMed]
22. Huang, Q.M.; Liu, S.M.; Wang, G.; Wu, B.; Zhang, Y.Z. Coalbed methane reservoir stimulation using guar-based fracturing fluid: A review. *J. Nat. Gas Sci. Eng.* **2019**, *66*, 107–125. [CrossRef]
23. Zhang, K.Y.; Zhang, G.D.; Li, Z.J.; Hou, J.R. Laboratory Evaluation of a Low pH and Low Polymer Concentration Zirconium-CMHPG Gel System for Hydraulic Fracturing. *Energy Fuels* **2019**, *33*, 9720–9735. [CrossRef]
24. Zhang, C.B.; Wang, Y.L.; Xu, N.; Gong, J.C.; Liang, S.A. Synthesis and Crosslinking Mechanism of Colloidal Graphene Oxide Crosslinker for Crosslinking Low-Concentration Hydroxypropyl Guar Gum Fracturing Fluids. *Energy Fuels* **2022**, *36*, 14760–14770. [CrossRef]
25. Xin, H.; Fang, B.; Yu, L.Y.; Lu, Y.J.; Xu, K.; Li, K.J. Rheological Performance of High-Temperature-Resistant, Salt-Resistant Fracturing Fluid Gel Based on Organic-Zirconium-Crosslinked HPAM. *Gels* **2023**, *9*, 151. [CrossRef] [PubMed]
26. Hurnaus, T.; Plank, J. Behavior of Titania Nanoparticles in Cross-linking Hydroxypropyl Guar Used in Hydraulic Fracturing Fluids For Oil Recovery. *Energy Fuels* **2015**, *29*, 3601–3608. [CrossRef]
27. Ma, Z.F.; Zhao, M.W.; Yang, Z.T.; Wang, X.Y.; Dai, C.L. Development and Gelation Mechanism of Ultra-High-Temperature-Resistant Polymer Gel. *Gels* **2023**, *9*, 726. [CrossRef]
28. Wang, J.Y.; Holditch, S.A.; McVay, D.A. Effect of gel damage on fracture fluid cleanup and long-term recovery in tight gas reservoirs. *J. Nat. Gas Sci. Eng.* **2012**, *9*, 108–118. [CrossRef]
29. Li, X.; Zhang, Q.; Liu, P.; Li, T.; Liu, G.; Liu, Z.; Zhao, H. Investigation on the microscopic damage mechanism of fracturing fluids to low-permeability sandstone oil reservoir by nuclear magnetic resonance. *J. Pet. Sci. Eng.* **2022**, *209*, 109821. [CrossRef]
30. Lu, Y.Y.; Yang, F.; Ge, Z.L.; Wang, Q.; Wang, S.Q. Influence of viscoelastic surfactant fracturing fluid on permeability of coal seams. *Fuel* **2017**, *194*, 1–6. [CrossRef]
31. Donmoyer, S.; Agrawal, V.; Sharma, S.; Hakala, J.A. Effect of oxidative breakers on organic matter degradation, contaminant mobility and critical mineral release during shale-fracturing fluid interactions in the Marcellus Shale. *Fuel* **2023**, *331*, 125678. [CrossRef]
32. Chivers, P.R.A.; Smith, D.K. Shaping and structuring supramolecular gels. *Nat. Rev. Mater.* **2019**, *4*, 463–478. [CrossRef]
33. Meazza, L.; Foster, J.A.; Fucke, K.; Metrangolo, P.; Resnati, G.; Steed, J.W. Halogen-bonding-triggered supramolecular gel formation. *Nat. Chem.* **2013**, *5*, 42–47. [CrossRef] [PubMed]
34. Dai, X.; Zhang, Y.; Gao, L.; Bai, T.; Wang, W.; Cui, Y.; Liu, W. A Mechanically Strong, Highly Stable, Thermoplastic, and Self-Healable Supramolecular Polymer Hydrogel. *Adv. Mater.* **2015**, *27*, 3566–3571. [CrossRef] [PubMed]
35. Rossow, T.; Seiffert, S. Supramolecular Polymer Networks: Preparation, Properties, and Potential. In *Supramolecular Polymer Networks and Gels*; Seiffert, S., Ed.; Springer: Berlin/Heidelberg, Germany, 2015; pp. 1–46.
36. Sun, X.; Liang, X.B.; Wang, S.Z.; Lu, Y. Experimental study on the rheology of CO_2 viscoelastic surfactant foam fracturing fluid. *J. Pet. Sci. Eng.* **2014**, *119*, 104–111. [CrossRef]
37. Li, C.; Huang, Y.; Sun, X.; Gao, R.; Zeng, F.; Tontiwachwuthikul, P.; Liang, Z. Rheological properties study of foam fracturing fluid using CO_2 and surfactant. *Chem. Eng. Sci.* **2017**, *170*, 720–730. [CrossRef]
38. Othman, A.; Aljawad, M.S.; Mahmoud, M.; Kamal, M.S.; Patil, S.; Bataweel, M. Chelating Agents Usage in Optimization of Fracturing Fluid Rheology Prepared from Seawater. *Polymers* **2021**, *13*, 2111. [CrossRef]
39. Lv, K.H.; Zhang, G.D.; Bai, Y.R.; Yang, J.B. Preparation of Encapsulated Breakers for Polymer Gels and Evaluation of Their Properties. *Gels* **2023**, *9*, 387. [CrossRef]
40. Lin, X.; Zhang, S.C.; Wang, Q.; Feng, Y.; Shuai, Y.Y. Improving the fracturing fluid loss control for multistage fracturing by the precise gel breaking time design. *J. Nat. Gas Sci. Eng.* **2015**, *25*, 367–370. [CrossRef]
41. Zhang, Q.F.; Zuo, M.M.; Li, G.H.; Sha, J.P.; Zuo, X.B. Synthesis of ammonium persulfate microcapsule with a polyaniline shell and its controlled burst release. *J. Appl. Polym. Sci.* **2021**, *138*, 49695. [CrossRef]
42. DeMella, K.C.; Raghavan, S.R. Catalyst-Loaded Capsules that Spontaneously Inflate and Violently Eject their Core. *Langmuir* **2019**, *35*, 13718–13726. [CrossRef] [PubMed]
43. Meng, F.; Li, M.; Wang, S.; Liu, X.; Gao, W.; Ma, Z.; Kong, C.; Ma, X.; Li, J. Encapsulation of potassium persulfate with ABS via coacervation for delaying the viscosity loss of fracturing fluid. *J. Appl. Polym. Sci.* **2019**, *136*, 47734. [CrossRef]
44. Zou, C.; Dai, C.; Liu, Y.; You, Q.; Ding, F.; Huang, Y.; Sun, N. A novel self-degradable gel (SDG) as liquid temporary plugging agent for high-temperature reservoirs. *J. Mol. Liq.* **2023**, *386*, 122463. [CrossRef]

45. Zhang, L.F.; Zhou, F.J.; Zhang, S.C.; Li, Z.; Wang, J.; Wang, Y.C. Evaluation of permeability damage caused by drilling and fracturing fluids in tight low permeability sandstone reservoirs. *J. Pet. Sci. Eng.* **2019**, *175*, 1122–1135.
46. Huang, Q.M.; Liu, S.M.; Cheng, W.M.; Wang, G. Fracture permeability damage and recovery behaviors with fracturing fluid treatment of coal: An experimental study. *Fuel* **2020**, *282*, 118809. [CrossRef]

gels

MDPI

Article

Alkaline Hydrolysis of Waste Acrylic Fibers Using the Micro-Water Method and Its Application in Drilling Fluid Gel Systems

Wenjun Long, Zhongjin Wei, Fengshan Zhou *, Shaohua Li, Kang Yin, Yu Zhao, Siting Yu and Hang Qi

Beijing Key Laboratory of Materials Utilization of Nonmetallic Minerals and Solid Wastes, National Laboratory of Mineral Materials, School of Materials Science and Technology, China University of Geosciences (Beijing), No. 29 Xueyuan Road, Haidian District, Beijing 100083, China; longwj@cugb.edu.cn (W.L.); 3003190032@cugb.edu.cn (Z.W.); 2103210080@email.cugb.edu.cn (S.L.); 2103220083@email.cugb.edu.cn (K.Y.); 2103210079@email.edu.cn (Y.Z.); 2103210078@email.cugb.edu.cn (S.Y.); 2003220043@email.cugb.edu.cn (H.Q.)

* Correspondence: zhoufs@cugb.edu.cn

Abstract: Filtrate reducer is a drilling fluid additive that can effectively control the filtration loss of drilling fluid to ensure the safe and efficient exploitation of oilfields. It is the most widely used treatment agent in oilfields. Due to its moderate conditions and controllable procedure, alkaline hydrolysis of high-purity waste polyacrylonitrile has been utilized for decades to produce filtrate reducer on a large scale in oilfields. However, the issues of long hydrolysis time, high viscosity of semi-finished products, high drying cost, and tail gas pollution have constrained the development of the industry. In this study, low-purity waste acrylic fiber was first separated and purified using high-temperature hydroplastization, and the hydrolyzed product was obtained using alkaline hydrolysis with the micro-water method, which was called MW−HPAN. The hydrolysis reaction was characterized using X-ray diffraction, scanning electron microscopy, infrared spectroscopy, and thermogravimetric analysis, and the elemental analysis showed a hydrolysis degree of 73.21%. The experimental results showed that after aging at 180 °C for 16 h, the filtration volume of the freshwater base slurry with 0.30% dosage and 4% brine base slurry with 1.20% dosage was 12.7 mL and 18.5 mL, respectively. The microstructure and particle size analysis of the drilling fluid gel system showed that MW−HPAN could prevent the agglomeration of clay and maintain a reasonable particle size distribution even under the combined deteriorating effect of high temperature and inorganic cations, thus forming a dense filter cake and achieving a low filtrate volume of the drilling fluid gel system. Compared with similar commercially available products, MW−HPAN has better resistance to temperature and salt in drilling fluid gel systems, and the novel preparation method is promising to be extended to practical production.

Keywords: waste acrylic fiber; hydroplastization; micro-water method; alkaline hydrolysis; filtrate reducer

Citation: Long, W.; Wei, Z.; Zhou, F.; Li, S.; Yin, K.; Zhao, Y.; Yu, S.; Qi, H. Alkaline Hydrolysis of Waste Acrylic Fibers Using the Micro-Water Method and Its Application in Drilling Fluid Gel Systems. *Gels* **2023**, *9*, 974. https://doi.org/10.3390/gels9120974

Academic Editor: Jordi Puiggali

Received: 6 November 2023
Revised: 1 December 2023
Accepted: 8 December 2023
Published: 13 December 2023

1. Introduction

The trend toward fast fashion and increased fiber consumption in the apparel industry has led to more frequent iterations of apparel products and shorter life cycles for textiles, generating large amounts of textile waste [1]. Textile waste is mainly generated during the textile production process, as well as during the use and disposal by consumers, and has a complex composition, being a mixture of different fibers, which is difficult to degrade in the natural environment, and coarse landfilling and incineration will pollute the atmosphere [2]. At present, the global annual production of used textiles is more than 40 million tons, and China's annual production is more than 26 million tons, accounting for about 6% of the total municipal solid waste [3,4]. However, due to the lack of a recycling system for used textiles and public awareness of environmental protection, as well as the relative backwardness of China's recycling technology, the overall recycling rate of used textiles in China is less than 10% [5], which is far below the level of the most developed countries [6].

Acrylic fiber is one of the most widely used textile raw materials due to its softness, thermal stability, and other advantages. China's acrylic fiber production ranks fourth among synthetic fibers, accounting for about 0.79%, with an average annual production of about 650,000 tons in the past ten years, in addition to more than 100,000 tons of imported acrylic fiber products every year (data from the website of the National Bureau of Statistics of China); thereby, a considerable amount of waste acrylic fiber is produced every year. The main chemical composition of waste acrylic fibers is polyacrylonitrile, where the polyacrylonitrile content varies from different sources. Waste acrylic fibers are difficult to degrade and cannot be naturally degraded or thermoformed. The incineration of these fibers produces toxic gases such as HCN and NH_3, which pollute the environment. Under the effect of acid, alkali, heat, and pressure, the hydrophobic cyano functional groups on the polyacrylonitrile molecular chain with high chemical reactivity can be hydrolyzed to generate amide and carboxyl groups to obtain functional polymer materials [7].

Hydrolyzed polyacrylonitrile products (HPAN) have been used as oilfield additives for several decades, mainly as a filtrate reducer for drilling fluid gel systems. The raw material for the preparation was changed from pure polyacrylonitrile to waste acrylic fibers to reduce the cost, and alkaline hydrolyzed polyacrylonitrile was used for large-scale industrial production due to the advantages of moderate conditions and controllable process. Over the past sixty years, scientists have conducted a number of studies on the mechanism underlying the alkaline hydrolysis reaction of polyacrylonitrile [8–13]. It is now generally accepted that the alkaline hydrolysis of polyacrylonitrile produces six-membered ring reaction intermediates (Figure 1) and that the properties of the hydrolysis products are closely related to the hydrolysis degree.

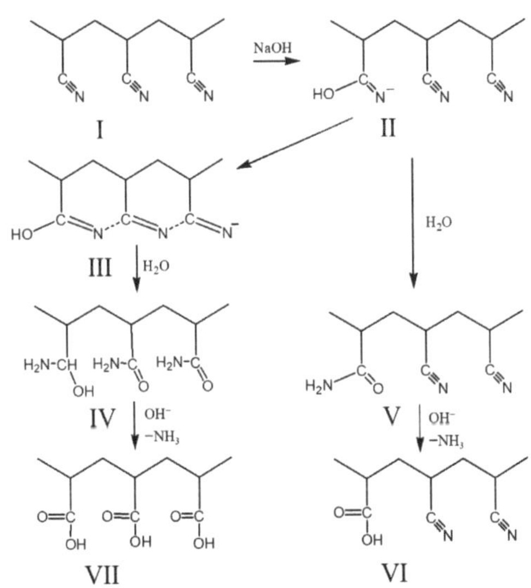

Figure 1. Mechanism of alkaline hydrolysis of polyacrylonitrile [14].

Solvents for the reaction of water-soluble polymers have a great effect on the efficiency and stability of the reaction. Take the starch modification reaction as an example, many research studies have been carried out in the field of the chemical modification of starch. Most chemical modifications of starch, such as esterification and carboxymethylation, are wet modifications, i.e., carried out in sufficient aqueous or organic solvents, including ethanol [15,16], N,N-dimethylformamide [17], pyridine [18], etc. The aqueous solution reaction system has the disadvantages of poor reaction efficiency, unstable product quality, and increasing side reactions. The organic solvent method can improve the reaction efficiency, but the cost increases and it is toxic and easily causes environmental pollution [19]. Scientists have developed dry and semi-dry starch modification methods on this basis [20–22]. Zhu et al. [23] investigated the effect of different starch moistures on solvent-free esterification reactions. At low water contents, the esterification reaction is difficult due to strong hydrogen bonding, whereas at too high moisture, the degree of substitution and the efficiency of the reaction decrease due to side reactions and dilution. Karic et al. [24] modified starch without solvent in the moist/semi-moist state using betaine hydrochloride (BHC) as a cationic reagent, resulting in starch derivatives with sufficiently high cationicity. Compared with aqueous and organic solvent methods, solvent-free modification keeps starch in a solid state during processing, resulting in a significantly different contact interface with chemical reagents. Due to the low solvent content, the occurrence of side reactions is suppressed to the maximum extent, and the microenvironment of the reaction system is made different from that of liquid-phase reactions, resulting in a localized high concentration at the reaction site, which improves the efficiency of the reaction. In addition, the solvent-free reaction is also an energy-saving and environmentally friendly process, and it is necessary to extend it to other chemical modification methods for water-soluble polymer materials [25].

Due to the agglomeration and hydrophobicity of polyacrylonitrile fibers, the hydrolysis reaction is an inhomogeneous reaction in the initial stage, resulting in a slow reaction speed, and the high-water retention of hydrolyzed semi-finished products makes them difficult to dry. In addition, the reaction process continues to emit irritating ammonia gas. The development of polyacrylonitrile hydrolysis technology that reduces environmental pollution, improves production efficiency, and lowers production costs has become essential. Thus, there is a need to improve the hydrolysis process to meet the requirements of being an environmentally friendly and low-cost production process.

In this study, polyacrylonitrile was hydrolyzed using a novel micro-water method process, and the effects of different reaction conditions on the filter loss reduction properties of the samples were investigated to determine the optimum hydrolysis reaction conditions. The occurrence of the hydrolysis reaction was characterized using analytical techniques such as FT-IR and elemental analysis. The filtration loss performance of drilling fluid gel systems with samples after aging at different temperatures was investigated by testing filtration loss volume, and the morphology of the filter cake, the rheological property, and particle size distribution of the drilling fluid gel system were analyzed. Compared with commercially available conventional HPAN products, the samples show better temperature and salt resistance. This research can improve the utilization and alkaline hydrolysis rate of waste acrylic fibers. Furthermore, the hydrolysis reaction of polyacrylonitrile with the micro-water method is simple in operation, and it can avoid the problems of high energy consumption and high cost, which is a more environmentally friendly process than the conventional process. This novel micro-water hydrolysis process is expected to replace the conventional process and be scaled up to practical production.

2. Results and Discussion

2.1. Structure Characterization of MW−HPAN

The structures of the used acrylic powder (LN-PAN) and MW−HPAN were tested, analyzed, and compared using an organic elemental analyzer, X-ray diffractometer, scanning electron microscope, Fourier transform infrared spectrometer, and thermogravimetry. MW−HPAN samples used for the tests were prepared under optimum reaction conditions: m_{PAN}:m_{NaOH}:m_{H2O} = 1:0.5:0.6, the reaction temperature was 160 °C, and the reaction time was 1 h.

2.1.1. Elemental Analysis

The organic elemental content of the samples before and after hydrolysis was determined using an organic elemental analyzer. According to the results shown in Table 1, after the hydrolysis reaction took place, there was a very significant decrease in the nitrogen content of the sample and a large increase in the C/N ratio, corresponding to the ammonia emitted by the cyano hydrolysis process. In addition, there was a large increase in the oxygen content and, therefore, a decrease in the C/O ratio, corresponding to the carboxyl groups produced during the hydrolysis process [26]. Based on the results of the elemental analysis and the conservation of the mass of carbon atoms, it was calculated that the mass of MW−HPAN after the reaction was 33.4 g. Therefore, 0.2221 mol NH_3 was released during the hydrolysis of polyacrylonitrile, which corresponds to the molar amount of carboxyl groups produced. If all the cyano groups were hydrolyzed to carboxyl groups, theoretically, 0.3034 mol of carboxyl groups would be produced, resulting in 73.21% hydrolysis of MW−HPAN [27]. This suggests that a large proportion of the cyano groups are hydrolyzed to carboxyl groups, which is consistent with the subsequent FTIR spectrogram.

Table 1. Elemental content of LC−PAN and MW−HPAN.

Sample	Elemental Content (wt%)			C/N	C/O
	C	N	O		
LC−PAN	61.60	21.24	9.31	2.90	6.62
MW−HPAN	36.89	3.41	42.66	10.82	0.86

2.1.2. XRD Analysis

The changes in the crystallinity of the polyacrylonitrile powder before and after the hydrolysis reaction were determined using an X-ray diffractometer, and the results are shown in Figure 2. The XRD diffraction pattern of the original polyacrylonitrile powder shows a strong diffraction peak at 2θ = 17.0° and a weaker diffraction peak at 2θ = 29.3°, corresponding to the (100) and (101) plane, respectively [28]. Between the two diffraction peaks, there is a large and diffuse amorphous diffraction peak at about 2θ = 25.0°, which is the characteristic diffraction peak of polyacrylonitrile, corresponding to the coexistence of relatively ordered quasi-crystalline regions and disordered amorphous regions [29]. After hydrolysis, the XRD diffraction peaks of MW−HPAN at 2θ = 17.0° and 29.3° disappeared completely, indicating that the crystallinity of MW−HPAN was reduced. A new characteristic peak appeared at around 2θ = 8.0°, corresponding to the appearance of the amide peak from the hydrolysis of PAN [29]. During the hydrolysis reaction, part of the cyano group reacted with sodium hydroxide, resulting in the destruction of the relatively regular structure of the polyacrylonitrile molecule and leading to a reduction in crystallinity, which was conducive to the improvement of hydrophilicity.

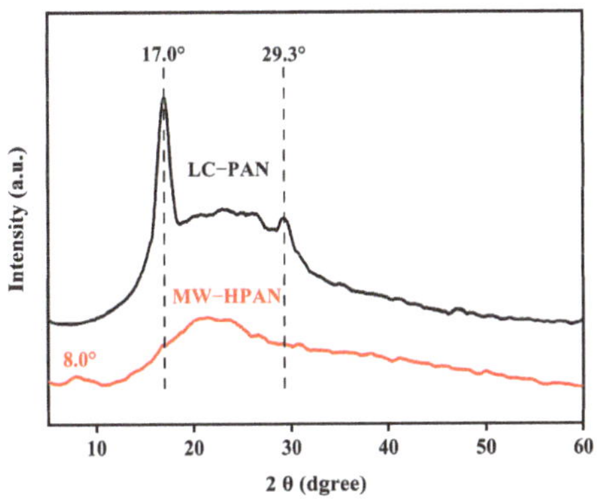

Figure 2. X-ray diffraction spectrograms of LC−PAN and MW−HPAN.

2.1.3. SEM Characterization

The morphology of the polyacrylonitrile powder before and after the hydrolysis reaction is shown in Figure 3. Before the reaction, the polyacrylonitrile powder still had a fibrous structure with a relatively smooth surface and a small number of micropores

(Figure 3a_1–a_3). After the reaction, the MW−HPAN powder was irregularly shaped, the particle size was reduced, the surface flatness was reduced and became rougher, and, at the same time, many cracks were formed (Figure 3b_1–b_3). The change in powder morphology confirms the decrease in crystallinity and the occurrence of the hydrolysis reaction.

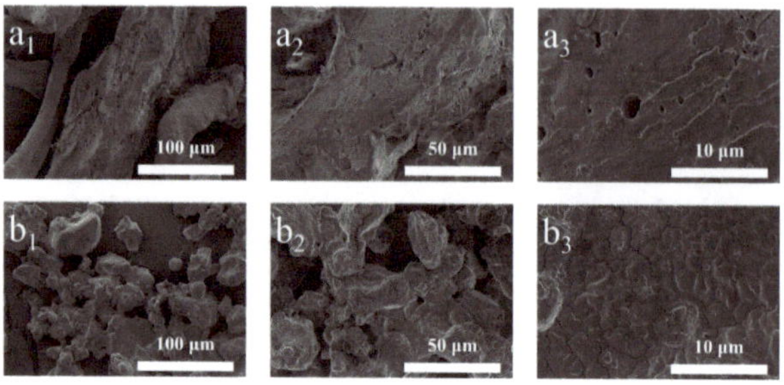

Figure 3. Scanning electron microscope images of LC−PAN and MW−HPAN. (**a₁**) LC−PAN × 500, (**a₂**) LC−PAN × 1.0 k, (**a₃**) LC−PAN × 5.0 k, (**b₁**) MW−HPAN × 500, (**b₂**) MW−HPAN × 1.0 k, and (**b₃**) MW−HPAN × 5.0 k.

2.1.4. FT-IR Characterization

Figure 4 shows the obvious changes in the infrared spectra of LC−PAN and MW−HPAN. The peaks at 2940 cm^{-1} and 2860 cm^{-1} on the spectra correspond to the asymmetric and symmetric stretching vibrations of the methylene group, respectively, and the peak at 1450 cm^{-1} is assigned to the metastable vibration of the methylene group [30]. A band at 2244 cm^{-1} is observed in the spectra of LC−PAN, which is the characteristic stretching vibration peak of −CN. The peak appears at 1736 cm^{-1}, and 1634 cm^{-1} is assigned to the stretching vibration of carbonyl and -C=C, which is derived from copolymer monomer, and the peaks at 1376 cm^{-1} correspond to the bending vibration of the methylene group [31]. Compared with LC−PAN, the broader peak at 3430 cm^{-1} in the MW−HPAN spectrum corresponds to the N-H stretching vibration of the amide group generated by the hydrolysis reaction, and the peak at 2244 cm^{-1} almost completely disappeared, indicating the consumption of the cyano group [32,33]. The peak at 1067 cm^{-1} is the stretching vibration peak of the amide group, and the peaks at 1573 cm^{-1} and 1410 cm^{-1} are typical carboxylate vibration peaks corresponding to antisymmetric and symmetric stretching vibrations, respectively. The widening and intensification of the peak at 1047 cm^{-1} is due to the generation of the carboxylic acid group [34,35]. The changes in the infrared spectra indicate that the alkaline hydrolysis process consumes the ester group and most of the cyano group on the molecular chain of polyacrylonitrile and generates an amide group and a sodium carboxylate group.

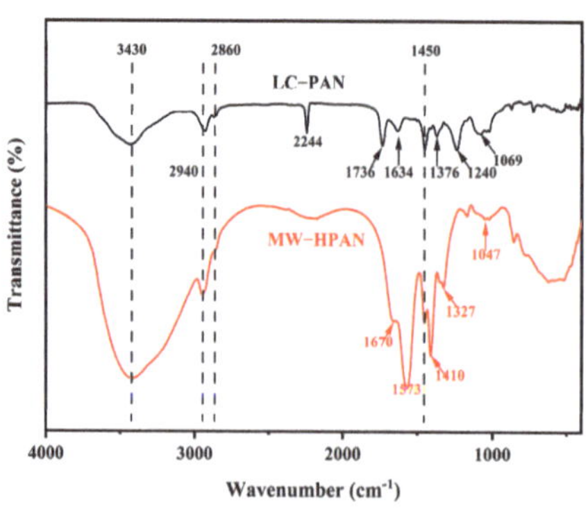

Figure 4. FT-IR spectrograms of LC−PAN and MW−HPAN.

2.1.5. Thermogravimetric Analysis

The thermal stability of LC−PAN and MW−HPAN was compared using thermogravimetric analysis. The TGA and DTG curves are displayed in Figure 5. The thermal decomposition process of both LC−PAN and MW−HPAN can be divided into three stages. The first mass loss stage of LC−PAN occurs before 268 °C and is attributed to the escape of HCN and NH_3, which is produced by the splitting of side and end groups not involved in the cyclization reaction between adjacent cyanide groups. This stage transforms PAN from a linear into a trapezoidal structure, losing only 0.81% of its mass [36]. The first stage of MW−HPAN occurs before 172 °C, resulting in a mass loss of about 5.69%, which corresponds to the evaporation of bound water. The second stage of PAN occurs at 268 °C~375 °C and is due to the decomposition of the cyclic structure, which releases a large amount of hydrogen and methane, reaching the maximum rate of thermal decomposition at 326 °C, and the mass loss of the substance is 33.33% [37,38]. The second stage of MW−HPAN occurs at 172 °C~400°C, during which the cyclization and thermal decomposition of anhydride take place. MW−HPAN forms a more stable trapezoidal structure than the polyacrylonitrile molecule due to the additional ionic mechanism, resulting in a mass loss of 13.81%, which is lower than polyacrylonitrile [29]. The third stage of mass loss of LC−PAN and MW−HPAN occurs at 375 °C~528 °C and 400 °C~535 °C respectively, corresponding to the further oxidative decomposition of the residue [39]. The residual carbon amount of MW−HPAN at 600 °C is higher than that of LC−PAN, and the thermal decomposition behavior of MW−HPAN indicates its resistance to high temperatures.

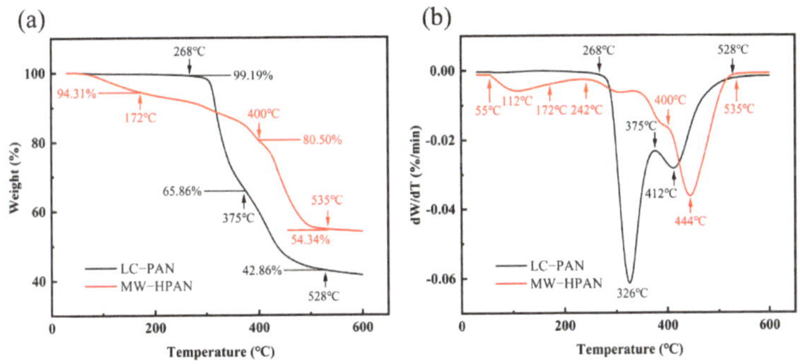

Figure 5. The thermal properties of LC−PAN and MW−HPAN. (**a**) Thermogravimetric (TG) analysis curves of LC−PAN and MW−HPAN (**b**) Differential thermogravimetric (DTG) curves of LC−PAN and MW−HPAN.

2.1.6. Particle Size Analysis

The particle size distribution of the sample was measured with a laser particle sizer. The particle size of a polymer affects its dissolution rate in the drilling fluid: the larger the particle size, the slower the dissolution rate. The particle size distribution of the sample is shown in Figure 6. Overall, 50% of the samples have a particle size of less than 107 μm, and 90% of the particle sizes are less than 230 μm, indicating that particle sizes are small and can dissolve quickly, which is consistent with the experimental phenomenon.

2.2. Optimization of the Hydrolysis Parameters

The experimental parameters were optimized by testing the filtration loss performance of the samples in freshwater and a 4% brine base slurry after being cured at room temperature for 16 h. The dosage of freshwater and 4% brine base slurry were 0.3% and 1.2%, respectively.

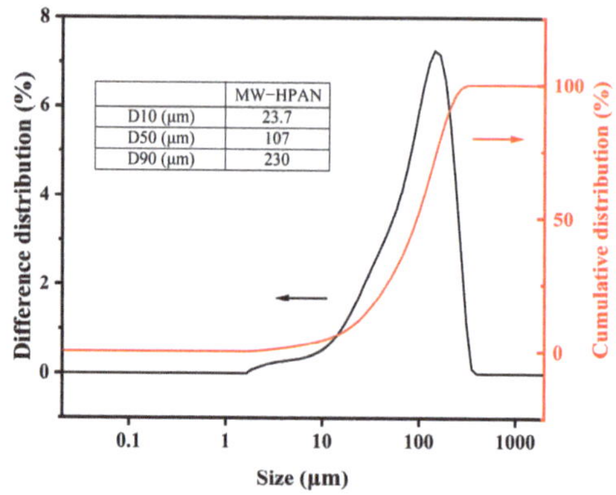

Figure 6. Particle size distribution of MW−HPAN.

2.2.1. Effect of Water Addition on the Filtrate Reduction Property

PAN was hydrolyzed with water addition dosages of 4 g, 8 g, 10 g, 12 g, 14 g, and 16 g, respectively, and the remaining experimental parameters remained unchanged. The results are shown in Figure 7a. The filter loss volume of the sample was reduced with the increase in water addition, indicating that the hydrolysis degree of the sample increased and the water solubility was improved, which suggests that the increase in water addition was beneficial to improving the filtration loss reduction performance of the sample. When the amount of water was 12 g, the filter loss volume of the sample in the fresh water and 4% brine base slurry was 14.7 mL and 9.6 mL, respectively, which could meet the actual requirements, and the sample could be pulverized without subsequent drying. When the water addition exceeded 12 g, the increase in water addition did not further improve the loss reduction performance of the sample, which would increase the cost of drying. Thus, 12 g was selected as the optimum water addition.

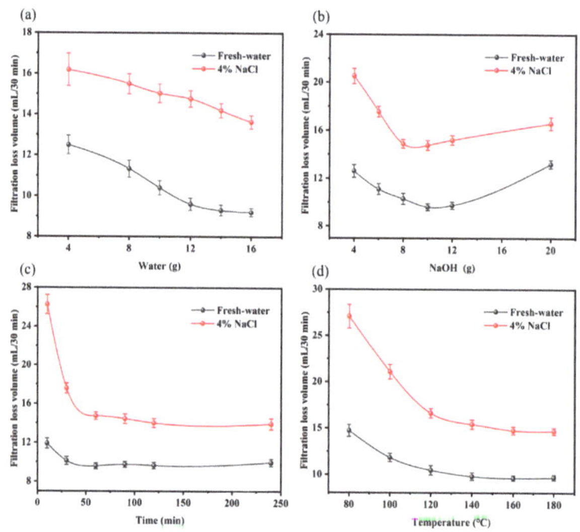

Figure 7. The room temperature filter loss volume of different experimental variable samples in the freshwater and 4% brine base slurry: (**a**) water addition, (**b**) NaOH addition, (**c**) reaction time, and (**d**) reaction temperature.

2.2.2. Effect of NaOH Addition on the Filtrate Reduction Property

PAN was hydrolyzed under the conditions of sodium hydroxide dosages of 4 g, 6 g, 8 g, 10 g, 12 g, and 20 g, respectively, and the remaining variables were kept constant. When the sodium hydroxide addition was less than 10 g, the filtration loss volume decreased with the increase in sodium hydroxide dosage, indicating the hydrolysis degree of PAN and the water solubility of the sample increased. When the amount of sodium hydroxide was higher than 10 g, excessive caustic soda caused the degradation of PAN and the decrease in molecular weight of the sample increased the filtration loss. Therefore, the optimum amount of sodium hydroxide was determined to be 10 g (Figure 7b).

2.2.3. Effect of Reaction Time on the Filtrate Reduction Property

PAN was hydrolyzed under the conditions of a reaction time of 10 min, 30 min, 60 min, 90 min, 120 min, and 240 min, respectively, and the other experimental parameters remained unchanged. As demonstrated in Figure 7c, when the reaction time was lower than 60 min, a significant reduction was shown in the filtration loss of the sample with the increase in the hydrolysis reaction time. This indicated that the hydrolysis degree of PAN increased, and the filtration loss reduction performance of the sample was improved. After the reaction time exceeded 60 min, the filtration loss of the drilling fluid gel system basically had no obvious change, which indicated that the increase in the reaction time at this time did not improve the hydrolysis degree of PAN, and the increase in the reaction time would increase the production cost. Therefore, the optimal reaction time was determined to be 60 min.

2.2.4. Effect of Temperature on the Filtrate Reduction Property

PAN was hydrolyzed under the conditions of reaction temperatures of 80 °C, 100 °C, 120 °C, 140 °C, 160 °C, and 180 °C, respectively, and the other parameters remained unchanged. As shown in Figure 7d, when the reaction temperature was lower than 100 °C, most of the water molecules in the system existed in liquid form, the reaction was not uniform, and the reaction speed was slow, leading to a low hydrolysis degree and a high filtration loss volume (higher residues of acrylic powder were observed in the product). As the temperature rose, the water molecules were transformed into a gaseous state, which was conducive to contact with acrylic powder and improved the uniformity of the reaction, thus increasing the reaction speed and hydrolysis degree and improving the filtration loss reduction performance of MW−HPAN. When the reaction temperature was higher than 160 °C, there was no significant difference in the properties of the products, and the optimum reaction temperature was determined to be 160 °C, according to the experimental results.

2.3. Evaluation of the Filtration Control Property of MW−HPAN

The samples were prepared under optimal experimental conditions in order to comprehensively evaluate the filtration loss performance and action mechanism, rheological properties, filtration reduction performance, particle size distribution, and morphology of the filter cake in the freshwater base slurry and the 4% NaCl brine base slurry before and after aging at 180 °C for 16 h were analyzed.

2.3.1. Rheological Properties of Drilling Fluid Gel Systems

A drilling fluid gel system is a kind of pseudoplastic fluid. The Bingham plastic model and the Herschel–Bulkely model were used to simulate and evaluate the multivariate fluid properties of drilling fluid gel systems scientifically [40].

Bingham plastic model:

$$\tau = \tau_0 + \mu_p \gamma \quad (1)$$

Herschel–Bulkely model:

$$\tau = \tau_y + K\gamma^n \quad (2)$$

where:
τ is the shear stress, γ is the shear rate, τ_0 is the yield point, μ_p is the plastic viscosity, τ_y is the yield point under this model, K is the flow consistency coefficient, and n is the fluid behavior index.

In general, a drilling fluid gel system is required to have a higher viscosity at a lower shear rate to carry or suspend the cuttings in the borehole, and a lower viscosity at a higher shear rate to pump quickly to the bottom of the borehole and release the cuttings easily [41]. First, the rheological properties of the drilling fluid gel system with the freshwater base slurry were studied at different aging temperatures. The fitting equations and curves of different rheological models are shown in Table 2 and Figure 8. It was found that the rheological curve of the drilling fluid gel system was more consistent with the Herschel–Barclay model. The linear correlation coefficients R^2 of the aging temperature at 25 °C, 120 °C, 150 °C, and 180 °C were 0.9986, 0.9949, 0.9980, and 0.9992, respectively.

Table 2. Rheological fitting equation of the drilling fluid at different aging temperatures.

Temperature °C	Flow Pattern			
	Bingham		Herschel–Bulkely	
	Equation	R^2	Equation	R^2
25	$\tau = 2.044 + 0.0100 \times \gamma$	0.9430	$\tau = 0.511 + 0.0392 \times \gamma^{0.8231}$	0.9986
120	$\tau = 2.044 + 0.0090 \times \gamma$	0.9247	$\tau = 0.511 + 0.0399 \times \gamma^{0.8074}$	0.9949
150	$\tau = 0.511 + 0.0050 \times \gamma$	0.9677	$\tau = 0 + 0.0131 \times \gamma^{0.8745}$	0.9980
180	$\tau = 0.511 + 0.0040 \times \gamma$	0.9595	$\tau = 0 + 0.0129 \times \gamma^{0.8480}$	0.9992

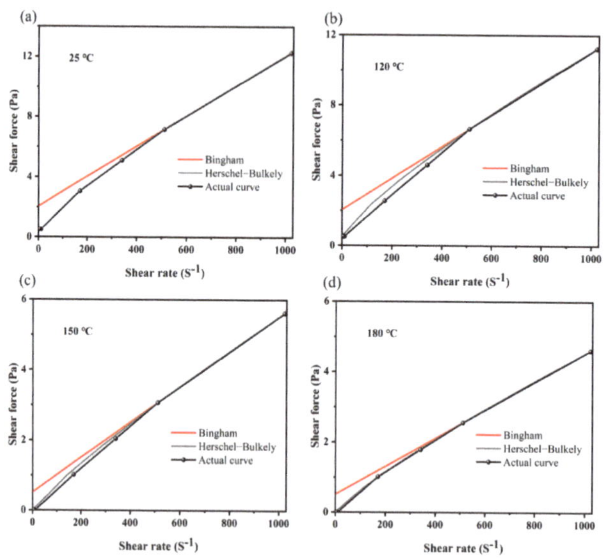

Figure 8. Rheological curve fitting for the drilling fluid gel system after aging at different temperatures: (**a**) 25 °C, (**b**) 120 °C, (**c**) 150 °C, and (**d**) 180 °C.

In addition, the effects of MW−HPAN on the rheological parameters of the drilling fluid gel system at different aging temperatures, such as AV, PV, YP, and RYP, were investigated. Temperature has a very important influence on the stability of a drilling fluid gel system, which is reflected in the interaction between clay particles and the polymer additive. First, the clay particles will disperse, agglomerate, and passivate at high temperatures [42,43]. Second, a high temperature will lead to degradation, a change in the side chain functional groups, or high-temperature cross-linking of the polymer [44]. Furthermore, the high-temperature effect will weaken the adsorption capacity of the polymer treatment agent on the clay surface, which is called high-temperature desorption [45]. At the same time, a high temperature will also reduce the hydration ability of the treatment agent on the clay surface, reduce the thickness of the hydration film, and weaken the adhesion protection performance, which is called high-temperature dehydration [46,47].

As shown in Figure 9, the AV and PV of the drilling fluid gel system increased the overall after aging at different temperatures and decreased slightly at 150 °C without the sample. The YP first increased to a maximum value at 120 °C, gradually decreased and reached a minimum value at 150 °C, and then increased with temperature. The RYP decreased overall. These phenomena were related to the dynamic equilibrium of high-temperature dispersion, agglomeration, and passivation of clay particles [48].

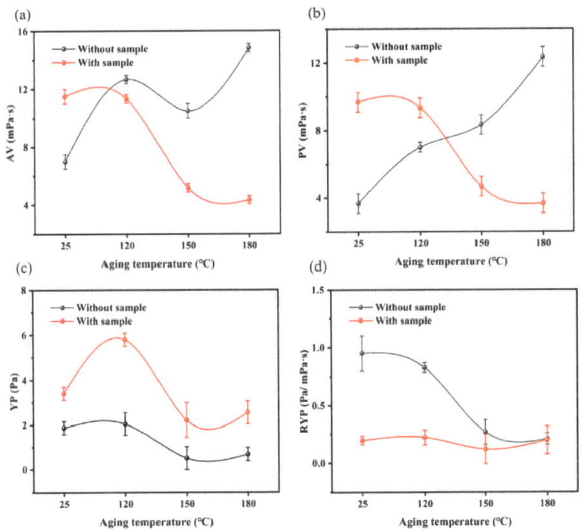

Figure 9. The influence of samples on rheological parameters of drilling fluid gel system after aging at different temperatures: (**a**) apparent viscosity, (**b**) plastic viscosity, (**c**) yield point, and (**d**) the ratio of yield point to plastic viscosity.

The AV of the drilling fluid gel system at room temperature was increased with the sample, which was because the MW−HPAN molecules in the drilling fluid gel system were in an irregular nematic conformation, the friction within the macromolecule was greater, and, at the same time, it adsorbed water molecules and increased the internal friction between polymers and water molecules. In addition, the cyano and amide functional groups of the MW−HPAN molecular side chain were adsorbed on the clay surface by hydrogen bonding and formed a spatial network structure. Under the combined effect of these factors, the AV, PV, YP, and RYP of the drilling fluid gel system increased. As the aging temperature increased, the AV and PV of the drilling fluid gel system gradually decreased. This was because the high temperature broke the MW−HPAN molecular chain. In addition, the cyano and amide functional groups underwent hydrolysis at high temperatures, even leading to the breakage of the functional groups and the main chain [49]. The adsorption and hydration capacity of MW−HPAN on clay surfaces weakened with the temperature, resulting in a reduction in the thickness of the hydration film on clay particles. These two effects negatively impact the adhesive protection ability of MW−HPAN [44,50]. The RYP of the drilling fluid gel system remained within a certain range with increasing aging temperature, indicating a certain level of high-temperature resistance of MW−HPAN. It was found that the drilling fluid gel system with MW−HPAN had good rheological properties.

2.3.2. Filtration Performance of the Drilling Fluid Gel System

In the drilling process, the key to controlling the filtration performance of the drilling fluid gel system is to improve the quality of the mud cake, which involves an enhancement in the compactness, shear resistance, and lubricity of the mud cake [51]. Bentonite, a commonly used slurry material for water-based drilling fluid gel systems, is beneficial for forming dense mud cakes that prevent water from penetrating into the formation. However, the instability and aggregation of bentonite can occur in high temperature and high salt content environments, leading to increased porosity and permeability of the filter cake and an increasing filtration loss [52].

The filtration performance of the drilling fluid gel system with MW−HPAN was measured under different aging temperatures and salt concentrations. First, the filtration performance of freshwater base slurry before and after aging at 120 °C, 150 °C, and 180 °C was evaluated, and the sample dosage was 0.3%. As shown in Figure 10a, the filtration loss of the freshwater base slurry without samples before aging is 24.9 mL/30 min, while after aging at 120 °C, 150 °C and 180 °C, the filtration loss increases to 25.2 mL, 27.1 mL, and 28.3 mL, respectively. The increase in filtration loss means that aging will affect the filtration performance of the drilling fluid gel system. However, the final filter loss with samples before and after aging is significantly reduced. When the sample addition is 0.3%, the filtration loss before and after aging at 120 °C, 150 °C, and 180 °C is reduced to 9.6 mL, 8.8 mL, 11.6 mL, and 12.2 mL, respectively, which meets the API recommended requirements. It can be observed that filter loss increases slightly with the temperature, which means the increasing aging temperature does not affect the filtration reduction performance of the sample.

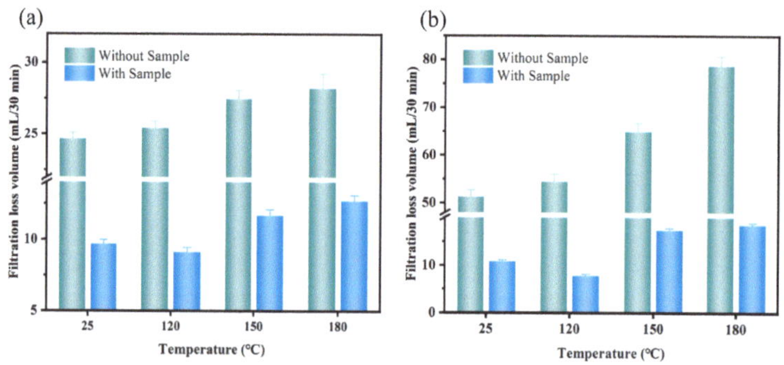

Figure 10. The filtration loss of the gel after aging at different temperatures. (**a**) Freshwater base slurry: the sample dosage was 0.3%; and (**b**) 4%NaCl brine base slurry: the sample dosage was 1.2%.

Second, in order to evaluate the salt resistance of the sample, the filtration performance of the 4% NaCl brine base slurry before and after aging at 25 °C, 120 °C, 150 °C and 180 °C was tested, and the sample dosage was 1.2%. As shown in Figure 10b, the filtration loss volume of 4% NaCl brine base slurry before aging was 51.0 mL, and it increased continuously with the increase in aging temperature. The filtration loss increased to 75.6 mL after aging at 180 °C, which was much higher than that of the freshwater base slurry, indicating that the filtration performance of the drilling fluid gel system became worse after the synergistic damage effect of high temperature and inorganic salt pollution. However, after adding 1.2% of the sample as a filtrate reducer to the 4% NaCl brine base slurry, the drilling fluid gel system can maintain a low filtration loss before and after hot rolling. After aging at 180 °C, the filtration loss is 17.8 mL, and the viscosity of the drilling fluid gel system is not high. It can be seen that viscosity may not be the main factor affecting the filtration performance of the drilling fluid gel system after aging. There are many active functional groups on the MW−HPAN molecular chain, of which the adsorption groups (cyano and amide groups) can be firmly adsorbed on the surface of clay particles by the formation of hydrogen bonds with the oxygen atoms on bentonite, and the presence of carboxyl groups improves the salt resistance of sample [53]. Furthermore, the rigidity of the C-C and cyano group in the MW−HPAN molecule enhances its high-temperature resistance. These factors are responsible for the good temperature and salt resistance performance of MW−HPAN.

2.3.3. Microstructural Analysis of the Drilling Fluid Gel System

The DLVO theory is the classical colloidal stability theory that explains the stability of charged colloidal particles in the liquid phase, showing that the stability of colloidal particles is determined by the combined effect of van der Waals' gravitational and electrostatic repulsive forces between the particles. The diffusion double-layer of colloidal particles is divided into an adsorption layer, formed by the attraction between the colloidal particles and the hetero ions, and a diffusion layer, formed by the thermal movement of the hetero ions. According to the DLVO colloidal stability theory, high temperature and cationic contamination will compress the diffuse double layer of negatively charged clay in a drilling fluid gel system, destroy the rheological and filtration properties and reducing the stability of the drilling fluid gel system, which easily produce a thick and loose filter cake that leads to an increase in filtration loss. Filtrate reducer can prevent the negative effect by forming thin and dense filter cake, and thus achieve the effect of reducing filtration loss [52,54].

The thickness of the fresh filter cake was measured, and the microstructure of the filter cake after freeze-drying was observed using SEM to evaluate the effect of MW−HPAN on the quality of the filter cake. The thickness of the filter cake from the freshwater base slurry without the sample before aging was 1.78 mm (Figure 11a_1); under the joint destruction of cation and high temperature, the structure of the filter cake became loose, and the thickness of the filter cake from the 4% NaCl brine base slurry was 7.72 mm (Figure 11c_1), indicating that the stability of the suspension system decreased and the filter loss increased. However, the thickness of the filter cake with the sample was significantly reduced before and after aging. The thickness of the filter cake of the 4% NaCl brine base slurry was reduced from 7.72 mm to 1.60 mm (Figure 11d_1), indicating that MW−HPAN could hinder the destruction effect of the electrolyte ion and high temperature on the filter cake.

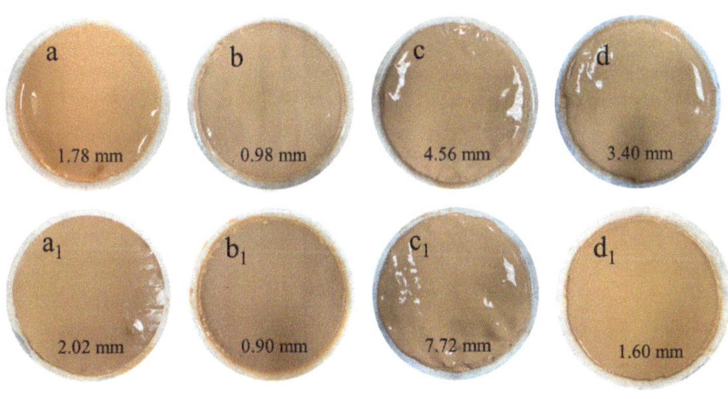

Figure 11. Images of the filter cake with and without the sample after aging (at 180 °C for 16 h) and before aging: (**a**) without the sample before aging (freshwater base slurry), (**b**) with the sample before aging (freshwater base slurry), (**c**) without the sample before aging (4% NaCl base slurry), (**d**) with the sample before aging (4% NaCl base slurry), (**a_1**) without the sample after aging (freshwater base slurry), (**b_1**) with the sample after aging (freshwater base slurry), (**c_1**) without the sample after aging (4% NaCl base slurry), and (**d_1**) with the sample after aging (4% NaCl base slurry).

The SEM images were consistent with the above findings (Figure 12). The agglomeration of clay appeared on the surface of the filter cake obtained from the base slurry, and a large number of wrinkles, cracks, and pores were formed after aging (Figure 12a$_1$). These phenomena were more significant with the effect of cations (Figure 12c$_1$). The increase in the permeability of the filter cake led to a rapid increase in filter loss. However, the surface of the filter cake after adding the sample to the base slurry was smooth and dense, and it looked like the polymer film covered it (Figure 12b$_1$). Under the destructive effect of high temperature and cation, MW−HPAN could reduce the agglomeration of clay, prevent the formation of wrinkles and cracks, reduce the permeability of filter cake, and obviously improve the filtration performance of the drilling fluid gel system. The law of variation in filter cake quality was consistent with the variation in filter loss in Figure 10.

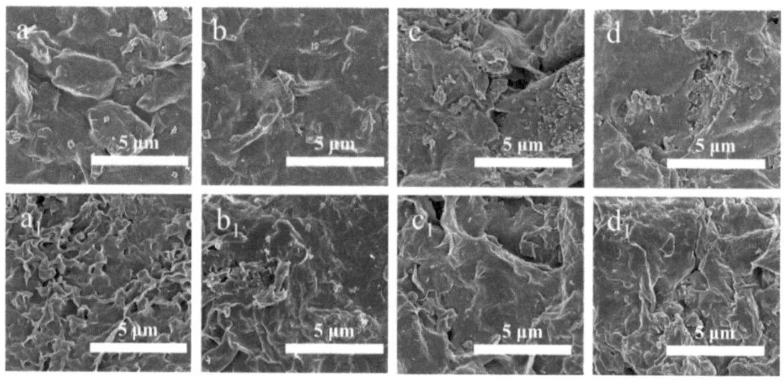

Figure 12. SEM images of the filter cake with and without the sample after aging (at 180 °C for 16 h) and before aging: (**a**) without the sample before aging (freshwater base slurry), (**b**) with the sample before aging (freshwater base slurry), (**c**) without the sample before aging (4% NaCl base slurry), (**d**) with the sample before aging (4% NaCl base slurry), (**a$_1$**) without the sample after aging (freshwater base slurry), (**b$_1$**) with the sample after aging (freshwater base slurry), (**c$_1$**) without the sample after aging (4% NaCl base slurry), and (**d$_1$**) with the sample after aging (4% NaCl base slurry). (×10.0 k magnification).

2.3.4. Particle Size Analysis of the Drilling Fluid Gel System

A reasonable size distribution of clay particles in a drilling fluid gel system is required to form a low-permeability mud cake. Larger clay particles accumulate with each other to form a relatively strong bridging structure, while small-sized clay particles enter into the cracks and pores, acting as a sealer in the structure of the filter cake, which is conducive to the formation of a stable and dense filter cake and reduces the penetration of water [55,56].

As shown in Figure 13, the particle size distribution of the freshwater base slurry without the sample before aging was relatively reasonable: about 53.67% of the clay particles were smaller than 10 µm, and the average particle size of the system was about 8.71 µm. The particle size distribution of the freshwater base slurry was narrower after aging at 180 °C, and the average particle size of the system increased to 10.64 µm, indicating the aggregation of clay particles at high temperatures. However, the drilling fluid gel system still maintained a suitable particle size distribution after aging with the addition of MW−HPAN. Specifically, 60.02% of the particle sizes in the system were less than 10 µm, and the average particle size was reduced to 6.90 µm, which was due to the firm adsorption between the sample and the clay surface through hydrogen bonding, preventing the clay particles from flocculating and becoming larger. In addition, MW−HPAN stabilized the fine particles that were not bound into large particles by adsorption, which increased the proportion of fine particles and was conducive to the formation of a dense filter cake, called the protective effect of colloid. The experimental results showed that the high temperature could hardly destroy the colloid stability (colloid stability refers to the invariability in the viscosity, particle size, and other properties of the colloidal solution system to a certain extent) of the drilling fluid gel system with MW−HPAN, and the results were consistent with the observations of the filter cake (Figures 11 and 12).

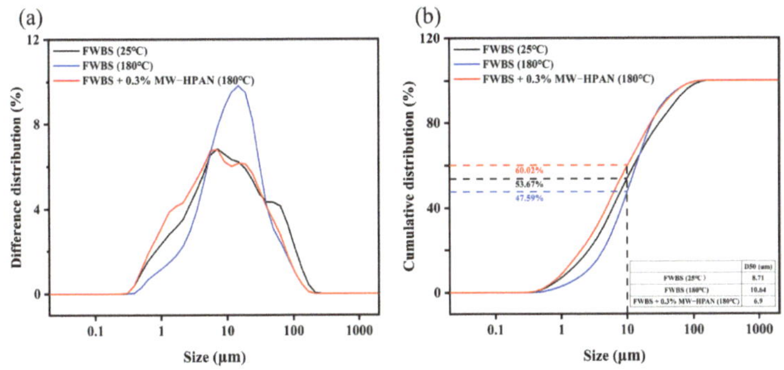

Figure 13. Particle size distribution of freshwater base slurry (FWBS) before and after aging. (**a**) The difference distribution and (**b**) the cumulative distribution.

2.4. Comparison with Commercially Available Products

The obtained sample, MW−HPAN, is essentially a hydrolyzed polyacrylonitrile sodium salt, and we denoted it as MW−HPAN since it was prepared using a novel micro-water method. Its properties were compared with hydrolyzed polyacrylonitrile sodium salt products synthesized with the conventional process. The filtration loss performance of MW−HPAN and commercially available hydrolyzed polyacrylonitrile sodium salt (ST-NaPAN was selected as a representative product for comparison) in the freshwater- and 4% brine-based slurries before and after aging at different temperatures was evaluated. As shown in Figure 14a, the filtration loss of ST–NaPAN was slightly lower than that of MW–HPAN when the aging temperature was lower than 120 °C in the freshwater base slurry, but the filtration loss reduction effect of MW−HPAN was obviously better than ST–NaPAN when the aging temperature was higher than 120 °C. This variation trend was more significant in the brine base slurry (Figure 14b): the filtration loss volume of MW–HPAN was 10.8 mL before aging, which was higher than ST–NaPAN, but it still met the requirements of API recommendations. The filtration loss reduction performance of ST–NaPAN decreased drastically even after aging at a lower temperature (120 °C). However, the filtration loss value of MW–HPAN first decreased and then increased with an increase in the aging temperature, and the filtration loss reduction performance could still maintain a better level after aging at 180 °C.

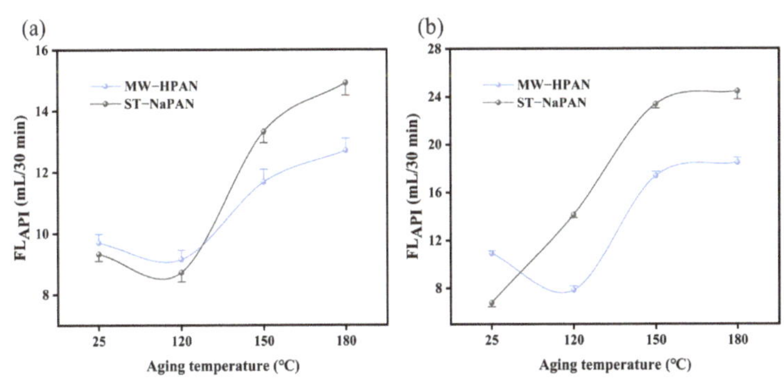

Figure 14. Filtration performance of MW–HPAN and ST–NaPAN in drilling fluid gel systems. (**a**) The freshwater base slurry and (**b**) the 4% NaCl brine base slurry.

These results indicate that MW–HPAN has a significantly better anti-temperature and anti-salt performance than ST–NaPAN as a filtrate reducer for drilling fluid gel systems. This is due to the difference between the two hydrolysis reaction processes. First, sufficient water is required in the traditional hydrolysis process as a solvent. Due to the agglomeration of acrylic fiber and the heterogeneity of the reaction, the hydrolysis process takes a long time, during which PAN undergoes more severe degradation in this highly alkaline environment. Second, the drying time is long due to the high water retention of the semi-finished product, during which further hydrolysis will occur. Therefore, a low molecular weight and high hydrolysis degree of the hydrolyzed polyacrylonitrile product (HPAN) will be obtained, and the reduction in molecular weight will greatly reduce the temperature and salt resistance of HPAN. Furthermore, the stability of the performance of the samples will vary from batch to batch due to the difference in the purity of the waste fibers. In this study, the hydrolysis of polyacrylonitrile with hydroplastization followed by the micro-water hydrolysis method could solve the above problems. Hydroplastization can separate and purify low-purity waste polyacrylonitrile fibers into a high-purity polyacrylonitrile powder, and the HPAN with a moderate hydrolysis degree can be obtained in a short time using the micro-water "vapor-solid" reaction at high temperature. There is no degradation of molecular weight in this novel process, and the drying cost is low. The performance of the HPAN can be significantly improved with this new hydrolysis process.

2.5. Mechanism of Filtration Loss Reduction

Clay particles can form stable ionic and hydrogen bonds with the hydration group and adsorption group of the polymer, respectively. These two effects contribute to the polymer being firmly adsorbed on clay particles and the formation of a stable spatial lattice structure, which is conducive to the formation of a dense mud cake and reducing the filtration loss volume [57]. As shown in Figure 15, the filter loss-reducing agent MW−HPAN obtained with the micro-water hydrolysis method contains carboxyl, amide, and unhydrolyzed cyano groups on the molecular chain, in which the adsorption groups (cyano and amide) can form a hydrogen bond with the oxygen atoms on bentonite, so that the MW–HPAN is firmly adsorbed on the surface of the clay, generating a good colloid protective effect on the suspension. A reasonable particle distribution of the drilling fluid gel system after aging at high temperatures can be obtained with MW−HPAN. A large number of small particles are more conducive to filling the cracks and pores, forming a dense filter cake and reducing the permeability of the filter cake. Furthermore, the molecular chains of MW–HPAN can be stretched in the drilling fluid gel system with the presence of a hydration group (carboxyl group) and adsorb free water into bound water, forming a hydration layer around the molecular chain. This large and thick adsorption of the hydration film can still be effective in preventing the flocculation of clay particles even after aging at high temperatures, achieving the effect of reducing filtration loss.

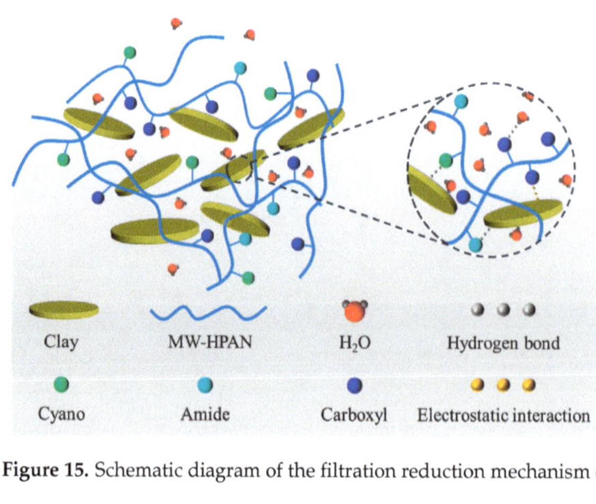

Figure 15. Schematic diagram of the filtration reduction mechanism of MW–HPAN [57].

3. Conclusions

The experimental results indicate that the reaction of hydrolysis of polyacrylonitrile using the micro-water method was successful. MW–HPAN obtained with the reaction at m_{PAN}:m_{NaOH}:m_{H2O} = 1:0.5:0.6 at 160 °C for 1 h can be used as a filtrate reducer for drilling fluid gel systems with excellent performance. The hydrolysis of polyacrylonitrile is a process in which hydrophobic cyano groups react under alkaline conditions to form hydrophilic amide and carboxyl groups. The addition of a small number of water molecules makes the microenvironment of the hydrolysis reaction system different from that of the liquid phase reaction, which results in a localized high concentration at the reaction site and greatly improves the efficiency of the reaction. The results of the elemental analysis showed that its hydrolysis degree could reach 73.21% in a short time, which was unattainable in traditional processes under the same conditions. MW–HPAN showed better temperature and salt resistance properties than commercially available HPAN products in the water-based drilling fluid gel system. The method of preparing MW–HPAN using micro-water hydrolysis has obvious economic benefits and a broad market prospect.

4. Materials and Methods

4.1. Materials

Waste acrylic fibers (industrial grade) used in the experiment were purchased from a textile factory in Luancheng City (Shijiazhuang, China), which consisted of polyacrylonitrile and other blended fibers (denoted as LC–PAN). Sodium hydroxide, sulphonic acid, and anhydrous sodium carbonate, used without any further purification, were analytical grade and purchased from Sinopharm Chemical Reagents Co., Ltd. (Beijing, China). Industrial grade hydrolyzed polyacrylonitrile sodium salt (Na-HPAN, Na-HPAN is an abbreviation for the product of alkaline hydrolysis of polyacrylonitrile) was obtained from Baoding Santuo Chemical Products Co., Ltd. (Baoding, China). Calcium bentonite for the base slurry was purchased from Drilling & Exploration Engineering Co., Ltd. (Tianjin, China). The England evaluation clay for the drilling fluid gel system test was purchased from the Beijing Institute of Exploration Engineering (Beijing, China).

4.2. Synthesis of Na-HPAN with the Micro-Water Method

4.2.1. Hydroplastization of Waste Acrylic Fibers

After opening, the waste acrylic fibers (40 g) and water (40 g) were added to a digestion tank and then placed in a hot-roller furnace. The hydroplastization reaction was carried out at 180 °C for 2 h. During the reaction, the structure of the acrylic fibers was destroyed, while the structure of the non-acrylic blended fibers was unchanged. The dry plasticized acrylic fibers were crushed with a grinder and separated from the other blended fibers with a 60-mesh screen to obtain a high-purity polyacrylonitrile powder (Figure 16, Phase I).

4.2.2. Alkaline Hydrolysis of the Polyacrylonitrile Powder with the Micro-Water Method

The obtained polyacrylonitrile powder (20 g) and sodium hydroxide (10 g) were mixed firmly in a grinder and poured into a reaction kettle with an outlet valve stem, to which water (12 g) was added in two batches at the same time (Figure 16, Phase II). After sealing, the kettle was placed in the hot-roller furnace for hydrolysis at 160 °C for 1 h, and then the kettle was removed. The ammonia gas and water vapor in the kettle were immediately absorbed with a dilute sulphonic acid solution through the valve stem. The sample was dried with residual heat, and the hydrolyzed polyacrylonitrile sodium salt was obtained using the micro-water method (MW–HPAN) after crushing (Figure 16, Phase III). The product was washed three times alternately with water and anhydrous ethanol and dried at 70 °C for 24 h. The dried product was ground to powder for subsequent characterization.

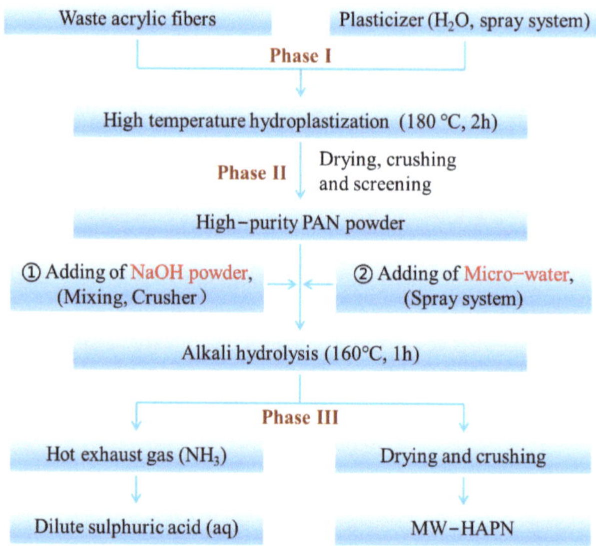

Figure 16. Schematic diagram of polyacrylonitrile hydrolysis using the micro-water method.

4.3. Structure Characterization Techniques

4.3.1. Elemental Analysis (EA)

The organic elemental content of the samples before and after hydrolysis was determined using a UNICUBE organic elemental analyzer (Elementar Trading Co., Ltd.; Shanghai, China). The limit of detection (LOD) was 50 ppm with an accuracy of 0.01%.

4.3.2. X-ray Diffraction (XRD)

The crystal structures of the samples were analyzed using a Bruker D8 Advance X-ray diffractometer (Hongkong, China) equipped with a Cu-Kα radiation source (λ = 0.154 nm), and the diffraction patterns from 0° to 60° were collected at a scan rate of 2°/min.

4.3.3. Scanning Electron Microscope (SEM)

The dry samples were mounted on an aluminum holder and an ion sputtering device (Model E-1010, Hitachi; Tokyo, Japan) was used for gold sputter coating to make it conductive. Then, the morphologies of the samples were characterized using a Hitachi SU8020 scanning electron microscope (Tokyo, Japan) operating at an accelerating voltage of 3 kV.

4.3.4. Fourier Transform Infrared Spectra (FT-IR)

The sample testing tablets were carried out using the KBr compression method and characterized with a Nicolet IS10 FT-IR spectrometer (Madison, WI, USA) in the range of 4000 cm^{-1} to 500 cm^{-1} with the resolution of 4 cm^{-1} and the signal-noise ratio was 50,000:1. All spectrums were obtained by accumulating 64 scans.

4.3.5. Thermogravimetric (TG)

The thermal decomposition behaviors of the samples were investigated using a Netzsch STA449F3 thermogravimetric analyzer (Selb, Germany) under a nitrogen atmosphere and a nitrogen flow rate of 50 mL/min. The samples were placed in a clean crucible and heated from 30 °C to 600 °C at a heating rate of 10 °C/min. The derivative thermogravimetric analysis (DTG) curves were obtained by applying the first-order derivative to the temperature with the TGA data.

4.3.6. Particle Size Analysis

The particle size distribution of the sample was measured using a Mastersizer 2000 laser particle sizer (Malvern Instruments Co., Ltd., Shanghai, China). Before each test, the samples were ultrasonically dispersed in ethanol solution for 5 min, and three parallel experimental tests were carried out. The shading rate was set at 14%, and the measuring range of the instrument was 0.02–2000 μm.

4.4. Comprehensive Performance Evaluation of the Water-Based Drilling Fluid Gel System

The performance of water-based drilling fluid gel systems, including filtration properties, temperature, salt resistance, and rheological properties were evaluated according to the American Petroleum Institute Recommended Practices (API RP 13B-1-2019). All performance evaluation experiments were carried out at least three times, and the mean and standard deviation of the parallel experiments were calculated. The apparatus used for the test is shown in Figure 17.

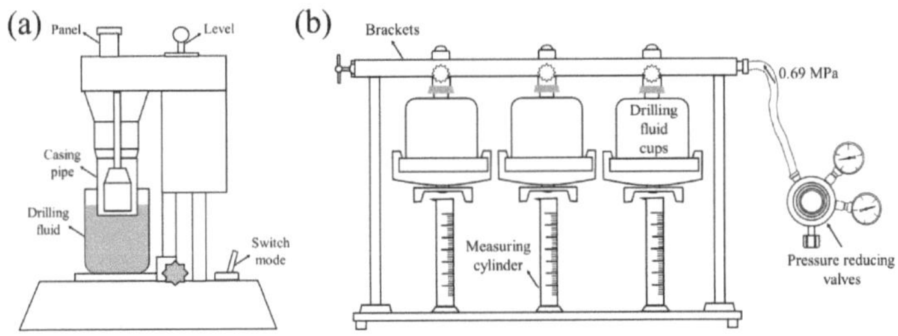

Figure 17. Schematic diagram of drilling fluid gel system performance testing instruments. (**a**) Six-speed rotational viscometer and (**b**) medium-pressure filter loss meter.

4.4.1. Preparation of the Freshwater Base Slurry

In total, 0.525 g of anhydrous sodium carbonate and 15.0 g of bentonite were added to a high stirring cup containing 350 mL of deionized water and stirred at high speed for 20 min, with at least two stops during this time to scrape off clay adhering to the wall of the cup. The mixture was then cured at room temperature for 24 h as a freshwater base slurry.

4.4.2. Preparation of the 4% NaCl Brine Base Slurry

In total, 4 g of sodium chloride, 2.1 g of anhydrous sodium carbonate, 60.0 g of drilling fluid gel system slurry with bentonite were added to a high stirring cup containing 350 mL of deionized water and stirred at high speed for 20 min, with at least two pauses to scrape off the clay attached to the wall of the cup. Then, the above dispersion was cured at room temperature for 24 h as a 4% NaCl brine base slurry.

4.4.3. Measurement of API Filtration Performance

The API filtration loss of the prepared drilling fluid gel systems was determined using a medium-pressure filter loss meter (SD3, Qingdao Tongchun Petroleum Instrument Co., Ltd., Qingdao, China) and standard Fann filter paper. A certain amount of the sample was added to the well-cured 350 mL base slurry (freshwater or 4% brine) and stirred at high speed for 20 min. Then, the mixture was poured into an airtight container and conditioned at room temperature. After 16 h, the conditioned drilling fluid gel systems were stirred at high speed for 5 min and poured into a filter loss cup until the level of the fluid was tangent to the graduated line. A pressure of 100 psi (0.69 MPa) was applied with a nitrogen cylinder. The 30-minute filtrate volume was collected to indicate the API filtration loss performance of the drilling fluid gel system. In addition, the filter cake was carefully removed from the loss of the filtration cup and slowly washed off the flowing mud on its surface with water, and its thickness was then determined using a vernier caliper.

4.4.4. Determination of Rheological Properties of Drilling Fluid Gel Systems

A certain amount of the sample was added to the well-conditioned 350 mL base slurry (freshwater or 4% brine) and stirred at high speed for 20 min before being conditioned at room temperature for 16 h. After 16 h, the sample was stirred at high speed for 5 min and then poured into the drilling fluid gel system beaker of a Fanns model 351 six-speed rotational viscometer (ZNN-D6, Qingdao Haitongda Special Instrument Co., Ltd., Qingdao, China) until the surface of the fluid was tangent to the scale of the viscometer, and the stable readings at different rotational speeds (600 rpm, 300 rpm, 200 rpm, 100 rpm, 6 rpm, 3 rpm) were then recorded. The apparent viscosity (AV), plastic viscosity (PV), yield point (YP), and the ratio of yield point to plastic viscosity (RYP) of the drilling fluid gel system were calculated using the following equations.

$$\mathrm{AV} = \frac{1}{2}\Phi 600 \tag{3}$$

$$\mathrm{PV} = \Phi 600 - \Phi 300 \tag{4}$$

$$\mathrm{YP} = \frac{1}{2}\Phi 300 - \mathrm{PV} \tag{5}$$

$$\mathrm{RYP} = \mathrm{YP}/\mathrm{PV} \tag{6}$$

where:

AV is the apparent viscosity (mPa·s); PV is the plastic viscosity (mPa·s); YP is the yield point (Pa); RYP is the ratio of the yield point to the plastic viscosity; Φ600 is the dial reading of the 6-speed rotational viscometer at 600 r/min (dia); and Φ300 is the dial reading of the 6-speed rotational viscometer at 300 r/min (dia);

4.4.5. Measurement of the Temperature Resistance of Drilling Fluid Gel Systems

A certain amount of the sample was added to the well-maintained 350 mL base slurry (freshwater or 4% brine) and stirred at high speed for 20 min. Then, the mixture was poured into a high-temperature aging tank, sealed, and placed in a high-temperature aging oven (XGRL-5, Qingdao Haitongda Special Instrument Co., Ltd., Qingdao, China), and aging was carried out at different temperatures for 16 h. After 16 h, the drilling fluid gel systems were allowed to cool to room temperature, high-speed stirring was carried out for 5 min, and then the filtration loss properties and rheological properties were determined.

4.4.6. Particle Size Analysis of Drilling Fluid Gel Systems

The particle size distribution of bentonite particles in the drilling fluid gel system before and after aging was measured using a Bettersize 2000 laser particle sizer (Dandong Bettersize Instruments Co., Ltd., Dandong, China). Water was chosen as the dispersion medium, sodium hexametaphosphate was used as the dispersant, the shading rate was set at 12%, and the measuring range of the instrument was 0.02–2000 μm.

4.4.7. Morphologies of Drilling Fluid Gel System Mud Cakes

The mud cake obtained from the API filtration loss test was immersed in liquid nitrogen and frozen for 2–3 min. Then, it was removed and rapidly transferred to a vacuum freeze dryer (ZLGJ-18, Zhengzhou Huachen Instrument Co., Ltd., Zhengzhou, China) and vacuum lyophilized at −30 °C for 48 h. The surface of the mud cake was then gold plated, and the morphology was characterized using a Hitachi SU8020 scanning electron microscope at an accelerating voltage of 3 kV.

Author Contributions: Data curation, W.L., S.L., K.Y., Y.Z., S.Y. and H.Q.; formal analysis, Z.W., S.L. and Y.Z.; funding acquisition, F.Z.; investigation, K.Y.; methodology, W.L.; project administration, F.Z.; software, W.L. and Z.W.; supervision, F.Z.; visualization, W.L.; writing—original draft, W.L., Z.W., S.L., K.Y., S.Y. and H.Q.; writing—review and editing, W.L. All authors have read and agreed to the published version of the manuscript.

Funding: This research was funded by "Science and Technology Achievement Transformation Guidance Fund of China University of Geosciences Beijing"; Funding number: 4-2-2022-002.

Institutional Review Board Statement: Not applicable.

Informed Consent Statement: Not applicable.

Data Availability Statement: All data and materials are available on request from the corresponding author. The data are not publicly available due to ongoing researches using a part of the data.

Conflicts of Interest: We declare that this is original research that has not been published previously and has not been under consideration for publishing elsewhere. There were no conflict of interest in the submission of this manuscript, and this manuscript was approved by all authors for publication.

References

1. Shirvanimoghaddam, K.; Motamed, B.; Ramakrishna, S.; Naebe, M. Death by waste: Fashion and textile circular economy case. *Sci. Total Environ.* **2020**, *718*, 137317. [CrossRef]
2. Pensupa, N.; Leu, S.; Hu, Y.; Du, C.; Liu, H.; Jing, H.; Wang, H.; Lin, C. Recent Trends in Sustainable Textile Waste Recycling Methods: Current Situation and Future Prospects. *Top. Curr. Chem.* **2017**, *375*, 76. [CrossRef] [PubMed]
3. Yousef, S.; Tatariants, M.; Tichonovas, M.; Sarwar, Z.; Jonuškienė, I.; Kliucininkas, L. A new strategy for using textile waste as a sustainable source of recovered cotton. *Resour. Conserv. Recycl.* **2019**, *145*, 359–369. [CrossRef]
4. Hu, Y.; Du, C.; Leu, S.; Jing, H.; Li, X.; Lin, C. Valorisation of textile waste by fungal solid state fermentation: An example of circular waste-based biorefinery. *Resour. Conserv. Recycl.* **2018**, *129*, 27–35. [CrossRef]
5. Qu, X.; Li, Z.; Xie, X.; Sui, Y.; Yang, L.; Chen, Y. Survey of composition and generation rate of household wastes in Beijing, China. *Waste Manag.* **2009**, *29*, 2618–2624. [CrossRef] [PubMed]
6. Barbero-Barrera, M.D.M.; Pombo, O.; Navacerrada, M.d. Textile fibre waste bindered with natural hydraulic lime. *Compos. Part B* **2016**, *94*, 26–33. [CrossRef]
7. Guo, J.; Xiang, H.; Gong, X.; Zhang, Y. Preparation and Performance of the Hydrolyzate of Waste Polyacrylonitrile Fiber/Poly (Ethylene Glycol) Graft Copolymerization. *Energy Sour. Part A* **2011**, *33*, 1067–1075. [CrossRef]
8. Bajaj, P.; Chavan, R.; Manjeet, B. Saponification Kinetics of Acrylonitrile Terpolymer and Polyacrylonitrile. *J. Macromol. Sci. Chem.* **1985**, *A22*, 1219–1239. [CrossRef]
9. Sanli, O. Homogeneous Hydrolysis of Polyacrylonitrile by Potassium Hydroxide. *Eur. Polym. J.* **1990**, *26*, 9–13. [CrossRef]
10. Ermakov, I.; Rebrov, A.; Litmanovich, A.; Plate, N. Alkaline hydrolysis of polyacrylonitrile, 1—Structure of the reaction products. *Macromol. Chem. Phys.* **2000**, *201*, 1415–1418. [CrossRef]
11. Truong, N.; Galin, J.; Francois, J.; Pham, Q. Microstructure of Acrylamide Acrylic-Acid Copolymers. 1. As Obtained by Alkaline-Hydrolysis. *Polymer* **1986**, *27*, 459–466. [CrossRef]
12. Litmanovich, A.; Plate, N. Alkaline hydrolysis of polyacrylonitrile. On the reaction mechanism. *Macromol. Chem. Phys.* **2000**, *201*, 2176–2180. [CrossRef]
13. Dyatlov, V.; Grebeneva, T.; Rustamov, I.; Koledenkov, A.; Kolotilova, N.; Kireev, V.; Prudskov, B. Hydrolysis of polyacrylonitrile in aqueous solution of sodium carbonate. *Polym. Sci. Ser. B* **2012**, *54*, 161–166. [CrossRef]
14. Hu, X.; Xiao, C. Characterization and Hydrolization of PAN By Sodium Hydroxide. *Adv. Mater. Res.* **2011**, *332–334*, 1032–1035. [CrossRef]
15. Long, W.; Luo, H.; Yan, Z.; Zhang, C.; Hao, W.; Wei, Z.; Zhu, X.; Zhou, F.; Cha, R. Synthesis of filtrate reducer from biogas residue and its application in drilling fluid. *Tappi. J.* **2020**, *19*, 151–158. [CrossRef]
16. Long, W.; Zhu, X.; Zhou, F.; Yan, Z.; Evelina, A.; Liu, J.; Wei, Z.; Ma, L. Preparation and Hydrogelling Performances of a New Drilling Fluid Filtrate Reducer from Plant Press Slag. *Gels* **2022**, *8*, 201. [CrossRef]
17. Huang, L.; Xiao, C.; Chen, B. A novel starch-based adsorbent for removing toxic Hg(II) and Pb(II) ions from aqueous solution. *J. Hazard. Mater.* **2011**, *192*, 832–836. [CrossRef]
18. Ojogbo, E.; Blanchard, R.; Mekonnen, T. Hydrophobic and Melt Processable Starch-Laurate Esters: Synthesis, Structure-Property Correlations. *J. Polym. Sci. Part A Polym. Chem.* **2018**, *56*, 2611–2622. [CrossRef]
19. Kweon, M.; Sosulski, F.; Bhirud, P. Cationization of waxy and normal corn and barley starches by an aqueous alcohol process. *Starch Stärke* **1997**, *49*, 59–66. [CrossRef]
20. Zuo, Y.; Gu, J.; Yang, L.; Qiao, Z.; Tan, H.; Zhang, Y. Synthesis and characterization of maleic anhydride esterified corn starch by the dry method. *Int. J. Biol. Macromol.* **2013**, *62*, 241–247. [CrossRef]

21. Zuo, Y.; Gu, J.; Qiao, Z.; Tan, H.; Cao, J.; Zhang, Y. Effects of dry method esterification of starch on the degradation characteristics of starch/polylactic acid composites. *Int. J. Biol. Macromol.* **2015**, *72*, 391–402. [CrossRef]
22. Fashi, A.; Fallah Delavar, A.; Zamani, A.; Noshiranzadeh, N. Solid state malic acid esterification on fungal alpha-amylase treated corn starch: Design of a green dual treatment. *Food Chem.* **2023**, *410*, 135439. [CrossRef]
23. Zhu, J.; Lu, K.; Liu, H.; Bao, X.; Yang, M.; Chen, L.; Yu, L. Influence of Moisture Content on Starch Esterification by Solvent-Free Method. *Starch Stärke* **2021**, *73*, 2100009. [CrossRef]
24. Karic, N.; Vukcevic, M.; Ristic, M.; Peric-Grujic, A.; Marinkovic, A.; Trivunac, K. A green approach to starch modification by solvent-free method with betaine hydrochloride. *Int. J. Biol. Macromol.* **2021**, *193 Pt B*, 1962–1971. [CrossRef]
25. Chen, Q.; Yu, H.; Wang, L.; ul Abdin, Z.; Chen, Y.; Wang, J.; Zhou, W.; Yang, X.; Khan, R.; Zhang, H.; et al. Recent progress in chemical modification of starch and its applications. *RSC Adv.* **2015**, *5*, 67459–67474. [CrossRef]
26. Perez-Alvarez, L.; Ruiz-Rubio, L.; Moreno, I.; Vilas-Vilela, J. Characterization and Optimization of the Alkaline Hydrolysis of Polyacrylonitrile Membranes. *Polymers* **2019**, *11*, 1843. [CrossRef] [PubMed]
27. Zhang, H.; Zhang, T.; Qiu, Q.; Qin, X. Quaternary ammonium salt–modified polyacrylonitrile/polycaprolactone electrospun nanofibers with enhanced antibacterial properties. *Text. Res. J.* **2021**, *91*, 2194–2203. [CrossRef]
28. Kozicki, M.; Sąsiadek, E.; Karbownik, I.; Maniukiewicz, W. Doped polyacrylonitrile fibres as UV radiation sensors. *Sens. Actuators B* **2015**, *213*, 234–243. [CrossRef]
29. Zhou, X.; Bai, L.; Liu, X.; Ren, Y.; Song, Z.; Yang, X. Preparation of halogen-free flame retardant polyacrylonitrile via hydrolyzing and grafting with diphenylphosphinyl chloride. *J. Macromol. Sci. Part A Pure Appl. Chem.* **2019**, *56*, 1097–1103. [CrossRef]
30. Wang, F.; Wang, X.; Jiang, Y.; Niu, Z.; Wu, W.; Zhang, H. Study of adsorption performance and adsorption mechanism for U(VI) ion on modified polyacrylonitrile fibers. *J. Radioanal. Nucl. Chem.* **2019**, *323*, 365–377. [CrossRef]
31. Fang, J.; Liu, G.; Chen, C.; Lin, C.; Zhang, B.; Jin, H.; Chen, Y.; Lu, J.; Zhu, L. Intrinsically antibacterial thin film composite membranes with supramolecularly assembled lysozyme nanofilm as selective layer for molecular separation. *Sep. Purif. Technol.* **2021**, *254*, 117585. [CrossRef]
32. Jin, S.; Kim, M.; Jeong, Y.; Yoon, Y.; Park, W. Effect of alkaline hydrolysis on cyclization reaction of PAN nanofibers. *Mater. Des.* **2017**, *124*, 69–77. [CrossRef]
33. Yu, P. Modification of waste polyacrylonitrile fiber and its application as a filtrate reducer for drilling. *Pet. Sci.* **2015**, *12*, 325–329. [CrossRef]
34. Austria, H.; Lecaros, R.; Hung, W.; Tayo, L.; Hu, C.; Tsai, H.; Lee, K.; Lai, J. Investigation of salt penetration mechanism in hydrolyzed polyacrylonitrile asymmetric membranes for pervaporation desalination. *Desalination* **2019**, *463*, 32–39. [CrossRef]
35. Awad, F.; AbouZied, K.; Bakry, A.; Abou El-Maaty, W.; El-Wakil, A.; El-Shall, M. Polyacrylonitrile modified partially reduced graphene oxide composites for the extraction of Hg(II) ions from polluted water. *J. Mater. Sci.* **2021**, *56*, 7982–7999. [CrossRef]
36. Ren, Y.; Zhang, Y.; Gu, Y.; Zeng, Q. Flame retardant polyacrylonitrile fabrics prepared by organic-inorganic hybrid silica coating via sol-gel technique. *Prog. Org. Coat.* **2017**, *112*, 225–233. [CrossRef]
37. Chae, H.; Minus, M.; Rasheed, A.; Kumar, S. Stabilization and carbonization of gel spun polyacrylonitrile/single wall carbon nanotube composite fibers. *Polymer* **2007**, *48*, 3781–3789. [CrossRef]
38. Peng, H.; Wang, D.; Li, M.; Zhang, L.; Liu, M.; Fu, S. N-P-Zn-containing 2D supermolecular networks grown on MoS2 nanosheets for mechanical and flame-retardant reinforcements of polyacrylonitrile fiber. *Chem. Eng. J.* **2019**, *372*, 873–885. [CrossRef]
39. Yan, X.; Zhou, W.; Zhao, X.; Xu, J.; Liu, P. Preparation, flame retardancy and thermal degradation behaviors of polyacrylonitrile fibers modified with diethylenetriamine and zinc ions. *J. Therm. Anal. Calorim.* **2015**, *124*, 719–728. [CrossRef]
40. Dargahi-Zaboli, M.; Sahraei, E.; Pourabbas, B. Hydrophobic silica nanoparticle-stabilized invert emulsion as drilling fluid for deep drilling. *Pet. Sci.* **2016**, *14*, 105–115. [CrossRef]
41. Guo, J.; Yan, J.; Wang, L.; Fan, W. Influence of composition of KCl/silicate drilling fluids on rheological properties. *J. Cent. S. Univ. Technol. Engl. Ed.* **2007**, *14* (Suppl. S1), 426–429. [CrossRef]
42. Arabi, A.; Dewu, B.; Funtua, I.; Oladipo, M.; Tukur, M.; Bilal, S.; Kurowska, E.; Kasim, S.; Yakasai, A.; Babale, S. Morphology, rheology and thermal stability of drilling fluid formulated from locally beneficiated clays of Pindiga Formation, Northeastern Nigeria. *Appl. Clay Sci.* **2018**, *161*, 90–102. [CrossRef]
43. Magzoub, M.; Mahmoud, M.; Nasser, M.; Hussein, I.; Elkatatny, S.; Sultan, A. Thermochemical Upgrading of Calcium Bentonite for Drilling Fluid Applications. *J. Energy Resour. Technol.* **2019**, *141*, 042902. [CrossRef]
44. Liu, J.; Zhou, F.; Deng, F.; Zhao, H.; Wei, Z.; Long, W.; Evelina, A.; Ma, C.; Chen, S.; Ma, L. Improving the rheological properties of water-based calcium bentonite drilling fluids using water-soluble polymers in high temperature applications. *J. Polym. Eng.* **2022**, *42*, 129–139. [CrossRef]
45. Du, M.; Liu, P.; Clode, P.; Liu, J.; Haq, B.; Leong, Y. Impact of additives with opposing effects on the rheological properties of bentonite drilling mud: Flow, ageing, microstructure and preparation method. *J. Pet. Sci. Eng.* **2020**, *192*, 107282. [CrossRef]
46. Huo, J.; Peng, Z.; Ye, Z.; Feng, Q.; Zheng, Y.; Zhang, J.; Liu, X. Investigation of synthesized polymer on the rheological and filtration performance of water-based drilling fluid system. *J. Pet. Sci. Eng.* **2018**, *165*, 655–663. [CrossRef]
47. Ben Azouz, K.; Bekkour, K.; Dupuis, D. Influence of the temperature on the rheological properties of bentonite suspensions in aqueous polymer solutions. *Appl. Clay Sci.* **2016**, *123*, 92–98. [CrossRef]
48. Bu, H.; Xu, T.; Sun, J.; Wang, H. The effect of high temperature on clays in drilling fluids. *Drill. Fluid Complet. Fluid* **2010**, *27*, 23–25. (In Chinese)

49. Guecioue r, A.; Benmounah, A.; Sekkiou, H.; Kheribet, R.; Safi, B. Valorization of KCl/PHPA system of water-based drilling fluid in presence of reactive clay: Application on Algerian field. *Appl. Clay Sci.* **2017**, *146*, 291–296. [CrossRef]
50. Liu, J.; Dai, Z.; Xu, K.; Yang, Y.; Lv, K.; Huang, X.; Sun, J. Water-Based Drilling Fluid Containing Bentonite/Poly(Sodium 4-Styrenesulfonate) Composite for Ultrahigh-Temperature Ultradeep Drilling and Its Field Performance. *SPE J.* **2020**, *25*, 1193–1203. [CrossRef]
51. Yang, L.; Jiang, G.; Shi, Y.; Lin, X.; Yang, X. Application of ionic liquid to a high-performance calcium-resistant additive for filtration control of bentonite/water-based drilling fluids. *J. Mater. Sci.* **2017**, *52*, 6362–6375. [CrossRef]
52. Liu, F.; Jiang, G.; Peng, S.; He, Y.; Wang, J. Amphoteric Polymer as an Anti-calcium Contamination Fluid-Loss Additive in Water-Based Drilling Fluids. *Energy Fuels* **2016**, *30*, 7221–7228. [CrossRef]
53. Li, J.; Sun, J.; Lv, K.; Ji, Y.; Ji, J.; Bai, Y.; Wang, J.; Jin, J.; Shi, S.; Huang, X.; et al. A zwitterionic copolymer as fluid loss reducer for water-based drilling fluids in high temperature and high salinity conditions. *Geoenergy Sci. Eng.* **2023**, *222*, 111200. [CrossRef]
54. Desbrieres, J. Cement cake properties in static filtration-Influence of polymeric additives on cement filter cake permeability. *Cem. Concr. Res.* **1993**, *23*, 347–358. [CrossRef]
55. Liu, F.; Yao, H.; Liu, Q.; Wang, X.; Dai, X.; Zhou, M.; Wang, Y.; Zhang, C.; Wang, D.; Deng, Y. Nano-silica/polymer composite as filtrate reducer in water-based drilling fluids. *Colloids Surf. A* **2021**, *627*, 127168. [CrossRef]
56. Gumaste, S.; Iyer, K.; Sharma, S.; Channabasavaraj, W.; Singh, D. Simulation of fabric in sedimented clays. *Appl. Clay Sci.* **2014**, *91–92*, 117–126. [CrossRef]
57. Sun, J.; Chan, X.; Lv, K.; Wang, J.; Zhang, F.; Jin, J.; Zhou, X.; Dai, Z. Environmentally friendly and salt-responsive polymer brush based on lignin nanoparticle as fluid-loss additive in water-based drilling fluids. *Colloids Surf. A* **2021**, *621*, 126482. [CrossRef]

gels

Article

Effect of Polymer and Crosslinker Concentration on Static and Dynamic Gelation Behavior of Phenolic Resin Hydrogel

Wenjuan Ji [1], Bei Chang [2], Haiyang Yu [3,*], Yilin Li [4] and Weiqiang Song [3]

[1] School of Petroleum Engineering, China University of Petroleum (East China), Qingdao 266580, China; jwj1224@126.com
[2] Chuanqing Drilling Engineering Company Limited, China National Petroleum Corporation Limited, Chengdu 610051, China; changb_cqyyq@cnpc.com.cn
[3] College of Energy and Mining Engineering, Shandong University of Science and Technology, Qingdao 266590, China; skd996020@sdust.edu.cn
[4] Drilling & Production Technology Research Institute of Jidong Oilfield, China National Petroleum Corporation Limited, Tangshan 063000, China; zcy_liyl@petrochina.com.cn
* Correspondence: yhy_skd@sdust.edu.cn

Abstract: The application results of profile control and water plugging technology are highly related to the gelation time and strength of phenolic resin hydrogel. In this work, a hydrogel solution was prepared by fully mixing the prepared polymer solution with a crosslinker. The static gelation process of PFR hydrogel in ampoule bottles and porous media was analyzed by changes in the viscosity and residual resistance coefficient. Then, the dynamic gelation of the PFR hydrogel in porous media was tested using a circulating flow device, and the changes in viscosity and injection pressure were analyzed during the dynamic gelation process. Finally, the effects of the polymer concentration and crosslinker concentration on dynamic gelation were analyzed. The initial gelation time and final gelation time in porous media were 1–1.5 times and 1.5–2 times those in ampoule bottles under static conditions, respectively. The initial dynamic gelation time in porous media was 2–2.5 times and 1.5–2 times the initial static gelation times in ampoule bottles and porous media, respectively. The final dynamic gelation time was four times and two times the initial static gelation times in ampoule bottles and porous media, respectively. The production after dynamic gelation in porous media comprised hydrogel aggregates and water fluid, leading to a high injection pressure and low viscosity of the produced liquid. As the concentration of polymer and crosslinker increased, the dynamic gelation time was shortened and the gel strength was increased. In the dynamic gelation process in porous media, the phenol resin hydrogel could migrate deeply, but it was limited by the concentrations of the polymer and crosslinker. The results of subsequent water flooding showed that the polymer hydrogel had a good plugging ability after dynamic gelation. The deep reservoir could only be blocked off in the subsequent water flooding process when the migration of hydrogel happened in the dynamic gelation process.

Keywords: PFR hydrogel; dynamic gelation; gelation time; partially hydrolyzed polyacrylamide; plugging ability

check for updates

Citation: Ji, W.; Chang, B.; Yu, H.; Li, Y.; Song, W. Effect of Polymer and Crosslinker Concentration on Static and Dynamic Gelation Behavior of Phenolic Resin Hydrogel. *Gels* **2024**, *10*, 325. https://doi.org/10.3390/gels10050325

Academic Editor: Rama Bansil

Received: 9 April 2024
Revised: 5 May 2024
Accepted: 7 May 2024
Published: 9 May 2024

1. Introduction

Polymer hydrogel is widely used as a profile control agent in worldwide oil fields [1–4], and its application results are highly related to gelation time and gel strength. The gelation time is divided into static gelation time and dynamic gelation time. Many polymer hydrogels can form whole structures with adjustable gelation times and controllable strength on the ground. However, it is difficult to predict the gelation situation after injection into the formation, which leads to application problems in profile control and water plugging [5]. A field application case showed that polymer hydrogel injected into a deep formation would not show an obvious blockage after extending the static gelation time by 20 days.

When subjected to shear in the rock core, whether the polymer gelation solution can form hydrogel or not is related to the dynamic gelation of polymer hydrogels. This directly affects the success of profile control and water plugging.

Recently, many scholars have studied the differences between the dynamic and static gelation of hydrogel. Ki et al. studied the fractal nature of the backbone network for the irreversible kinetic gelation model. They suggested that the static percolation model might be adequate to describe the sol–gel transition and related phenomena of the irreversible growth model [6]. Dolan et al. described the gelation time and strength of the hydrogel using small-amplitude oscillation under zero shear rate and stable shear conditions [7]. Kolnes et al. studied the effects of shear and temperature on the gelation of xanthan gum/chromium hydrogel, and the results showed that shear disrupted the cross-linking structure of the polymer and crosslinker [8]. Bhaskar et al. investigated the effect of different shear modes on the gelation process of redox systems [9]. Josset et al. studied the gelation process of cross-linked polymer hydrogel under a high shear rate and high shear conditions in the formation near the wellbore [8]. Carvalho et al. studied the cross-linking reaction process of cross-linked polymers during the shearing process. The results showed that shearing could promote or delay the cross-linking reaction with a critical shear rate, switching between these two effects [10]. McCool et al. studied the gelation behavior of chromium acetate weak hydrogel in porous media flow using a 1036 feet long stainless-steel conduit [11]. Series et al. used a 100 feet long thin iron pipe to simulate the flow of hydrogel in cracks. The results showed that the residual resistance coefficient reached its peak at 20 feet and then began to decrease, while the middle and rear ends (60–100 feet) of the thin iron pipe remained unchanged [12]. Abete et al. studied the structure and dynamics in the formation of irreversible gels by molecular dynamics simulations, and suggested that gelation transition was due to the random percolation of permanent bonds between neighboring particles [13]. Sun, Z. et al. studied the influences of polymer molecular weight and concentration on gelation time, gel strength, and adhesion ability, and indicated that increasing the molecular weight and concentration could shorten the gelation time and enhance the gel strength, and that adhesion strength is mainly related to the number of hydrogen bonds [14]. Khurshid, I. and Afgan, I. analyzed the effect of polymer drive composition on surfactant retention and new surface complexation reactions [15]. Our research group has also studied the dynamic gelling behavior of polymer hydrogels in recent years. The static gelation time and the main factors of the gel strength of hydrogel at high temperatures were studied, resulting in the conclusion that temperature is an important factor affecting the gelation process of hydrogel [16]. The effect of the injected speed on dynamic gelation was analyzed in reference [17], showing that the injection speed has little effect on the dynamic gelation time, but has a great effect on the gel strength. Mechanical shear rate can also have an important influence on the gelling behavior of hydrogels [18]. The static gelation and dynamic gelation of hydrogels are quite different due to many factors, such as adsorption and shear [19]. The results showed that the dynamic gelling behavior of polymer hydrogels was greatly affected by the shear rate, which was directly related to the reservoir permeability and injection rate.

Previous studies have clarified the static and dynamic gelation processes of hydrogel. However, the migration characteristics and main control factors after dynamic gelation in porous media are not clear, especially the influence of the mass concentration of hydrogel on the migration characteristics. When the polymer hydrogel migrates in the formation, it is subjected to core shear and its structure is destroyed. During the migration process, it forms a whole hydrogel or a dispersed hydrogel particle structure. What are the changes in the gelation time and gel strength in formation and on the ground? These issues directly affect the success of profile control and water plugging by polymer hydrogel. Therefore, it is of great importance to study the gelation time and gel strength during the dynamic gelation process of polymer hydrogels [20,21]. Based on the above issues, this article studied static gelation in ampoule bottles and porous media, and the dynamic gelation of a hydrogel system composed of ordinary partially hydrolyzed polyacrylamide HPAM and phenol formaldehyde resin (PFR). The relationship between gelation time and gel strength under these three different conditions was investigated, and a quantitative relationship between the static gelation time and dynamic gelation time was established. The plugging ability of static gelation and dynamic gelation in porous media for subsequent water flooding was comparatively analyzed.

2. Results and Discussion

2.1. Static Gelation in Ampoules

A water-soluble phenolic resin crosslinker was obtained by the poly-condensation of phenol and formaldehyde (excess) under the catalyst action of sodium hydroxide [22]. The reaction mechanism of the crosslinker is shown in Figure 1. The hydroxymethyl groups (-CH_2OH) on the crosslinker and the amide groups (-$CONH_2$) on the polymer were dehydrated and condensed to form a bulk gel. The formation mechanism of the bulk gel is shown in Figure 2.

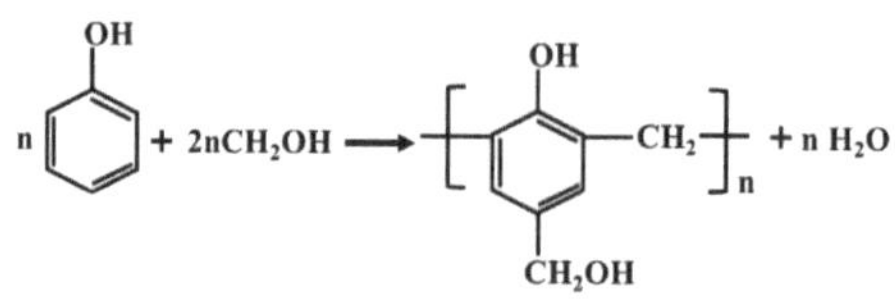

Figure 1. Reaction mechanism of water-soluble phenolic resin crosslinker.

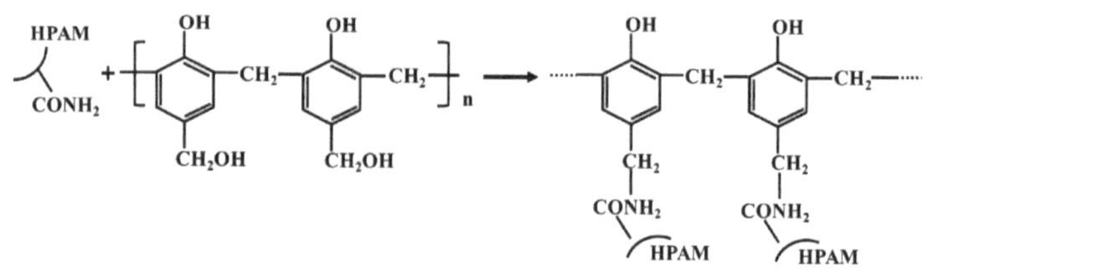

Figure 2. Formation mechanism of the bulk gel.

The commonly used methods for determining the gelation time of hydrogel include the visual strength code method and viscosity method [23,24]. This article used the viscosity method to determine the relationship between the viscosity of the hydrogel system and time under different gelation times (see Figure 3). It could be seen that the viscosity did not change significantly with time before the cross-linking reaction began. During the gelation process, the viscosity rapidly increased with time and then stabilized [18]. The probability of contact between the amide groups (-$CONH_2$) and hydroxymethyl groups (-CH_2OH) in the system increased with increasing concentrations of polymer and crosslinker, which made the formed spatial network structure denser. The gelation time of the hydrogel was shortened and the gel strength was increased [19]. According to the research results of Mokhtari et al., the gelation time can be divided into the initial gelation time (IGT) and final gelation time (FGT). The initial gelation time refers to the moment when the viscosity of the system begins to significantly increase at the beginning of the cross-linking reaction. The final gelation time is the time when the system viscosity reaches stability at the end of the cross-linking reaction [25]. The initial gelation time and final gelation time with different formulations of hydrogels can be obtained from the curves in Figure 3. The micro morphology of the PFR hydrogel after static gelation in ampoule bottles is shown in Figure 4.

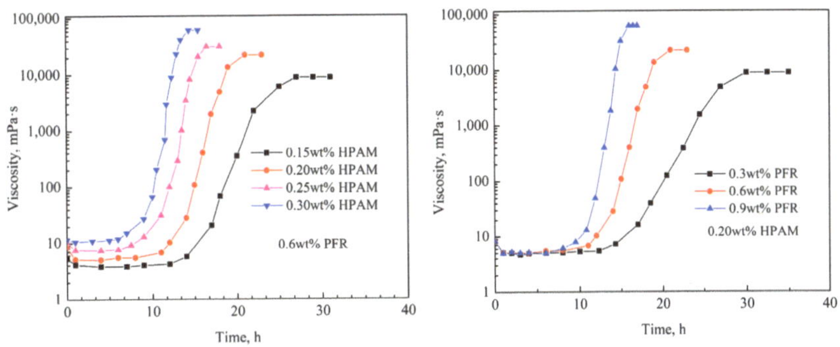

Figure 3. The change in viscosity of PFR hydrogel with time.

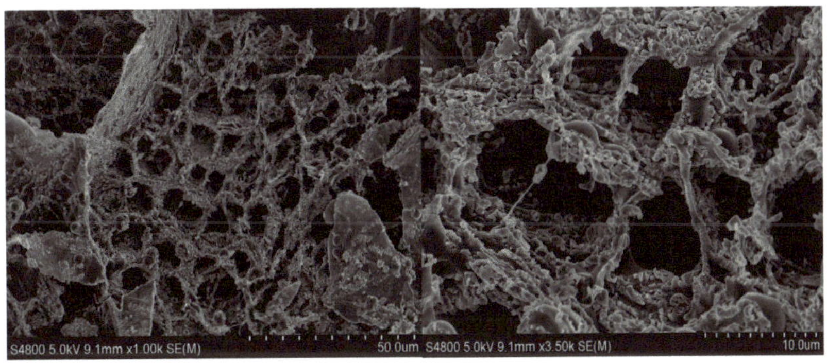

(**a**) (**b**)

Figure 4. The micro morphology of PFR hydrogel after static gelation in ampoule bottles; (**a**) micro morphology at 50 μm; (**b**) micro morphology at 10 μm.

As shown in Figure 4, the microscopic morphology of the phenolic resin hydrogel can be clearly seen at a lower magnification, and the phenolic resin hydrogel formed a regular network structure after static gelation in the ampoule bottle. The regular network structure was formed by the cross-linking of the amide groups in the polymers with the hydroxymethyl groups in the phenolic resin prepolymers. At higher multiples, it was observed that there were pores of similar size in the middle of the network structure, with pore sizes distributed around 3–5 μm.

2.2. Static Gelation in Porous Media

The amide groups on the polymer and the hydroxymethyl groups on the crosslinker could be cross-linked to form a three-dimensional network structure, which had a certain plugging ability on porous media. The strength of three-dimensional network structure was related to the concentration of the hydrogel, which directly affected the plugging efficiency of the porous media. The pressure change characteristics of the cross-linked polymer hydrogel when flowing in porous media reflect the degree of the hydrogel cross-linking reaction. As the cross-linking reaction proceeds, the plugging ability of the hydrogel for porous media gradually increases until the end of the reaction [17]. Generally, indicators such as the resistance coefficient and residual resistance coefficient or transition pressure are used to describe the flow characteristics of cross-linked polymer hydrogel [26,27]. The residual resistance coefficient is calculated by measuring the breakthrough pressure gradient of hydrogel in a series of sand pack pipes to characterize the process of the cross-linking reaction of hydrogel. The values reflect the viscosity of the hydrogel in porous media (see Figure 5).

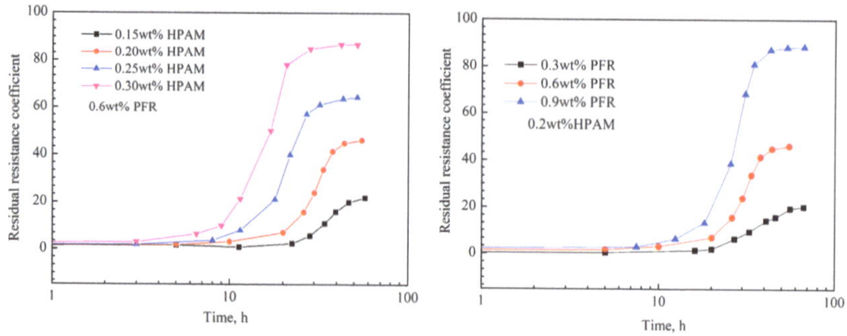

Figure 5. The change in static gelation of PFR hydrogel with time in porous media.

From Figure 5, it can be seen that the residual resistance coefficient first showed no significant change with an increasing placement time, and then rapidly increased until it stabilized. The reaction process of hydrogel can be divided into three stages: induction, gelation, and stabilization. The initial gelation time and final gelation time were both shortened with increasing concentrations of polymer and crosslinker. The micro morphology of the PFR hydrogel after static gelation in porous media is shown in Figure 6.

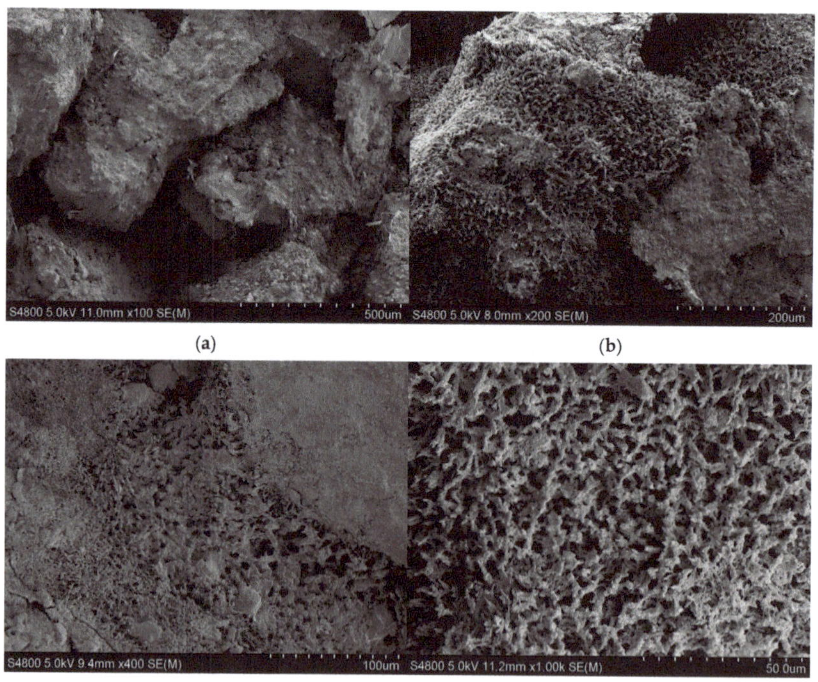

(a) (b) (c) (d)

Figure 6. The micro morphology of PFR hydrogel after static gelation in porous media (**a**) micro morphology at 500 μm; (**b**) micro morphology at 200 μm; (**c**) micro morphology at 100 μm; (**d**) micro morphology at 50 μm.

As shown in Figure 6, the existence form of phenolic resin hydrogel after static gelation in porous media can be clearly reflected at a lower magnification. The hydrogels were mainly adsorbed on the surface of porous media and trapped at smaller pore throats. At a higher magnification, it can be observed that the structure of phenolic resin hydrogel was composed of a network structure formed by the winding and cementation of some thicker chains and larger pores. Under the action of shear failure and adsorption retention, the phenolic resin hydrogel network structure formed in the process of static gelation in porous media was not as clear and complete as that formed by static gelation in ampoule bottles.

The results of the comparison of static gelation in ampoule bottles and porous media are shown in Table 1. During the hydrogel injection into the sandpack pipes, the polymer and crosslinker were subjected by the shear of porous media, which led to prolonging the gelation time and reducing the gel strength. Due to the different molecular sizes of the polymer and crosslinker, there was a significant difference in their migration speed in the porous media. A change in the concentration ratio of the polymer and crosslinker can affect the gelation time. The adsorption of the polymer and crosslinker on the surface of the porous media directly led to a decrease in the number of molecules involved in the cross-linking reaction, thereby prolonging the gelation time [28]. Therefore, the static gelation time of the PFR hydrogel in porous media was longer than that in ampoules, with an initial gelation time of 1–1.5 times and a final gelation time of 1.5–2 times.

Table 1. The IGTs and FGTs of PFR hydrogel in the process of static gelation in ampoule bottles and porous media.

No	HPAM, %	PFR, %	Static Gelation in Ampoule Bottles		Static Gelation in Porous Media	
			IGT, h	FGT, h	IGT, h	FGT, h
1	0.15	0.6	14	27	25	45
2	0.2	0.6	12	21	17	40
3	0.25	0.6	9	16.5	10	29
4	0.3	0.6	7	14.4	8	23
5	0.2	0.3	14.3	30	20	45
6	0.2	0.9	9.5	18	10	35

2.3. Dynamic Gelation in Porous Media

2.3.1. Analysis of Dynamic Gelation Process

In the static gelation of the hydrogel in porous media, there was no shear force affecting the cross-linking process, ultimately forming a three-dimensional network structure. However, the hydrogel could be subjected to the shear force from the porous media in the dynamic gelation process. When the hydrogel network structure size was increased to larger than the pore size, the network structure could be destroyed by the shear force, forming dispersed hydrogel particles rather than a whole bulk hydrogel. The phenolic resin hydrogel 0.2 wt% HPAM+0.3 wt% PFR was analyzed using a circulating flow experimental device. The change in the injection pressure (ΔP_{ad}, ΔP_{bd}, and ΔP_{cd}) was recorded with time (see Figure 7). During the dynamic gelation process, ΔP_{ad} is the pressure difference between the injection end a and the outlet end d, ΔP_{bd} is the pressure difference between the measuring point b and the outlet end d, and ΔP_{cd} is the pressure difference between the measuring point b and the outlet end d. The permeability was 7.07 μm^2 and the injected speed was 0.5 mL/min.

From Figure 7, in the dynamic gelation process, the pressure difference ΔP_{ad} firstly changed smoothly, then rapidly increased before finally stabilizing, which indicates that the hydrogel underwent the induction stage, gelation stage, and stabilization stage. During the induction stage, the polymers that had not been cross-linked still maintained a certain degree of viscoelasticity. When passing through a smaller core throat, they were compressed and deformed to generate a certain degree of elasticity, which required a certain pressure difference to pass through [29]. At this stage, the polymer molecules still formed as individual particles rather than a network structure [30]. Macroscopically, there was no significant change in pressure. During the gelation stage, polymer hydrogels were subjected by two types of forces: cross-linking reaction force and shear fragmentation force. The former could enhance the three-dimensional network structure of the hydrogel and increase the apparent viscosity, while the latter could destroy the network structure and reduce the apparent viscosity. When the size of the phenolic resin hydrogel increased with time until its cohesion was overcome by the shear of the porous media, the hydrogel was sheared and destroyed to form a dispersed gel particle system, instead of becoming a whole hydrogel, as shown in Figure 8. The formed hydrogel particles remained at the pore throat of the porous media to play a plugging role, thus leading to the subsequent pressure rise.

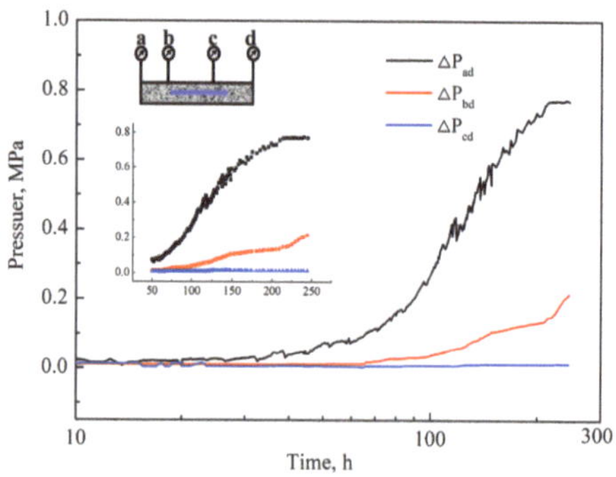

Figure 7. The change in pressure of PFR hydrogel with time in the dynamic gelation process.

Figure 8. The micro morphology of PFR gel after dynamic gelation in porous media.

The microstructure of the phenolic resin hydrogel after dynamic gelation in porous media was dispersed hydrogel particles rather than a whole bulk hydrogel. The network structure of the phenolic resin hydrogel was not found at a higher magnification. These hydrogel particles were adsorbed on the surface of the porous media and aggregated at the pore throat, reducing the seepage capacity of the porous media and playing a plugging role. Compared to the static gelation of the phenolic resin hydrogel in porous media, the main form of the hydrogel particles after dynamic gelation was trapping. The viscosity results of the hydrogel solution from the middle container after the experiment are shown in Table 2. It could be seen that, after the dynamic gelation process in the porous media, the viscosity of the hydrogel solution was lower than the viscosity of the initial polymer gelation solution. This confirmed that the polymer was cross-linked, but hydrogel particles were formed under the shear action of the porous media. The adsorption and migration of the hydrogel particles in the porous medium kept the displacement pressure gradually increasing. The output liquid was free water, which was generated by the bound water inside the hydrogel under the action of shear force. The pressure difference ΔP_{bd} saw a small increase, and the start time of this increase was significantly delayed compared to the pressure difference ΔP_{ad}. This shows that hydrogel particles formed by dynamic gelation can realize deep migration in porous media.

Table 2. Viscosity of PFR hydrogel production after dynamic gelation in porous media.

HPAM, wt%	PFR, wt%	Speed, mL/min	K, μm^2	Viscosity, mPa·s
0.15	0.6	0.5	7.22	2.1
0.2	0.6	0.5	8.08	3.7
0.25	0.6	0.5	8.28	6.3
0.3	0.6	0.5	8.99	4.2
0.2	0.3	0.5	7.07	5.7
0.2	0.9	0.5	8.49	1.6

2.3.2. The Effect of Polymer and Crosslinker Concentration on Dynamic Gelation

The dynamic gelation experiments of the PFR hydrogel in porous media under different formulations were carried out by changing the concentrations of the polymer and crosslinker. The injection rate of 0.5 mL/min was kept unchanged (see Figure 9). Due to the small change in the pressure difference ΔP_{bd} and almost no change in the pressure difference ΔP_{cd}, the dynamic gelation process was characterized by the pressure difference ΔP_{ad}. The dynamic initial gelation time of PFR hydrogel in porous media refers to the time that the cross linking reaction begins, and viscosity increases significantly during the flow process [31]. This is shown in the figure as the first inflection point on the inlet pressure difference curve ΔP_{ad}. The final gelation time refers to the moment when the inlet pressure difference ΔP_{ad} no longer increases and tends to stabilize, which is expressed as the second inflection point on the inlet pressure difference ΔP_{ad} curve. In order to eliminate the influence of permeability, $F = \Delta P \times K$ is defined as the seepage resistance of fluid flowing in a porous medium.

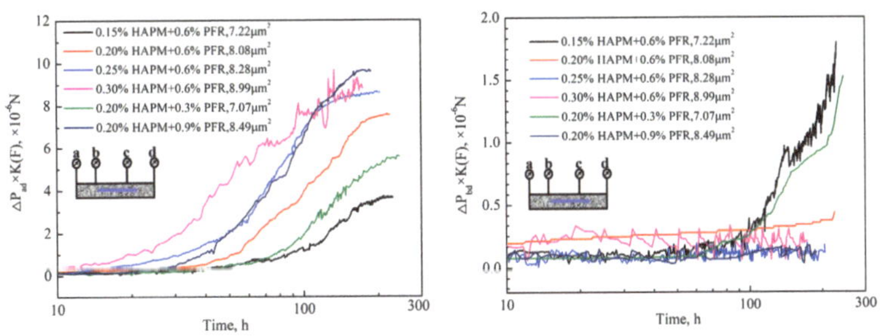

(a) (b)

Figure 9. The change in dynamic gelation of PFR hydrogel with time in porous media under different concentrations of polymer and crosslinker; (**a**) $\Delta P_{ad} \times K$ vs. time; (**b**) $\Delta P_{bd} \times K$ vs. time.

The cross-linking active points increased with increasing concentrations of polymer and crosslinker. Under the same shear conditions, the cross-linking reaction rate accelerated and the gelation time was shortened. Therefore, the time for the inflection point to appear on the inlet pressure difference curve was advanced, indicating that the initial gelation time and final gelation time of the dynamic gelation in porous media were both shortened (Figure 9a). From Figure 9b, it can be seen that, as the concentrations of polymer and crosslinker agent increased, especially to a polymer concentration greater than 0.2 wt%, the change amplitude of $\Delta P_{bd} \times K$ decreased. This indicates that the migration ability of the PFR hydrogel during dynamic gelation was limited by the concentrations of the polymer and crosslinker. According to the experimental method of dynamic gelation in porous media, it can be seen that the polymer hydrogel solution with two times the pore volume (2 PV) flowed alternately in the porous media during the entire cycle. When 1 PV hydrogel solution flowed in the porous media at 75 °C, the other PV hydrogel solution flowed in the intermediate container at room temperature (25 °C). The hydrogel solution of 0.2 wt% HPAM+0.6wt% with an initial viscosity value of 8.4 mPa·s aged at room temperature for 30 days. The viscosity was 10.9 mPa·s, indicating that the PFR hydrogel solution was not cross-linked at room temperature. Therefore, the dynamic gelation time in porous media should be half the time required for the entire gelation process. The initial and final gelation time were obtained through dynamic gelation experiments in porous media with different polymer and crosslinker concentrations, and were compared with static gelation in ampoule bottles and porous media. The IGTs and FGTs of the PFR hydrogel in the process of dynamic gelation are shown in Table 3.

Table 3. The IGTs and FGTs of PFR hydrogel in the process of dynamic gelation.

No	HPAM, %	PFR, %	Dynamic Gelation in Porous Media	
			IGT, h	FGT, h
1	0.15	0.6	40	105
2	0.2	0.6	25	95
3	0.25	0.6	20	80
4	0.3	0.6	14	46
5	0.2	0.3	35	106
6	0.2	0.9	20	80

According to Table 3, the initial and final gelation times of dynamic gelation in PFR hydrogel porous media were longer than those in ampoule bottles and porous media. The dynamic initial gelation time was 2–2.5 times the static initial gelation time in ampoule bottles and 1.5–2 times the static initial gelation time in porous media. According to Formula (1), the shear rate of the PFR hydrogel during dynamic gelation in porous media is calculated [32–34], as shown in Table 4.

$$\gamma = \frac{3\text{n}+1}{\text{n}} \cdot \frac{v}{\sqrt[0.5]{8C'K\phi}} \quad (1)$$

Table 4. Shear rate of PFR hydrogel under the process of dynamic gelation in porous media.

HPAM, wt%	PFR, wt%	Speed, mL/min	K, μm^2	Porosity	n	C′	Shear Rate, s^{-1}
0.2	0.3	0.5	7.07	0.352	0.573	2.29	4.34
0.2	0.6	0.5	8.08	0.367	0.440	2.29	3.71
0.2	0.9	0.5	8.48	0.371	0.445	2.29	3.61
0.15	0.6	0.5	7.22	0.353	0.645	2.29	4.43
0.25	0.6	0.5	8.28	0.369	0.427	2.29	3.63
0.3	0.6	0.5	8.99	0.373	0.346	2.29	3.28

γ—shear rate, s^{-1}; n—viscosity index of fluid, mPa·s^{n}; v—injection rate, cm/s; C'—the coefficient related to tortuosity, usually ranging from 25/12 to 2.5; K—permeability, μm^2; and Φ—porosity. The porosity value is the ratio of the saturated water volume to the total volume, and the tortuosity value is the average of the lower and upper limits.

According to Table 4, there was a certain shear rate during the dynamic gelation process of the PFR hydrogel in porous media. There were two forces in the process of PFR hydrogel formation: one was the cross-linking effect that facilitated the formation of the network structure, and the other was the shear effect that destroyed the network structure. Therefore, the initial gelation time and final gelation time of dynamic gelation in porous media were longer than those of static gelation in ampoule bottles and porous media.

2.3.3. Analysis of Water Flooding after Dynamic Gelation

After the dynamic gelation process of the phenolic resin hydrogel was completed, the changes in the pressure difference of subsequent water flooding were recorded at each point with the injection water volume (see Figure 10). According to the experimental method of the dynamic gelation of phenolic resin hydrogel, it can be seen that, during the entire dynamic gelation process, there were two PV phenolic resin hydrogel solutions circulating alternately in the porous media. Therefore, the residual resistance coefficient of the subsequent water flooding was generated by the two PV phenolic resin hydrogels.

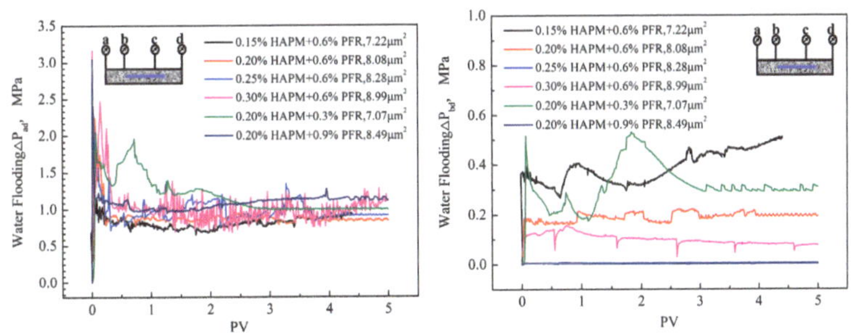

Figure 10. The change in water flooding after dynamic gelation with pore volume under different concentrations of polymer and crosslinker.

From Figure 10, it can be seen that, under different concentrations of polymer and crosslinker, the subsequent water flooding pressure difference ΔP_{ad} exhibited the same change trend after the dynamic gelation of the phenolic resin hydrogel. With an increase in the water flooding pore volume, ΔP_{ad} first rapidly increased to the maximum value, then decreased and stabilized. The change in ΔP_{bd} was related to the concentrations of the polymer and crosslinker. When the concentrations of the polymer and crosslinker were relatively small, ΔP_{bd} showed a certain value, especially if the polymer concentration was less than 0.2 wt%. When the pressure difference ΔP_{bd} in the process of dynamic gelation had a certain value, the subsequent water flooding pressure difference ΔP_{bd} showed significant changes. This indicates that only when the phenolic resin hydrogel migrated during the dynamic gelation process would it have a plugging effect on the deep formation during subsequent water flooding. The residual resistance coefficient of water flooding after dynamic gelation in porous media is calculated, as shown in Table 5.

Table 5. The residual resistance coefficient of water flooding after dynamic gelation of hydrogel with different concentrations of polymer and crosslinker.

HPAM, wt%	PFR, wt%	K, μm^2	Speed, mL/min	Residual Resistance Coefficient	
				ad	bd
0.15	0.6	7.22	0.5	98	77
0.2	0.6	8.08	0.5	101	33
0.25	0.6	8.28	0.5	112	-
0.3	0.6	8.99	0.5	150	3
0.2	0.3	7.07	0.5	104	46
0.2	0.9	8.49	0.5	140	-

According to Table 5, the residual resistance coefficient of the ad section gradually increased with increasing concentrations of polymer and crosslinker. Increases in the numbers of polymer and crosslinker molecules were conducive to the formation of a high-strength hydrogel, which was shown by the increase in the plugging strength of the porous media. Therefore, the residual resistance coefficient of subsequent water flooding increased. The difference was not significant compared with the residual resistance coefficient of water flooding after static gelation in porous media, indicating that the phenolic resin hydrogel after dynamic gelation had a good plugging ability. The residual resistance coefficient of the ad section decreased with increasing concentrations of polymer and crosslinker. When the concentration of the polymer was greater than 0.25 wt% and the concentration of the crosslinker was greater than 0.9 wt%, effective plugging could not be formed after the dynamic gelation of the bd section. The concentrations of polymer and crosslinker were positively correlated with the gel strength. The storage modulus and loss modulus of the hydrogel increased with increasing concentrations of polymer and crosslinker, which made the hydrogel difficult to deform and gave it a strong resistance to impact and local damage. Hydrogel with a high internal friction resistance finds it difficult to migrate in rock pores. Therefore, hydrogel with a high gel strength was mainly retained at the injection end after dynamic gel formation, which cannot produce effective plugging in deep formations.

3. Conclusions

The field implementation of hydrogel deep profile control can greatly improve oil recovery. It is affected by the dynamic gelation of hydrogel in porous media, especially the influence of the mass concentration of hydrogel components on the migration rule after dynamic gelation. This paper studied the quantitative relationship between dynamic and static gelling, the regulation of the subsequent water flooding of dynamic gelation, and the influence of the polymer concentration and crosslinker concentration on the deep migration of hydrogel. The main conclusions were as follows:

(1) The quantitative relationship between the static gelation of hydrogels in different environments was established. The static gelation time of PFR hydrogel in porous media was longer than that in ampoule bottles, where the initial gelation time increased to 1~1.5 times and the final gelation time increased to 1.5~2 times, respectively. The phenolic resin hydrogel could form a three-dimensional network structure after static gelation in ampoule bottles and porous media.

(2) The quantitative relationship between the dynamic gelation of hydrogels in different environments was established, and revealed the existing form of hydrogels after dynamic gelation. The dynamic initial and final gelation time of PFR hydrogel in porous media was longer than the static initial gelation time in ampoule bottles and porous media. The phenolic resin hydrogel could only form dispersed hydrogel particles after dynamic gelation in porous media.

(3) The subsequent water flooding experiment of dynamic gelation showed that the residual resistance coefficient of water flooding was not significantly different from that after static gelation in porous media, indicating that the polymer hydrogel after dynamic gelation had a good plugging ability.

(4) The phenolic resin hydrogel could realize deep migration during dynamic gelation in porous media, but it was limited by the concentrations of the polymer and crosslinker. The deep reservoir could only be blocked off in the subsequent water flooding process when the migration of the hydrogel happened in the dynamic gelation process.

4. Materials and Methods

4.1. Materials

The polymer employed in this research was the ordinary partially hydrolyzed polyacrylamide (HPAM), whose molecular weight was 1.2×10^7 and the degree of hydrolysis was 22%. The crosslinker was a water-soluble phenolic resin prepolymer, which was obtained by the poly-condensation of phenol and formaldehyde (excess) under the catalyst action of sodium hydroxide [20]. The used synthetic water (SW) contained 6921 ppm Na^+, 412 ppm Ca^{2+}, 148 ppm Mg^{2+}, and 11853 ppm Cl^-.

4.2. Experimental Methods

4.2.1. Static Gelation in Ampoule Bottles

The viscosity method was used to determine the relationship between the viscosity of the hydrogel system and time under different gelation times. The experimental method was as follows: the prepared polymer solutions were diluted with simulated water, then crosslinker agents with different concentrations were added. After thorough mixing, they were placed in a constant temperature oven at 75 °C. The viscosity of the system was measured at different gelation times using a DV-II viscometer.

4.2.2. Static gelation in Porous Media

The permeability and pore volume of sandpack pipe models (Φ 2.5 cm × 10 cm) were measured after being saturated with water. In total, 1 PV of hydrogel solution was injected into every pipe, and the sandpack pipe models were placed in a 75 °C oven. At each interval, a sandpack pipe was taken out to carry out the water flooding experiment. The relationship between the pressure and pore volume was measured and the residual resistance coefficient was calculated under 1 mL/min during water flooding. The determination of the gelation time was due to the relationship between the residual resistance coefficient and gelling time.

4.2.3. Dynamic Gelation in Porous Media

The dynamic gelation of HPAM/PFR hydrogel in porous media could be estimated with the circulating device, which comprised two piston containers at room temperature and a sandpack at 75 °C. Specific methods have been shown in the literature [15], as shown in Figure 11.

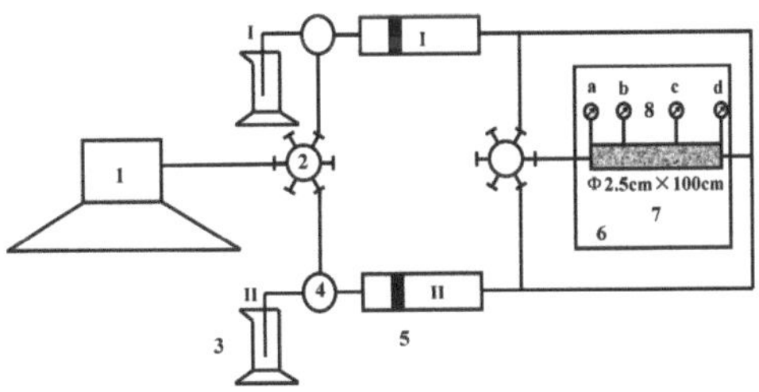

Figure 11. Schematic of circulated equipment for dynamic gelation in porous media.

4.2.4. Water Flooding after Dynamic Gelation in Porous Media

In order to confirm the residual resistance coefficient of the used hydrogel system after dynamic gelation, subsequent water flooding under 1 mL/min was conducted when the process of dynamic gelation was finished. The plugging ability could be determined by the curve of the pressure difference with the injected pore volume of water.

Author Contributions: Conceptualization, H.Y. and W.J.; methodology, Y.L.; validation, B.C.; formal analysis, W.J.; investigation, W.J.; writing—original draft preparation, H.Y.; writing—review and editing, W.J. and W.S. All authors have read and agreed to the published version of the manuscript.

Funding: This research was funded by National Key R&D Program of China (NO. 2022YFE0129900), National Natural Science Foundation of China (NO. 52174026 and NO. 52204009), CNPC Innovation Found (NO. 2021DQ02-0206).

Institutional Review Board Statement: The study did not require ethical approval.

Informed Consent Statement: Informed consent was obtained from all subjects involved in the study.

Data Availability Statement: The data presented in this study are openly available in article.

Acknowledgments: The authors acknowledge Jidong oilfield for financial support and experimental materials.

Conflicts of Interest: Author B.C. was employed by the Chuanqing Drilling Engineering Company Limited. The remaining authors declare that the research was conducted in the absence of any commercial or financial relationships that could be construed as a potential conflict of interest.

References

1. Al-Muntasheri, G.A.; Nasr-El-Din, H.A.; Peters, J.A.; Zitha, P.L.J. Investigation of a High-Temperature Organic Water-Shutoff Gel: Reaction Mechanisms. *SPE J.* **2006**, *11*, 497–504. [CrossRef]
2. Banerjee, R.; Patil, K.; Khilar, K.C. Studies on Phenol-Formaldehyde Gel Formation at a High Temperature and at Different pH. *Can. J. Chem. Eng.* **2006**, *84*, 328–337. [CrossRef]
3. Aldhaheri, M.; Wei, M.Z.; Zhang, N.; Bai, B.J. A Review of Field Oil-Production Response of Injection-Well Gel Treatments. *SPE Res. Eval. Eng.* **2019**, *22*, 597–611. [CrossRef]
4. Khurshid, I.; Afgan, I. Geochemical Investigation of Electrical Conductivity and Electrical Double Layer Based Wettability Alteration During Engineered Water Injection in Carbonates. *J. Petrol. Sci. Eng.* **2022**, *215*, 110627. [CrossRef]
5. Zhou, B.B.; Kang, W.L.; Yang, H.B.; Zhu, T.Y.; Zhang, H.W.; Li, X.X.; Sarsenbekuly, B.; Sarsenbek, T. Preparation and Properties of an Acid-Resistant Preformed Particle Gel for Conformance Control. *J. Petrol. Sci. Eng.* **2021**, *197*, 107964. [CrossRef]
6. Ki, D.Y.; Woo, K.Y.; Lee, S.B. Static and Dynamic Properties of the Backbone Network for the Irreversible Kinetic Gelation Model. *Phys. Rev. E* **2000**, *62*, 821. [CrossRef] [PubMed]
7. Dolan, D.M.; Thiele, J.L.; Willhite, G.P. Effects of pH and Shear on the Gelation of a Cr(III)-Xanthan System. *SPE Prod. Oper.* **1998**, *13*, 97–103. [CrossRef]
8. Kolnes, J.; Stavland, A.; Thorsen, S. The effect of temperature on the gelation time of xanthan/Cr(III) systems. In Proceedings of the SPE International Conference on Oilfield Chemistry, Anaheim, CA, USA, 20–22 February 1991. SPE 21001-MS.
9. Bhasker, R.; Stinson, J.; Willhite, G.; Thiele, J. The effects of shear history on the gelation of polyacrylamide/chromium (VI)/thiourea solutions. *SPE Reserv. Eng.* **1988**, *3*, 1251–1256. [CrossRef]
10. Carvalho, W.; Djabourov, M. Physical Gelation under Shear for Gelatin Gels. *Rheol. Acta* **1997**, *36*, 591–609. [CrossRef]
11. McCool, S.; Li, X.P.; Willhite, G.P. Flow of a Polyacrylamide/Chromium Acetate System in a Long Conduit. *SPE J.* **2009**, *14*, 54–66. [CrossRef]
12. Seright, R. Use of Preformed Gels for Conformance Control in Fractured Systems. *Old Prod. Facil.* **1997**, *12*, 59–65. [CrossRef]
13. Abete, T.; De Candia, A.; Del Gado, E.; Fierro, A.; Coniglio, A. Static and Dynamic Heterogeneities in a Model for Irreversible Gelation. *Phys. Rev. Lett.* **2007**, *98*, 088301. [CrossRef] [PubMed]
14. Sun, Z.; Wang, S.; Zhu, Q.; Cao, X.; Lv, K.; Feng, Y.; Yin, H. Insights into Polyacrylamide Hydrogels Used for Oil and Gas Exploration: Gelation Time, Gel Strength, and Adhesion Strength. *Energy Fuels* **2023**, *37*, 19548–19561. [CrossRef]
15. Khurshid, I.; Afgan, I. Novel insights into the geochemical evaluation of polymer drive composition on surfactant retention in carbonates using the surface complexation modeling. *Sci. Rep.* **2022**, *12*, 17542. [CrossRef] [PubMed]
16. Yu, H.Y.; Jiang, X.R.; Ji, W.J.; Song, W.Q.; Cao, Y.M.; Yan, F.; Luo, C.; Yuan, B. The New Low Viscosity and High-Temperature Resistant Composite Hydrogel. *Chem. Pap.* **2023**, *77*, 3561–3570. [CrossRef]
17. Yu, H.Y.; Ma, Z.F.; Tang, L.; Li, Y.S.; Shao, X.Z.; Tian, Y.X.; Qian, J.; Fu, J.; Li, D.; Wang, L.; et al. The Effect of Shear Rate on Dynamic Gelation of Phenol Formaldehyde Resin Gel in Porous Media. *Gels* **2022**, *8*, 185. [CrossRef] [PubMed]
18. Yu, H.Y.; Yu, J.F.; Ji, W.J.; Zheng, J.P.; Wang, Y.F. Dynamic Gelation of the HPAM/Phenol-Formaldehyde Resin Gel under Oscillatory Shear: Critical Gelation Shear Rate and Reformation. *Chem. Pap.* **2021**, *75*, 1313–1322. [CrossRef]
19. Yu, H.Y.; Ji, W.J.; Zheng, J.P. Dynamic and Static Gelation Behavior of Phenol Formaldehyde Resin Gel System in Ampoule Bottles and Porous Media. *Oil Gas Sci. Technol.* **2020**, *75*, 55. [CrossRef]
20. Albonico, P.; Bartosek, M.; Malandrino, A.; Bryant, S.; Lockhart, T.P. Studies on Phenol-Formaldehyde Crosslinked Polymer Gels in Bulk and in Porous Media. In Proceedings of the SPE International Symposium on Oilfield Chemistry, San Antonio, TX, USA, 14–17 February 1995. SPE 28983.
21. Bryant, S.L.; Rabaioli, M.R.; Lockhart, T.P. Influence of Syneresis on Permeability Reduction by Polymer Gels. *SPE Prod. Oper.* **1996**, *11*, 209–215. [CrossRef]
22. Shi, J.; Wei, X. Synthesis and Evaluation of Polymer Oil Displacement Agent Crosslinked by Phenolaldehyde Resin Prepolymer. *Adv. Fine Petrochem.* **2003**, *7*, 45–48.
23. Sydansk, R. A New Conformance-Improvement-Treatment Chromium(III) Gel Technology. In Proceedings of the DOE Symposium on Enhanced Oil Recovery, Tulsa, OK, USA, 17–20 April 1988. SPE/DOE 17329.
24. Reddy, B.R.; Eoff, L.; Dalrymple, E.D.; Brown, D. Natural Polymer-Based Compositions Designed for Use in Conformance Gel Systems. *SPE J.* **2005**, *10*, 385–393. [CrossRef]

25. Mokhtari, M.; Ozbayoglu, M.E. Laboratory Investigation on Gelation Behavior of Xanthan Crosslinked with Borate Intended to Combat Lost Circulation. In Proceedings of the SPE International Production and Operations Conference and Exhibition, Tunis, Tunisia, 8–10 June 2010. SPE 136094.
26. Smith, J.E. The Transition Pressure: A Quick Method for Quantifying Polyacrylamide Gel Strength. In Proceedings of the SPE International Conference on Oilfield Chemistry, Houston, TX, USA, 8–10 February 1989. SPE 18739.
27. Prada, A.; Civan, F.; Dalrymple, E.D. Evaluation of Gelation Systems for Conformance Control. In Proceedings of the SPE Improved Oil Recovery Conference, Tulsa, OK, USA, 3–5 April 2000. SPE 59322.
28. Park, P.J.; Sung, W. Polymer Translocation Induced by Adsorption. *J. Chem. Phys.* **1998**, *108*, 3013–3018. [CrossRef]
29. Delshad, M.; Kim, D.H.; Magbagbeola, O.A.; Huh, C.; Pope, G.; Tarahhom, F. Mechanistic Interpretation and Utilization of Viscoelastic Behavior of Polymer Solutions for Improved Polymer-Flood Efficiency. In Proceedings of the SPE Improved Oil Recovery Conference, SPE Improved Oil Recovery Conference, Tulsa, OK, USA, 20–23 April 2008. SPE 113620.
30. Chauveteau, G.; Tabary, R.; Renard, M.; Omari, A. Controlling In-Situ Gelation of Polyacrylamides by Zirconium for Water Shutoff. In Proceedings of the SPE International Symposium on Oilfield Chemistry, Houston, TX, USA, 16–19 February 1999. SPE 50752.
31. Yu, H.Y.; Wang, Y.F.; Zhang, J.; Lv, P.; Shi, S.L. Dynamic Gelation of HPAM/Cr(III) under Shear in an Agitator and Porous Media. *Oil Gas Sci. Technol.* **2015**, *70*, 941–949.
32. Savins, J.G. Non-Newtonian Flow Through Porous Media. *Ind. Eng. Chem.* **1969**, *61*, 18–47. [CrossRef]
33. Hirasaki, G.J.; Pope, G.A. Analysis of Factors Influencing Mobility and Adsorption in the Flow of Polymer Solution Through Porous Media. *SPE J.* **1974**, *14*, 337–346. [CrossRef]
34. Camilleri, D.; Engelson, S.; Lake, L.W.; Lin, E.C.; Ohno, T.; Pope, A.G.; Sepehrnoori, K. Description of an Improved Compositional Micellar/Polymer Simulator. *SPE Reserv. Eng.* **1987**, *2*, 427–432. [CrossRef]

gels

Article

Micro and Macro Flooding Mechanism and Law of a Gel Particle System in Strong Heterogeneous Reservoirs

Rongjun Ye [1,2,*], Lei Wang [1,2,*], Wenjun Xu [1,2], Jianpeng Zhang [1,2] and Zhengbang Chen [1,2]

1 School of Petroleum Engineering, Yangtze University, Wuhan 430100, China; xuwenjun@yangtzeu.edu.cn (W.X.); jianpzhang@163.com (J.Z.); chen03012024@163.com (Z.C.)

2 Key Laboratory of Drilling and Production Engineering for Oil and Gas, Wuhan 430100, China

* Correspondence: weng_yip@163.com (R.Y.); wang-lei@yangtzeu.edu.cn (L.W.); Tel.: +86-159-1508-0560 (R.Y.); +86-189-7116-8785 (L.W.)

Abstract: To address the issue of ineffective injection resulting from the consistent channeling of injected water through highly permeable channels in ultra-deep, high-temperature, high-salinity, and strongly heterogeneous reservoirs during the production process, a gel particle profile control agent suitable for high-temperature and high-salinity conditions was chosen. With the help of the glass etching visual microscopic model and the heterogeneous long core model, the formation mechanism of a water flooding channeling path and the distribution law of the remaining oil were explored, the microscopic profile control mechanism of the different parameters was clarified, and the profile control effect of macroscopic core displacement was analyzed. The research shows that the formation mechanism of a water flooding channeling path is dominated by the distribution law of the permeability section and the connection mode between different penetration zones. The remaining oil types after water flooding are mainly contiguous block, parallel throats, and multi-branch clusters. The profile control effect of gel particles on reservoir vertical heterogeneity is better than that of reservoir lateral heterogeneity. It was found that 10 wt% submicron particles with a median diameter of 600 nm play a good role in profiling and plugging pores of 5–20 μm. In addition, 10 wt% micron-sized particles with a median diameter of 2.63 μm mainly play a strong plugging role in the pores of 20–30 μm, and 5 wt% micron-sized particles with a median diameter of 2.63 μm mainly form a weak plugging effect on the pores of 10–20 μm. The overall profile control effect of 10 wt% submicro particles is the best, and changes in concentration parameters have a more significant effect on the profile control effect. In the macroscopic core profile control, enhanced oil recovery (EOR) can reach 16%, and the gel particles show plugging, deformation migration, and re-plugging. The research results provide theoretical guidance for tapping the potential of the remaining oil in strong heterogeneous reservoirs. To date, the gel particles have been applied in the Tahe oilfield and have produced an obvious profile control effect; the oil production has risen to the highest value of 26.4 t/d, and the comprehensive water content has fallen to the lowest percentage of 32.1%.

Keywords: strong heterogeneous reservoir; channeling path; gel particle; physical simulation; enhanced oil recovery (EOR)

Citation: Ye, R.; Wang, L.; Xu, W.; Zhang, J.; Chen, Z. Micro and Macro Flooding Mechanism and Law of a Gel Particle System in Strong Heterogeneous Reservoirs. *Gels* **2024**, *10*, 151. https://doi.org/10.3390/gels10020151

Academic Editor: Jordi Puiggali

Received: 17 January 2024
Revised: 7 February 2024
Accepted: 16 February 2024
Published: 19 February 2024

1. Introduction

With the development of China's oilfields gradually entering the middle and late stages, the long-term water injection development method has led to a significant increase in the water content of the produced liquid of oil wells. For heterogeneous strong reservoirs, this will further enhance their heterogeneity, resulting in excess dispersion of the remaining oil. Therefore, the remaining oil in a strong heterogeneous reservoir developed by long-term water injection is rich in potential space [1–4]. The reservoir which is the fourth Triassic member in the T block of the Tahe oilfield is an ultra-deep, high-temperature, and high-salinity reservoir [5,6]. There are two sets of interlayers with moderate heterogeneity in the longitudinal direction and strong heterogeneity in the plane. After long-term water injection development, the current oil recovery is still low. The main reason for this is that there is a local concentration of plane streamlines in the water injection block, and the injected water is easy to channel along the dominant channel, resulting in premature water breakthrough in the oil well and ineffective water injection. Behzad Vaferi et al. applied deep-learning alternative models to prove that the water injection pattern strongly affects the water channeling reduction and the ultimate oil recovery increase [7]. However, at present, studies on the formation law of micro-waterflood channels in real-model experiments are still insufficient, and it is difficult to take effective measures to deal with the problems in water channels [8,9].

As an important measure to improve the development effect of water flooding, control water, and stabilize oil, profile control and flooding technology has been widely used in oilfields at home and abroad [10,11]. The essence of this technology is to improve reservoir heterogeneity and improve the water flooding sweep efficiency and oil washing efficiency by injecting chemical agents with both profile control abilities and oil washing abilities into injection wells, so as to achieve the purpose of enhance oil recovery (EOR) [12,13]. For the harsh conditions of ultra-deep, high-temperature, and high-salt reservoirs, conventional profile control and flooding systems such as inorganic gel have good temperature and salt resistance but short gelation times, and it is difficult to control the gelation time [14,15]. Inorganic particles are less affected by temperature and salinity but are poor in selective plugging, and raw materials are greatly affected by the origin [16–18]. Foams have strong selective plugging abilities which enables them achieve high-temperature and high-salt resistance, but they have short validity periods and involve complex construction technology [19–22]. Resin has a high plugging strength and a long validity period but poor selective plugging and is difficult to remove after plugging [23–25]. Organic polymer particles are widely used because of their low cost, good selectivity, and ability to absorb water and expand to form deep profile control under high-temperature and high-salt conditions. At present, research into organic particles is relatively mature, which includes research regarding profile control colloidal dispersion gel (CDG), precrosslinked gel particles (PPGs), polymer microspheres, and gelled dispersion particles (DPGs) [26–30]. Considering the technical comparison above and factors such as temperature and salt resistance, deep migration, strong reservoir matching, and simple construction, the organic polymer particle control agent is still the best choice for the control of strong heterogeneous reservoirs.

In recent years, a kind of granular micro and nano expanded polymer gel micellar emulsion technology has been developed. The preparation process of this technology is simple, efficient, and environmentally friendly, and the particles at the nanometer to millimeter level can be prepared according to the requirements of the reservoir. However, the influence of the key injection parameters of the particles on the oil displacement effect is still unknown. Therefore, on the basis of this technology, this paper developed a kind of reservoir pore throat size particle suitable for block T in the Tahe oilfield. At the microscopic level, the visual microscopic model of glass etching was used to compare and analyze the oil displacement effect of the system under different key parameters, and the formation mechanism of a water drive channel and the distribution law of remaining oil were studied. On the macro level, the core displacement experiment was used to analyze the displacement effect of gel particles and verify the characteristics of the micro displacement mechanism.

2. Results and Discussion

2.1. Study on Water Flooding Microscopic Channeling Mechanism and Remaining Oil Distribution Law

2.1.1. Mechanism Analysis of Water Flooding Channeling Path

The water flooding process is shown in Figure 1. In Figures 1–9, blue is the simulated water, red is the simulated oil, white is the rock skeleton, and the displacement direction is from left to right. The red circle indicates remaining oil, the black circle represents gel particles, and the green arrow indicates the inlet and outlet. In the early stage of water flooding, the overall performance of the water drive front is finger breakthrough. This means that the high-permeability section advances rapidly, while the middle- and low-permeability sections on both sides progress slowly, leading to the formation of local finger phenomenon within each penetration section (Figure 1a). With the increase in displacement pressure and the difference of oil–water viscosity, the injected water not only advances forward, but also spreads to both sides of the model (Figure 1b). The analysis shows that the occurrence of this phenomenon depends on the degree of pore connectivity at the interface between the permeability zones. The connection of different permeability zones belongs to the intralayer connection, where the pores on both sides are in an almost interconnected state, with the large pores dominating. At this time, the degree of pore connection has improved, and the speed of the two sides is faster. If the connection belongs to the interlayer connection, meaning there is a certain degree of interlayer at the connection, which makes the displacement phase unable to spread to another permeability zone, then the degree of pore connection is poor, and the speed of the two sides is slow or almost zero. However, the formation of the overall fingering phenomenon of water flooding is still dominated by the pore size. The injected water is rapidly channeled along the middle high-permeability section to form the dominant channel, followed by the middle-permeability section above (Figure 1c).

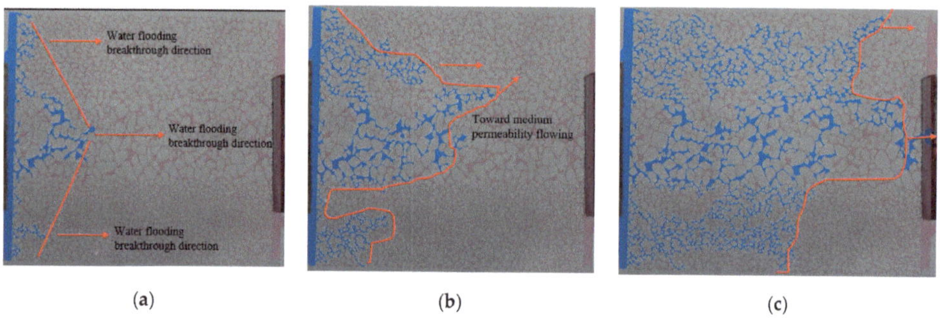

Figure 1. Water flooding phenomena in different periods: (**a**) Initial stage of water flooding; (**b**) Middle stage of water flooding; (**c**) Later stage of water flooding.

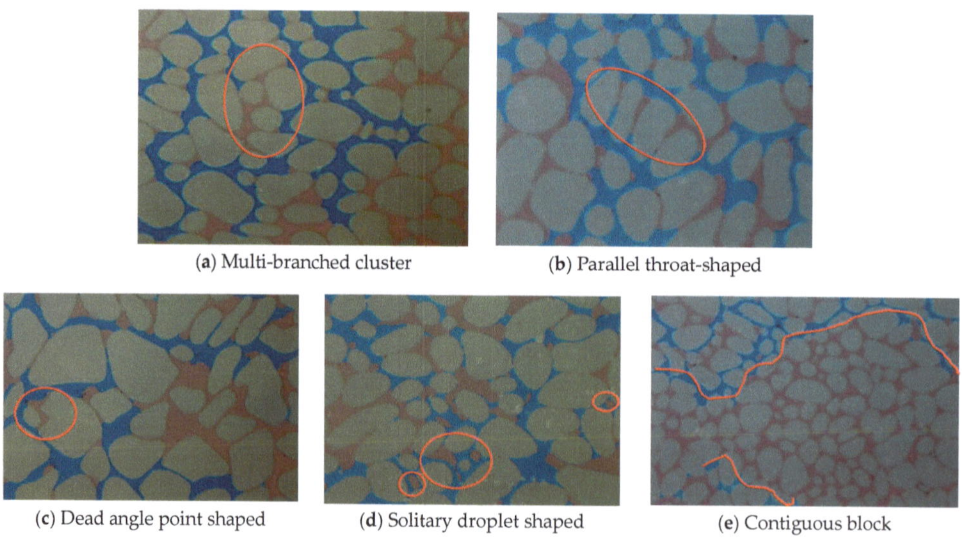

Figure 2. The microscopic local enlargement diagram of various remaining oil forms.

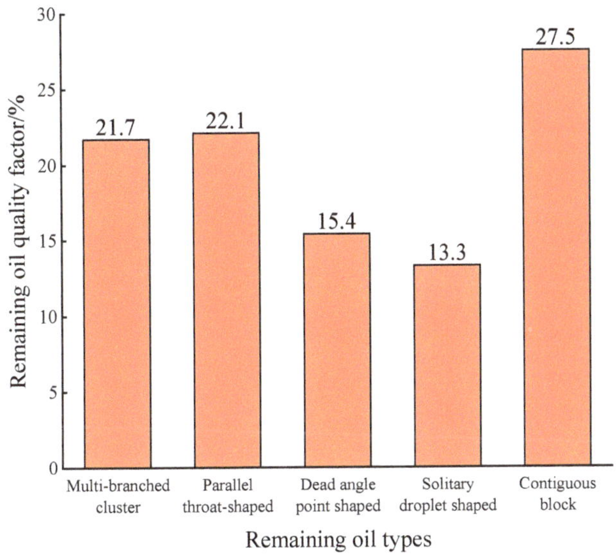

Figure 3. Pile-up histogram of various types of remaining oil after water flooding.

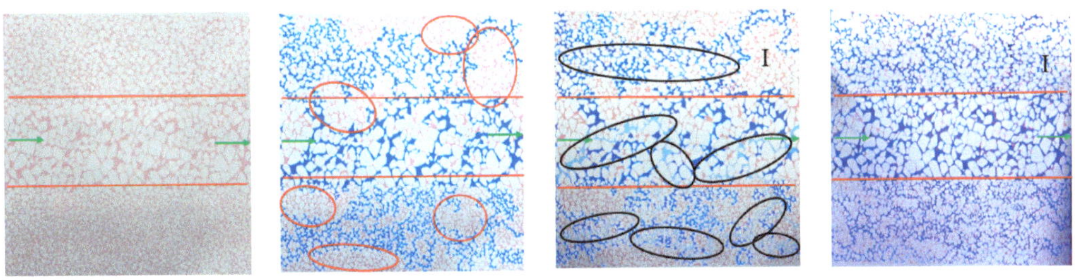

(**a**) (**b**) (**c**) (**d**)

Figure 4. Experiment 1: Remaining oil distribution map (vertical, 10 wt% submicron scale 100 μL): (**a**) oil flooding; (**b**) water flooding; (**c**) particle flooding; (**d**) subsequent water flooding.

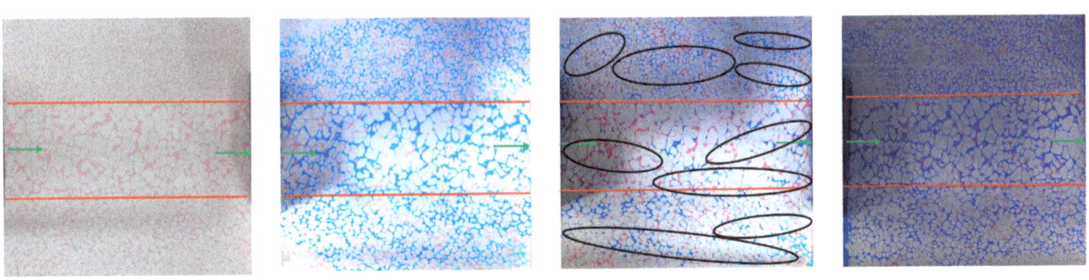

(**a**) (**b**) (**c**) (**d**)

Figure 5. Experiment 1: Remaining oil distribution map (horizontal, 10 wt% submicron scale 100 μL): (**a**) oil flooding; (**b**) water flooding; (**c**) particle flooding; (**d**) subsequent water flooding.

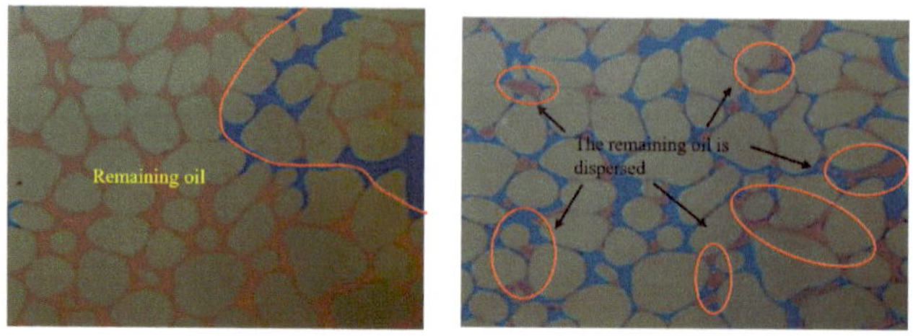

(a) (b)

Figure 6. Experiment 1: Local enlargement of Region I: (**a**) particle flooding; (**b**) subsequent water flooding.

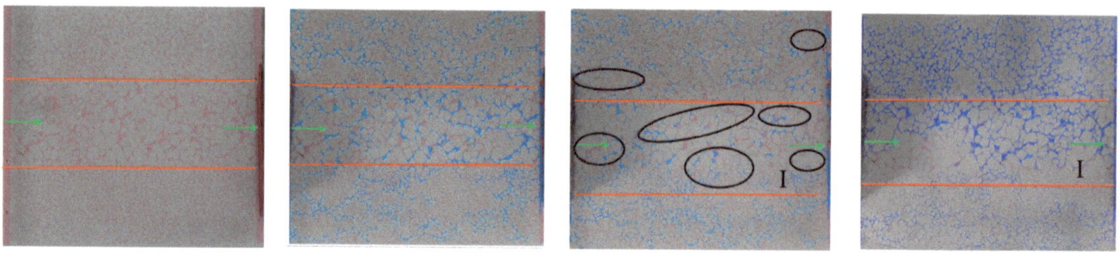

(a) (b) (c) (d)

Figure 7. Experiment 3: Remaining oil distribution map (vertical, 10 wt% micron 100 μL): (**a**) oil flooding; (**b**) water flooding; (**c**) particle flooding; (**d**) subsequent water flooding.

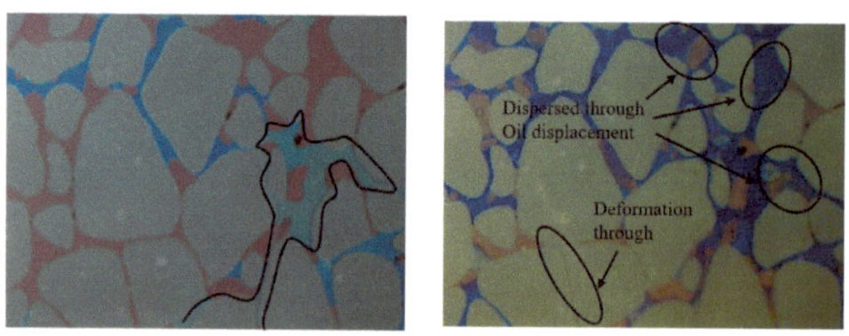

(a) (b)

Figure 8. Experiment 3: Local enlargement of Region I: (**a**) particle flooding; (**b**) subsequent water flooding.

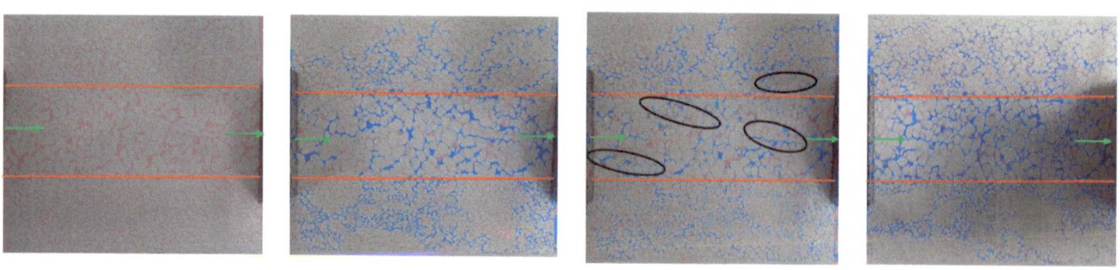

(a) (b) (c) (d)

Figure 9. Experiment 4: Remaining oil distribution map (vertical, 5 wt% micron 100 μL): (**a**) oil flooding; (**b**) water flooding; (**c**) particle flooding; (**d**) subsequent water flooding.

2.1.2. Remaining Oil Distribution

After water flooding, the oil—water distribution image of the model is locally enlarged, the distribution law of the remaining oil is analyzed, and the remaining oil is quantitatively assessed using image analysis software. As shown in Figure 2, the distribution types of the remaining oil in the model after water flooding are summarized as follows: (1) Large pores surrounded by multiple fine pore throats, making it difficult for injected water to overcome large seepage resistance and form multi-branch cluster residual oil (Figure 2a); (2) A small pore throat is formed in parallel with the large pore channel. The injected water tends to enter the larger pores, bypassing the smaller ones, and eventually establishes a consistent and stable flow pathway within the larger pores, resulting in the formation of parallel residual oil reservoirs. (Figure 2b); (3) In situations where there is only a single flow outlet outside the channel and the injected water flows vertically towards the outlet end, it becomes challenging for the injected water to alter the direction of its streamline and create a pressure differential within the channel. As a result, this leads to the formation of dead-angle points where residual oil accumulates. (Figure 2c); (4) When the rock surface is hydrophilic, the injected water tends to flow along the contours of the rock surface, thereby creating isolated droplets of residual oil in the central region of the larger channel. (Figure 2d); (5) In the place where a large number of small pores exists or the large pores are nearly enclosed by small pores, it becomes challenging for the injected water to create a local breakthrough and subsequently disperse the flake oil, thus forming a contiguous block of residual oil (Figure 2e) [31–34]. The distribution of various types of remaining oil after water flooding is statistically analyzed. It can be seen from Figure 3 that the contiguous block, parallel throat shape, and multi-branch cluster shape account for more than 20%, and the remaining oil types mainly consist of these three types.

2.2. *Study on Micro Flooding Mechanism of Key Parameters of Gel Particles*

In this study, we change the heterogeneous distribution form of the model, the size of the injected particles, and the concentration of the injected particles. We compare the microscopic profile control effects under different key parameters of the gel particle profile control agent.

2.2.1. Heterogeneous Distribution Form of the Model

Changing the form of the heterogeneous distribution, the vertical placement represents vertical heterogeneity, while the horizontal placement represents plane heterogeneity. Comparing the experimental results of Experiment 1 and Experiment 2, as shown in Figures 4 and 5, the colorless liquid area in the subsequent water flooding represents water after high-temperature fading. The red circle indicates oil, the black circle represents gel particles, and the green arrow indicates the inlet and outlet. Under the condition of longitudinal heterogeneity, the water flooding efficiency is highest in the permeability section. Simultaneously, the influence of gravity causes an excess accumulation of the remaining oil in the upper part of the permeable section. However, sizable pockets of remaining oil are still observed in the lowest part of the low-permeability section, highlighting that pore size remains a critical factor in the formation of residual oil. The type of residual oil after water flooding is mainly contiguous residual oil. A large number of unexpanded gel particles enter the hypertonic channel. After 72 h of thermal expansion of gel particles in the model, large pores can be effectively blocked. A large amount of contiguous residual oil is dispersed into residual droplets through subsequent water flooding and peristaltic gel particles, and then it flows out with the water [35–37] (Figure 6). The remaining oil distribution in the three permeable sections after water flooding under the plane heterogeneity is relatively uniform, mainly existing in parallel throats and multi-branch clusters. The presence of expansion particles enables the efficient utilization of residual oil in the three penetration sections. However, the effectiveness primarily depends on profile adjustments, and the oil displacement effect is not particularly significant. From an overall perspective of the profile control effect image, the gravitational impact of vertical heterogeneity affects the water flooding effect. When it comes to controlling displacement in a longitudinal heterogeneous reservoir, the impact of expansion particles surpasses that of a transverse heterogeneous reservoir. This observation is further supported by the analysis conducted on the enhanced oil recovery (EOR) below.

2.2.2. Different Injection Particle Sizes

By changing the injection particle size, the experimental results of Experiment 1 and Experiment 3 are compared. Figures 4 and 7 show that 10 wt% submicron particles (600 nm) can penetrate the middle/high/low-permeability sections. The subsequent water flooding can produce steering in these sections, leading to the effective utilization of the remaining oil. The 10 wt% micron particles (2.63 μm) primarily flow into the main channel (20–30 μm) of the high-permeability section, causing a significant plugging effect, while they hardly enter the medium- and low-permeability sections. Therefore, the subsequent water flooding mainly turns to the medium- and high-permeability remaining oil area and continues to advance along the larger pores in the low-permeability section. During subsequent water flooding process after 72 h of heat preservation, the particles disperse and deform from their original plugging position to other areas, creating an oil displacement effect. They deform through the fine pore throat to form re-displacement and plugging (Figure 8). From the pressure data in Table 1, it can be seen that the subsequent water flooding pressure in Experiment 1 and Experiment 3 is 0.11 MPa and 0.13 MPa, respectively. The better the pressure-lifting effect of large particles, the more effective the plugging effect, and the stronger the subsequent liquid flow steering ability. However, when combined with the experimental image comparison results mentioned above, it can be observed that its profile control effect in the low permeability section is inferior to that of small particles. It can be observed that increasing the size of injected particles can elevate the injection pressure and counteract the large capillary force of small pore throats. However, due to its large particle size, it fails to enter the large pore channels of the low-permeability section effectively. This limitation hinders the improvement the internal heterogeneity in the low permeability section, resulting in a poor outcome in enhancing oil recovery. Therefore, when selecting the injection particle size of the profile control agent, it is important not only to consider the pressurization effects but also to comprehensively consider the migration and plugging effect in different permeability sections.

Table 1. The pressure performance at different displacement stages under different parameters.

Number	Placement Mode	Injection Particle Sized	Injection Concentration/wt%	Injection Volume/μL	Water Flooding Pressure /MPa	Particle Flooding Pressure /MPa	Subsequent Water Flooding Pressure /MPa
1	vertical	submicro	10	100	0.005	0.006	0.016
2	horizontal	submicro	10		0.003	0.004	0.011
3	vertical	micron	10		0.0035	0.0075	0.0165
4	vertical	micron	5		0.0075	0.0075	0.0175

2.2.3. Different Injection Concentration

By changing the injection concentration, we compare the experimental results of Experiment 3 and Experiment 4. Figures 7 and 9 show that 5 wt% micron particles (2.63 μm) mainly form plugging in the middle- and high-permeability sections. However, due to the low concentration of gel particles, the plugging strength is weak, so the subsequent profile control effect is worse than that of the 10 wt% micron (Figure 9d). Moreover, due to the decrease in concentration, the likelihood of collisions between particles is reduced, making it challenging for particles to adsorb and coalesce with each other. It is difficult to achieve a state of aggregation and accumulation to block the large pores, which hinders the increase in displacement pressure. The particle displacement pressure in Experiment 4 is equal to the initial water displacement pressure (Table 1). However, since the diameter of large-particle-size and low-concentration particle aggregates is smaller than that of large-particle-size and high-concentration aggregates, the low-concentration particle solution will also partially enter the low-permeability section to play a role in profile control and flooding. It can be seen that the ability of particles to migrate to the low-permeability section to perform profile control is not only dependent on the particle size but also on the concentration of the particles.

2.2.4. Analysis of Enhanced Oil Recovery (EOR) under Different Parameters

By using image analysis software to quantitatively analyze the remaining oil in each stage of the model. Based on the results presented in Table 2, it is evident that when altering the heterogeneous distribution form and injection conditions, the enhanced oil recovery (EOR) during the particle flooding stage remains predominantly below 0.5%. The overall enhanced oil recovery (EOR) is mainly influenced by the subsequent water drive flow and the expansion and migration of particles. This indicates that the effect of adjustment is greater than that of flooding. On the other hand, Yuan Chengdong believes that enhancing the microscopic oil displacement efficiency is typically accomplished by increasing the capillary number. The lower the interfacial tension between the displacing phase and the displaced phase, the higher the capillary number [38]. According to the definition of the Newtonian capillary number, along with the interfacial tension mentioned above, the viscosity of the displacing phase, the velocity of displacement, and the contact angle, the capillary number can be calculated to be approximately on the order of 10^{-3} [39]. Therefore, the capillary number of the gel particle agent is smaller and the oil washing efficiency is poor.

Table 2. The EOR performance at different stages under different parameters.

Number	Experimental Parameters	Oil Recovery/%			Enhanced Oil Recovery (EOR)/%		
		Water Flooding	Particle Flooding	Subsequent Water Flooding	Particle Flooding	Subsequent Water Flooding	Particle + Subsequent Water Flooding
1	vertical, 10 wt% submicron	55.98	56.38	78.30	0.40	21.92	22.32
2	horizontal, 10 wt% submicron	59.43	59.84	78.36	0.41	18.51	18.92
3	vertical, 10 wt% micron	57.86	58.28	78.2	0.42	19.92	20.34
4	vertical, 5 wt% micron	56.28	56.6	74.27	0.32	17.67	17.99

The EOR in Experiment 1 is the highest. Compared with Experiment 2, the 3.4% increase in EOR indicates that the gel particles have a better effect on the vertical heterogeneity. The difference in enhanced oil recovery (EOR) between Experiment 1 and Experiment 3, utilizing pellet flooding followed by subsequent water flooding, is 1.98%. Meanwhile, the EOR difference between Experiment 3 and Experiment 4, employing particle flooding along with subsequent water flooding, is 2.35%. Upon comparing these differences, it becomes evident that the impact of concentration parameters on enhanced oil recovery outweighs that of particle size parameters. Comparing the EOR of particle flooding + subsequent water flooding with different particle sizes and concentrations, it can be observed that the profile control effect of 10 wt% submicron particles is the most effective.

2.3. Evaluation of Flooding Effect of Gel Particle Macro Core

Since the experimental conditions of the microscopic model involve normal temperature and pressure, they do not simulate the complex high-temperature and high-pressure environment in the actual reservoir. As a result, the enhanced oil recovery (EOR) is higher than that of the core flooding experiment. Due to the small pore volume of the model, accurately and quantitatively describing the change characteristics of the oil recovery factor and water content in the displacement process is challenging. Therefore, the actual production profile control and displacement effect of the gel particle should be based on the core displacement experiment.

Consistent with the design of micro-scale experimental parameters, the core displacement experiment compares the effects of various particle sizes and concentrations under the conditions of vertical heterogeneity. The results are shown in Table 3. Under similar flooding conditions, the final enhanced oil recovery of submicron particles is higher than that of micron particles. In the case of particles of the same size, the final enhanced oil recovery of the high-concentration 10 wt% particle solution is greater than that of the low-concentration 5 wt% particle solution, which aligns with the enhanced oil recovery (EOR) analysis of the microscopic model.

Table 3. The pressure and oil recovery at different stages under different parameters.

Condition	Parameters	Pressure/MPa			Oil Recovery/%			Ultimate Oil Recovery/%	EOR/%
		Water Flooding	Particle Flooding	Subsequent Water Flooding	Water Flooding	Particle Flooding	Subsequent Water Flooding		
Vertical heterogeneity, similar amount of profile control flooding	Micro particles	0.06	0.31	0.45	58.25	2.26	9.86	70.37	12.12
	Submicro particles	0.08	0.30	0.46	57.98	2.51	10.95	71.44	13.46
Vertical heterogeneity, submicron particles	5 wt%	0.07	0.31	0.55	58.02	2.78	11.25	71.44	14.03
	10 wt%	0.08	0.33	0.57	57.61	2.88	13.01	73.50	15.89

Analysis of the production characteristic curve of the gel particle profile control (Figure 10) shows that before the injection amount reaches about 0.3 pore volumes (pv) in the initial stage, it is necessary to first go through the anhydrous oil production period. During this period, all the produced liquid is oil, and the water content is almost zero. The water content of the produced liquid sharply increases to about 90% after the subsequent injection of approximately 0.2 pore volumes (pv) water. When the water content of the produced fluid reaches about 98%, the efficiency of enhanced oil recovery (EOR) is almost 0%, and it can be considered that the dominant channel of water flooding has been stably formed. The decrease in water content and the increase in the oil recovery factor during the gel particle flooding stage are not significant. This is because the gel particles first infiltrate into the core through the dominant channel, carried by high-salinity water initially. The remaining oil is not utilized during this stage, aligning with the aforementioned microscopic experiment. However, the gel particles are a type of particle agent with a specific diameter. As the injection amount increases, they tend to cause blockages, leading to a preference for flowing through narrower channels during particle flooding. This process helps to drive out the remaining oil. At this time, the water content will decrease and the oil recovery factor will increase gradually. When the particles are heated and expanded in the core, they exert a strong plugging effect on the dominant channel, leading to increased pressure and flow around the smaller pore throat during subsequent water flooding. Based on the pressure data values presented in Table 3, after the gel particles are kept warm for 72 h for subsequent water flooding, the maximum increase in subsequent water flooding pressure can reach 0.24 MPa. This suggests that the expanded particles exhibit a more effective plugging effect on larger pores. At present, the gel particles have been applied in Well Group H, Block T, Tahe Oilfield, and have produced a noticeable profile control effect. The oil production has increased to a peak of 26.4 t/d, and the comprehensive water cut has decreased to a minimum of 32.1%.

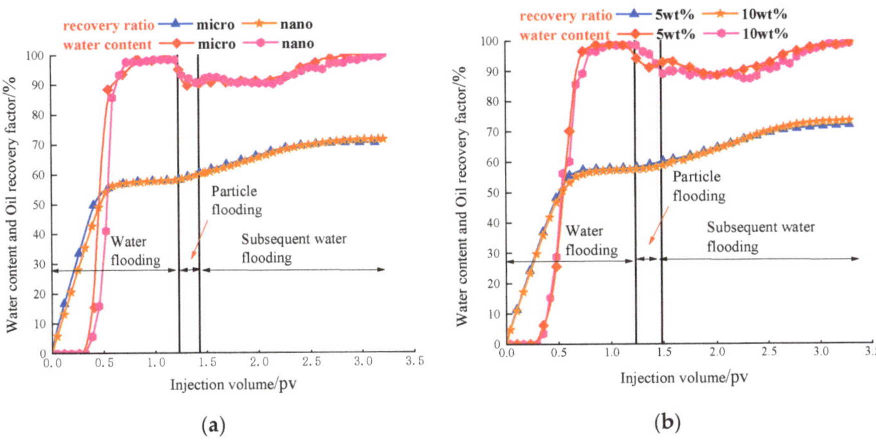

Figure 10. Comparison of the change in the oil recovery factor and water content in the core flooding experiment:(**a**) Change particle size; (**b**) Change particle concentration.

3. Conclusions

(1) From the results of primary water flooding, the formation law of water flooding channels is determined by the heterogeneity distribution of the model and the connection between different permeability sections. There are five types of remaining oil after primary water flooding: contiguous block, multi-branch cluster, parallel throat, corner concave, and dispersed solitary drop, mainly focusing on the first three types;

(2) The profile control effect of gel particles on reservoir vertical heterogeneity is better than that on reservoir lateral heterogeneity. The 10 wt% submicron particle (600 nm) can penetrate into the channel with a designed throat radius of 5–10 μm in the low-permeability section and the medium−high-permeability section to effectively plug it. The 10 wt% micron-sized particles (2.63 μm) can penetrate the channel with a designed throat radius of 20–30 μm in the high-permeability section to effectively plug it. However, it is nearly impossible for them to penetrate and cause plugging in sections with medium and low permeability. The 5 wt% micron-sized particles (2.63 μm) can penetrate into the channel with a designed throat radius of 10–20 μm in the medium-permeability section and high-permeability section, mainly forming weak plugging in the medium- and high-permeability section;

(3) Both microscopic experiments and macroscopic core displacement experiments show that 10 wt% submicron particles have the best effect on profile control and flooding. The concentration parameters have a more significant impact on the effectiveness of profile control and flooding. The mechanism of action of gel particles is primarily characterized by the agent initially entering the dominant channel and subsequently expanding under the influence of high-temperature water. This process results in the phenomenon of plugging, migration and deformation, and re-plugging. Therefore, it has an efficient profile control and flooding effect on heterogeneous reservoirs.

The effects of various key parameters of the expanded gel particles were investigated using an indoor physical model. Since the model could not simulate the actual conditions of the high-temperature and high-salt reservoir, as well as the changes in the pore throat after long-term water flooding, the experimental discussion may be inadequate. In the future, more attention should be paid to the similarity between the model and the actual field conditions so that the conclusions of laboratory experiments can better guide the field application. In addition, considering that the oil washing efficiency of the gel particle agent is not high, the use of nano-sized particles with smaller particle sizes or with surfactants can be considered to enhance collaborative oil flooding.

4. Materials and Methods

4.1. Visual Microscopic Glass Etching Experiment

Using the visual microscopic glass etching model, we designed various key parameters for particles. The aim was to explore the mechanism of profiling and flooding effects caused by the key parameters of the gel particle profile control agent at the micro level. Furthermore, we studied the mechanism of water flooding and the distribution of remaining oil. Equipment: high-grade GP−300C microscope system, polarizing microscope, micro-injection pump, model holder, etc., as shown in Figure 11. The internal pore size of the microscopic model is 45 mm × 45 mm. The penetration zone and three connected channels are etched at the inlet and outlet of the model. The upper, middle, and lower parts of the three penetration sections correspond to different permeability zones: the medium-permeability zone with a throat diameter of 10–20 μm, the high-permeability zone with a throat diameter of 20–30 μm, and the low-permeability zone with a throat diameter of 5–10 μm.

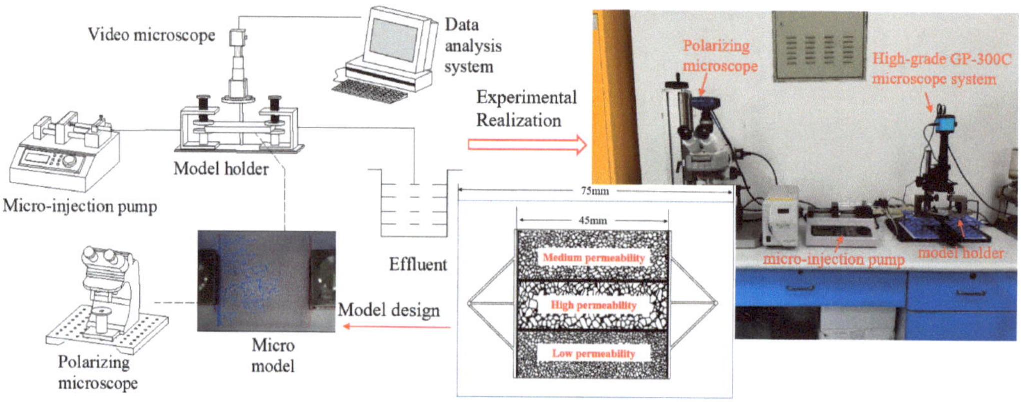

Figure 11. Visualization of microscopic model displacement experiment.

The simulated oil was a mixture of white oil and aviation kerosene in proportion to maintain a viscosity ratio of approximately 5.8:1 at room temperature. It was dyed red with Sudan III and filtered to prevent undissolved dyes from blocking the throat. The simulated water salinity was 210,000 mg/L, and the injected water was stained with methyl blue and then filtered. The gel particle profile control agent is a stable particle system formed by a high viscoelastic polymer gel through high-speed mechanical shear. The gel particle profile control agents used in this paper are composed of submicron particles with an initial median diameter of 600 nm and micron particles with an initial median diameter of 2.63 μm. Under the actual high-temperature and high-salt conditions of the reservoir (102 °C, 210,000 mg/L), the gel particle size can expand approximately four times within 72 h. Moreover, the viscosity of 10 wt% gel particles can be sustained at around 3 mPa·s after shear for 5, 10, 15, 20, and 30 min at 1000 rpm, respectively. At normal temperature, the interfacial tension of oil and water is 25.6 mN/m. This tension decreases by 2.3 mN/m and 1.2 mN/m, respectively, after the addition of submicron and micron profile control agents. By utilizing the sand-filled tube model, the system's plugging rate can remain above 80% after 30 days of aging, even under the storage environment of high temperature and high salt. The wettability of the model is hydrophilic, and the contact angle of the injected water in the model is 39.89°.

Experimental steps:

1. The software was connected to the microscope and the parameters adjusted;
2. The vacuum-saturated simulation of oil was modeled at room temperature;

3. The simulated water was injected at a constant speed of 50 μL/min until the pressure stabilized, and the injection volume reached approximately 320 μL;
4. According to the experimental scheme in Table 4, various particle sizes and concentrations of profile control and flooding agents were injected. Subsequently, the formation temperature was maintained for 72 h;
5. Subsequent water flooding. The experiment was carried out at room temperature, and the injection pressure, injection volume, and injection speed were recorded. The displacement process and local feature images were taken.

Table 4. Visualization microscopic experiment scheme.

Number	Placement Mode	Injection Particle Type	Injection Concentration/wt%	Injection Volume/μL
1	vertical	submicron	10	100
2	horizontal	submicron	10	
3	vertical	micron	10	
4	vertical	micron	5	

4.2. Heterogeneous Rectangular Core Displacement Experiment

It is used to analyze the profile control and flooding effects of gel particles and verify the characteristics of microscopic profile control and flooding mechanisms. Equipment: High-temperature and high-pressure core flow evaluation device includes an advection pump, thermostat, hand pump, core holder, intermediate container, etc., as shown in Figure 12. The simulated water and gel particle profile control agent is the same as that used in the microscopic experiment. The simulated oil is industrial white oil (102 °C, 1.64 mPa·s). The permeability design of the three-layer heterogeneous core model was based on the permeability distribution of the perforated interval of the control and flooding well group selected in the T block of the Tahe oilfield. Finally, the permeability design for the rectangular three-layer heterogeneous core used in the laboratory experiment was upper/middle/lower corresponding to the medium permeability/high permeability/low permeability = 140/280/40 × 10^{-3} μm², 30 × 4.5 × 4.5 cm.

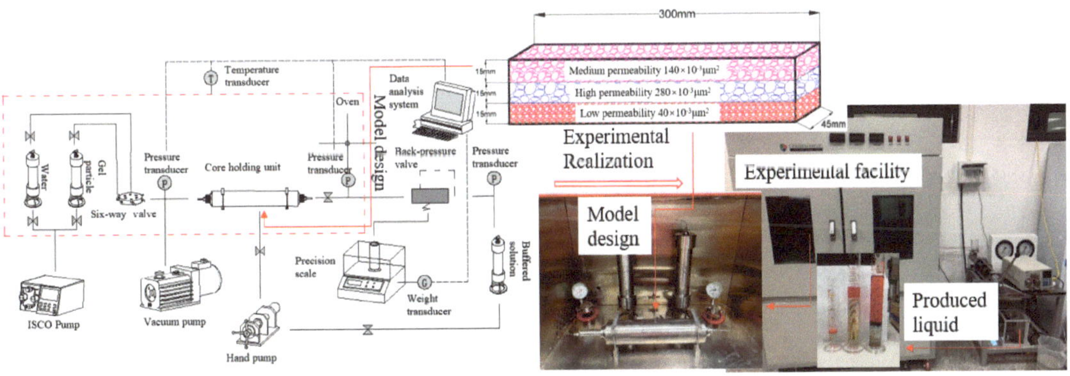

Figure 12. Heterogeneous three-layer long core displacement experiment.

Experimental steps:

1. The air tightness of the equipment was tested;
2. Core drying, weighing, and vacuum pressure saturation simulation of the oil was carried out;
3. Formation temperature aging for 24 h was conducted;
4. Water flooding of the core was stopped when the water content reached 98% at the outlet end;

5. According to the experimental scheme, the flooding agent was injected and maintained at the formation temperature for 72 h;
6. Subsequent water flooding to the outlet end of the water content reached 98%. The experimental formation temperature was carried out, the pressure, produced water, and oil were recorded, and the change trend in water content and the oil recovery factor were then calculated.

By comparing the permeability measured by the gas and liquid of each core, it can be seen from Table 5 that the gas permeability and liquid permeability measured by different cores show little difference, which accords with the experimental error range. The experiment completed by using this model is credible.

Table 5. Basic parameters of the core used in the experiment.

Condition	Parameters	Gas Permeability/10^{-3} μm^2	Liquid Permeability/10^{-3} μm^2	Porosity/%
Vertical heterogeneity, similar amount of profile control flooding	micro particles	154	140.55	20.44
	submicron particles	156	141.98	20.58
Vertical heterogeneity, submicron particles	5 wt%	156	140.36	20.24
	10 wt%	152	140.45	20.96

Author Contributions: Writing—original draft, R.Y.; revising—original draft, L.W.; methodology, W.X.; investigation, J.Z.; project resources, Z.C. All authors have read and agreed to the published version of the manuscript.

Funding: This research was funded by the National Natural Science Foundation of China (NSFC) project "Research on High Temperature Thixotropic Mechanism of Thermosensitive Polymer Nanofluids" (52074038).

Institutional Review Board Statement: Not applicable.

Informed Consent Statement: Not applicable.

Data Availability Statement: The data presented in this study are openly available in the article.

Acknowledgments: The authors gratefully acknowledge the support of Hubei Provincial Key Laboratory of Oil and Gas Drilling and Production Engineering for this work.

Conflicts of Interest: The authors declare no conflicts of interest.

References

1. Gao, Q.; Zhong, C.; Han, P.; Cao, R.; Jiang, G. Synergistic Effect of Alkali–Surfactant–Polymer and Preformed Particle Gel on Profile Control after Polymer Flooding in Heterogeneous Reservoirs. *Energy Fuels* **2020**, *34*, 15957–15968. [CrossRef]
2. Lu, X.; Cao, B.; Xie, K.; Cao, W.; Liu, Y.; Zhang, Y.; Wang, X.; Zhang, J. Enhanced oil recovery mechanisms of polymer flooding in a heterogeneous oil reservoir. *Pet. Explor. Dev.* **2021**, *48*, 169–178. [CrossRef]
3. Gong, W.; Liu, Y.; Xi, C.; Xi, C.; Yang, G.; Ju, Y.; Wang, M. Dynamic characterization of residual oil during long-term waterflooding experiments in heterogeneous porous structures. *Fuel* **2024**, *356*, 129567. [CrossRef]
4. Du, Q. Variation law and microscopic mechanism of permeability in sandstone reservoir during long-term water flooding development. *Acta Pet. Sin.* **2016**, *37*, 1159–1164.
5. Liao, H.; Xu, T.; Yu, H. Progress and prospects of EOR technology in deep, massive sandstone reservoirs with a strong bottom-water drive. *Energy Geosci.* **2023**, *28*, 100164. [CrossRef]
6. Liu, L.; Cao, F.; Liu, X.; Tan, T.; Zheng, X.; Liu, R. Development Characteristics and Potential Tapping Strategies of Massive Sandstone Reservoirs With Bottom Water in Tahe Oilfield. *Xinjiang Pet. Geol.* **2023**, *44*, 450–455.
7. Singh, V.; Ruwali, N.; Pandey, R.K.; Vaferi, B.; Wood, D.A. Applying a deep-learning surrogate model to simulate and compare achievable oil recovery by different waterflood scenarios. *Pet. Sci. Technol.* **2023**, 1–19. [CrossRef]

8. Peng, S.; Shi, Y.; Han, T.; Yao, T.; Huang, J.; Li, H.; Li, X.; Hu, H. A quantitative description method for channeling-path of reservoirs during high water cut period. *Acta Pet. Sin.* **2007**, *28*, 79–84.
9. Wang, Q.; Ma, X.; Bi, Y.; He, H.; Gu, X.; Liu, T.; Zhao, L. Analysis of Channeling-Path Phenomena in a Complex Fault-Block Reservoir with Low Recovery Factor and High Water-Cut Ratio. *Fluid Dyn. Mater. Process.* **2023**, *19*, 361–377.
10. Wei, P.; Zheng, L.; Yang, M.; Wang, C.; Chang, Q.; Zhang, W. Fuzzy-ball fluid self-selective profile control for enhanced oil recovery in heterogeneous reservoirs: The techniques and the mechanisms. *Fuel* **2020**, *275*, 117959. [CrossRef]
11. Jin, F.; Li, Q.; He, Y.; Luo, Q.; Pu, W. Experimental Study on Enhanced Oil Recovery Method in Tahe High-Temperature and High-Salinity Channel Sand Reservoir: Combination of Profile Control and Chemical Flooding. *ACS Omega* **2020**, *5*, 5657–5665. [CrossRef]
12. Cao, B.; Xie, K.; Lu, X.; Cao, W.; Cao, W.; He, X.; Xiao, Z.; Zhang, Y.; Wang, X.; Su, C. Effect and mechanism of combined operation of profile modification and water shutoff with in-depth displacement in high-heterogeneity oil reservoirs. *Colloids Surf. A Physicochem. Eng. Asp.* **2021**, *631*, 127673. [CrossRef]
13. Cao, W.; Xie, K.; Lu, X.; Liu, Y.; Zhang, Y. Effect of Profile-Control Oil-Displacement Agent on Increasing Oil Recovery and Its Mechanism. *Fuel* **2019**, *237*, 1151–1160. [CrossRef]
14. Cao, W.; Xie, K.; Cao, B.; Lu, X.; Tian, Z. Inorganic gel enhanced oil recovery in high temperature reservoir. *J. Petrol. Sci. Eng.* **2021**, *196*, 107691. [CrossRef]
15. Altunina, L.K.; Bokserman, A.A.; Kuvshinov, V.A.; Polkovnikov, V.V. Inorganic gels for enhanced oil recovery at high temperature. *Geol. Soc. Lond. Spec. Publ.* **1995**, *84*, 219–223. [CrossRef]
16. Zhao, X.; Chen, Z.; Chen, W.; Ma, H.; Zhai, D.; Ren, Z. The Current Research Situation And Development Trend Of Particle Profile-control Plugging Agents. *Oil Drill. Prod. Technol.* **2015**, *37*, 105–112.
17. Zhao, F.; Sun, G.; Sun, M.; Xie, T.; Wang, F. Studies On Channel Plugging In A Reservoir By Using Clay As A Profile Control Agent In A Double-fluid Method. *Acta Pet. Sin.* **1994**, *15*, 56–65.
18. Tran, T.V.; Civan, F.; Robb, I. Correlating Flowing Time and Condition For Perforation Plugging By Suspended Particles. *SPE Drill. Complet.* **2009**, *24*, 398–403. [CrossRef]
19. Li, S.; Cheng, H.; Wei, Y.; Li, M.; Wang, Z. Two-Dimensional Visual Analysis of Three-Phase Foam Flooding under Different Reservoir Conditions. *Energy Fuels* **2023**, *37*, 17277–17289. [CrossRef]
20. Wang, L.; Wang, T.; Wang, J.; Tian, H.; Chen, Y.; Song, W. Enhanced Oil Recovery Mechanism and Technical Boundary of Gel Foam Profile Control System for Heterogeneous Reservoirs in Changqing. *Gels* **2022**, *8*, 371. [CrossRef]
21. Tao, J.; Dai, C.; Kang, W.; Zhao, G.; Liu, Y.; Fang, J.; Gao, M.; You, Q. Experimental Study on Low Interfacial Tension Foam for Enhanced Oil Recovery in High-Temperature and High-Salinity Reservoirs. *Energy Fuels* **2017**, *31*, 13416–13426. [CrossRef]
22. Wang, Z.; Cao, Z.; Li, S.; Li, S. Investigation of the plugging capacity and enhanced oil recovery of flexible particle three-phase foam. *J. Mol. Liq.* **2023**, *386*, 122459. [CrossRef]
23. Wu, W.; Hou, J.; Qu, M.; Wu, W.; Zhang, W.; Wen, Y.; Liang, T.; Xiao, L. Application of Phenolic Resin Crosslinked Polymer Gel in Fractured-Vuggy Carbonate Reservoir with High Temperature and High Salinity. In Proceedings of the International Petroleum Technology Conference, Riyadh, Saudi Arabia, 12–14 February 2022.
24. Meng, X.; Zhang, G.; Wu, J.; Zhao, X.; Wang, L.; Zhang, F. Experimental Study on Phenol-Formaldehyde Resin Aggregates as In-Depth Conformance Control Agents Stabilized by Polymer. *Polymers* **2022**, *14*, 3159. [CrossRef]
25. Ran, Y.; Zhang, G.; Jiang, P.; Pei, H. Study on Water-Soluble Phenolic Resin Gels for High-Temperature and High-Salinity Oil Reservoir. *Gels* **2023**, *9*, 489. [CrossRef]
26. You, Q.; Tang, Y.; Dai, C.; Zhao, M.; Zhao, F.; Zhang, Y. A Study on the Morphology of a Dispersed Particle Gel Used as a Profile Control Agent for Improved Oil Recovery. *J. Chem.* **2014**, *2014*, 1–9. [CrossRef]
27. Yang, H.; Zhou, B.; Zhu, T.; Wang, P.; Zhang, X.; Wang, T.; Wu, F.; Zhang, L.; Kang, W.; Ketova, Y.A.; et al. Conformance control mechanism of low elastic polymer microspheres in porous medium. *J. Petrol. Sci. Eng.* **2021**, *196*, 107708. [CrossRef]
28. Zhao, G.; Dai, C.; You, Q. Characteristics and displacement mechanisms of the dispersed particle gel soft heterogeneous compound flooding system. *Pet. Explor. Dev.* **2018**, *45*, 481–490. [CrossRef]
29. Zhou, Y.; Wang, D.; Wang, Z.; Cao, R. The formation and viscoelasticity of pore-throat scale emulsion in porous media. *Pet. Explor. Dev.* **2017**, *44*, 111–118. [CrossRef]
30. Zhao, G.; Dai, C.; Zhao, M.; You, Q.; Chen, A. Investigation of preparation and mechanisms of a dispersed particle gel formed from a polymer gel at room temperature. *PLoS ONE* **2013**, *8*, e82651. [CrossRef]
31. Wang, D.; Luo, Y.; Lai, R.; Cui, K.; Li, H.; Zhang, Z.; Zhang, Y.; Shi, R. New Technique for Enhancing Oil Recovery from Low-Permeability Reservoirs: The Synergy of Silica Nanoparticles and Biosurfactant. *Energy Fuels* **2021**, *35*, 318–328. [CrossRef]
32. Kumar, G.; Mani, E.; Sangwai, J.S. Microfluidic Investigation of Surfactant-Assisted Functional Silica Nanofluids in Low-Salinity Seawater for Enhanced Oil Recovery Using Reservoir-on-a-Chip. *Energy Fuels* **2023**, *37*, 10329–10343. [CrossRef]
33. Su, Y.; Zhang, X.; Li, L.; Hao, Y.; Zhan, S.; Wang, W.; Wu, Z.; Zhang, W. Experimental study on microscopic mechanisms and displacement efficiency of N2 flooding in deep-buried clastic reservoirs. *J. Petrol. Sci. Eng.* **2022**, *208*, 109789. [CrossRef]
34. Li, J.; Liu, Y.; Gao, Y.; Cheng, B.; Meng, F.; Xu, H. Effects of microscopic pore structure heterogeneity on the distribution and morphology of remaining oil. *Pet. Explor. Dev.* **2018**, *45*, 1112–1122. [CrossRef]
35. Zhao, G.; You, Q.; Gu, C.; Lü, Y.; Dai, C. Preparation mechanism of multiscale dispersed particle gel. *Acta Pet. Sin.* **2017**, *38*, 821.

36. He, Y.; Liao, K.; Bai, J.; Fu, L.; Ma, Q.; Zhang, X.; Ren, Z.; Wang, W. Study on a Nonionic Surfactant/Nanoparticle Composite Flooding System for Enhanced Oil Recovery. *ACS Omega* **2021**, *6*, 11068–11076. [CrossRef] [PubMed]
37. Wang, J.; Liu, H.; Xu, J. Mechanistic Simulation Studies on Viscous-Elastic Polymer Flooding in Petroleum Reservoirs. *J. Dispers. Sci. Technol.* **2013**, *34*, 417–426. [CrossRef]
38. Yuan, C. Enhanced Oil Recovery Mechanism of Dispersion Gel-Surfactant Flooding for High Temperature and High Salinity Reservoirs. Ph.D. Thesis, Southwest Petroleum University, Nanchong, China, 2016.
39. Jamaloei, B.Y.; Asghari, K.; Kharrat, R. The investigation of suitability of different capillary number definitions for flow behavior characterization of surfactant-based chemical flooding in heavy oil reservoirs. *J. Pet. Sci. Eng.* **2012**, *90*, 48–55. [CrossRef]

gels

Review

Development and Applications of CO_2-Responsive Gels in CO_2 Flooding and Geological Storage

Yanxu Ding [1,2], Yang Zhao [1,2,*], Xin Wen [1,2], Yueliang Liu [1,2], Ming Feng [3] and Zhenhua Rui [1,2]

1 National Key Laboratory of Petroleum Resources and Engineering, China University of Petroleum (Beijing), Beijing 102249, China; 2021210398@student.cup.edu.cn (Y.D.); wenxin990825@163.com (X.W.); sdliuyueliang@163.com (Y.L.); ruizh@cup.edu.cn (Z.R.)
2 College of Petroleum Engineering, China University of Petroleum (Beijing), Beijing 102249, China
3 CNPC Engineering Technology R&D Company Limited, Beijing 102249, China; fengmingdri@cnpc.com.cn
* Correspondence: zhaoy@cup.edu.cn

Abstract: Gel systems are widely used as plugging materials in the oil and gas industry. Gas channeling can be mitigated by reducing the heterogeneity of the formation and the mobility ratio of CO_2 to crude oil. Cracks and other CO_2 leaking pathways can be plugged during the geological storage of CO_2 to increase the storage stability. By adding CO_2-responsive groups to the classic polymer gel's molecular chain, CO_2 responsive gel is able to seal and recognize CO_2 in the formation while maintaining the superior performance of traditional polymer gel. The application of CO_2 responsive gels in oil and gas production is still in the stage of laboratory testing on the whole. To actually achieve the commercial application of CO_2 responsive gels in the oil and gas industry, it is imperative to thoroughly understand the CO_2 responsive mechanisms of the various types of CO_2 responsive gels, as well as the advantages and drawbacks of the gels and the direction of future development prospects. This work provides an overview of the research progress and response mechanisms of various types of CO_2 responsive groups and CO_2 responsive gels. Studies of the CO_2 responsive gel development, injectivity, and plugging performance are comprehensively reviewed and summarized. The shortcomings of the existing CO_2 responsive gels system are discussed and the paths for future CO_2 responsive gel development are suggested.

Keywords: CCUS; gas channeling; CO_2 leakage; CO_2 responsive gel; comprehensive review

Citation: Ding, Y.; Zhao, Y.; Wen, X.; Liu, Y.; Feng, M.; Rui, Z. Development and Applications of CO_2-Responsive Gels in CO_2 Flooding and Geological Storage. *Gels* **2023**, *9*, 936. https://doi.org/10.3390/gels9120936

Academic Editor: Georgios Bokias

Received: 1 November 2023
Revised: 22 November 2023
Accepted: 24 November 2023
Published: 29 November 2023

1. Introduction

The demand for energy is growing worldwide, and the majority of the energy that people need is currently provided by fossil fuels [1]. This also leads to the release of a large amount of greenhouse gases such as CO_2, CH_4, N_2O, HFCs, PFCs, and ReF_6 into the atmosphere, causing the Earth's temperature to rise day by day. According to the National Weather Service, the earth's temperature has risen by 0.7 °C since the 19th century [2]. CO_2 is one greenhouse gas and makes up more than 60% of the greenhouse effect. The global cumulative emission reduction in CCUS is 5.5×10^{11} to 1.017×10^{12} t CO_2 to accomplish the goal of a 1.5 °C temperature rise control by the year 2100. The potential of CCUS to reduce global emissions has been assessed by the International Energy Agency and the Intergovernmental Panel on Climate Change (IPCC). To achieve near zero emissions globally by 2070, CCUS technology must accumulate an emission reduction of about 15% [3–6].

The development of CO_2 capture utilization and storage (CCUS) as an important technology for carbon neutrality is highly promising. CO_2-based enhanced oil recovery (CO_2-EOR) in hydrocarbon reservoirs is one of the most economically attractive means to achieve large-scale CO_2 utilization and geological storage. However, CO_2 channeling through high-permeability features seriously inhibits the performance of CO_2-EOR. In addition to gas channeling issues, in the process of CO_2 flooding, the injected CO_2 can also cause inorganic salt precipitation (such as metal carbonates) and asphaltene precipitation, leading to a decrease in reservoir permeability. Due to the fact that precipitation is also generated during water flooding, and the mechanism of precipitation generation is similar to CO_2 flooding It is possible to refer to the treatment method of water flooding and introduce inhibitors into the CO_2 flooding process to hinder or reduce the generation of precipitation. Also, there is an urgent need for CO_2 geological storage technology to solve the problem of potential CO_2 leakage via cracks, faults, and other high-permeability channels. Injecting CO_2 into hydrocarbon reservoirs is an attractive way to achieve the effective utilization and storage of CO_2 [7,8]. Current CCUS projects are mainly applied to conventional oil and gas reservoirs, in which 0.1 to 0.6 tons of crude oil can be produced for every 1 ton of CO_2 injected. The next stage of research mainly focuses on improving the recovery rate of unconventional oil and gas reservoirs with carbon dioxide. How to effectively inject CO_2 into shale or coal bed methane remains a key research and development direction for improving the recovery performance of such oil and gas reservoirs [9,10]. When the source and sink matching conditions are suitable, some CCUS projects in China have lower costs than the enhanced oil recovery (EOR) benefits, and have the potential for negative cost reduction [11,12]. In terms of CO_2-enhanced oil recovery and geological storage, it is a huge challenge to choose a suitable CO_2 sealing agent in order to control the migration of CO_2 in the formations. As a new type of plugging agent, CO_2-responsive gel can directionally identify and plug CO_2 in the formation, and has good acid resistance. It has high application potential in the CO_2 sealing of formations, and has received widespread attention from researchers [13,14].

2. Gas Channeling Control in CO_2-EOR

Gas channeling, as the primary issue that restricts the significant improvement of CO_2-EOR, requires an understanding of its generation mechanisms. The factors causing gas channeling can be attributed to the significant physical differences between CO_2 and crude oil and reservoir heterogeneities. The former can be divided into two situations (see Figure 1): (1) the gravity overriding phenomenon: during CO_2 flooding, due to the low density of CO_2, a large amount of CO_2 will gradually migrate above the crude oil, ultimately forming a gas channel at the top of the crude oil seepage channel [15]; (2) the viscous fingering phenomenon: the viscosity of CO_2 is lower than the viscosity of crude oil, which can cause uneven propulsion during the displacement process. When the flow rate of CO_2 in a local area is too fast, gas channeling will occur [16]. The gas channeling caused by reservoir heterogeneity can be divided into the following scenarios: (1) when there are high-permeability channels such as natural fractures, artificial fractures, wormholes, and conduits, CO_2 will bypass the matrix and cannot displace the crude oil inside [17,18]; (2) when there is a significant difference in the permeabilities of different layers or zones, CO_2 preferentially flows through high-permeability reservoirs [19]; (3) the presence of wormholes, ducts, and high-permeability cracks in the matrix leads to the ineffective flow of CO_2 [20].

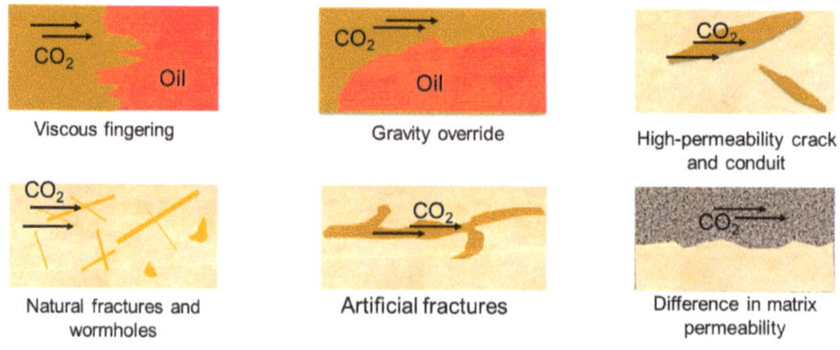

Figure 1. Formation mechanisms of gas channeling.

At present, there are several methods to mitigate oil and gas channeling in CO_2 flooding: water-alternating-gas (WAG) injection, the direct thickening of CO_2, hydrogel plugging, foam plugging, and nanoparticle plugging. These methods for suppressing CO_2 gas migration have been proven to be effective in practical applications and laboratory simulations. The main purpose of this section is to figure out the sealing mechanisms of different CO_2 channeling control methods and the parameters that affect their effectiveness.

2.1. Water-Alternating-Gas (WAG) Injection

WAG is considered a reliable method to suppress gas channeling during CO_2 flooding [21,22]. Awan et al. reported that, in the WAG process, CO_2 flooding provides high sweep efficiency at the micro level, while water flooding provides high sweep efficiency at the macro level. The organic combination of the two results in a significant improvement in WAG oil recovery overall [23]. According to Leeuwenburg's explanation, the injected water during WAG oil displacement can adjust the CO_2 flow to a certain extent. If CO_2 is suppressed from flowing to high-permeability zones, it is forced to displace oil left in low-permeability zones, thereby improving the overall oil recovery performance [24]. Kamali et al. also proposed a similar view, using three oil displacement methods: continuous CO_2 injection, simultaneous CO_2 and water injection, and WAG injection to conduct displacement experiments on sandstone cores. It was found that WAG had the best effect and continuous CO_2 injection had the worst effect. Then, numerical simulation experiments were conducted, and it was found that the presence of injected water during the WAG process effectively reduced the permeability of CO_2, thereby reducing the mobility ratio of CO_2 to crude oil [25]. Han explained the enhanced oil recovery mechanism of WAG from the perspective of miscible displacement, and they believe that an increased volume of injected water significantly increases the injection pressure of CO_2, making it easy for CO_2 to mix with crude oil. This reduces the interfacial tension between crude oil and CO_2, as well as the viscosity of crude oil, allowing more formation crude oil to be extracted from the ground [26].

It is worth noting that precipitation is often generated during the WAG process, which can hinder the flow of crude oil in the formation and affect the improvement of oil recovery. According to the type of precipitation, it can be divided into organic precipitation and inorganic precipitation. Organic precipitation usually refers to asphaltene precipitation [27]. Due to the injection of foreign fluids, changes in the thermodynamic parameters (temperature and pressure) and composition of the crude oil system will lead to aggregation reactions of asphaltene in the crude oil, resulting in solid asphaltene precipitation. The generated asphaltene precipitate will block the channel and seal the reservoir pores, ultimately leading to a decrease in reservoir permeability. Usually, a precipitation inhibitor is added to the injected fluid to hinder the flocculation of unstable asphaltene or make the generated asphaltene precipitate easy to wash out, thereby alleviating the blocking effect of precipitation on the formation. Inorganic precipitation mainly includes metal carbonates. During the process of alternating water and gas injection, some CO_2 dissolves in the injected water, converting it into acidic carbonated water and accumulating $CO_3{}^{2-}$ in the aqueous solution. When water contains metal scale ions such as Mg^{2+} and Ca^{2+}, $CO_3{}^{2-}$ will combine with metal scale ions. When the concentration of metal carbonate reaches a critical value, precipitation occurs. These sediments can block small pores, leading to a decrease in reservoir permeability. In addition, acidic carbonated water can also erode the reservoir, alter its permeability and pore structure, and cause metal scale ions in the reservoir to enter the injected water, further exacerbating the generation of precipitation [28]. Therefore, for the injected water in WAG, chelating agents should be used before injection into the formation to reduce the content of Ca^{2+} and Mg^{2+}. Alternatively, inhibitors can be added to inhibit the formation of precipitation.

The recovery efficiency of WAG oil displacement will be greatly impacted by the injection parameters, which include porosity, permeability, and other formation properties, as well as injection parameters like the water to gas slug ratio. Hao et al. first used a thin tube experiment to determine the minimum miscibility pressure of CO_2 and crude oil at 22.79 MPa, and then connected three different permeability cores in parallel [29]. After CO_2 displacement, it was changed to WAG displacement, and it was found that under both displacement conditions, the core with the highest permeability contributed the most to the recovery performance, both exceeding 90%. The injection pressures were kept at 15 MPa and 25 MPa, respectively, with oil recovery efficiencies of 33.01% and 39.42%, respectively. A higher injection pressure was beneficial for oil recovery improvement [30]. Hosseini and Wang et al. found that, after CO_2 and WAG displacement, the oil permeability and porosity of the core significantly decreased, and some areas of the core showed a reversal of wettability. They attributed this phenomenon to the presence of CO_2-caused precipitation and the accumulation of asphaltene in the crude oil, blocking some channels, resulting in a decrease in crude oil permeability and core porosity. A combination with surfactants or other types of chemical inhibitors during WAG flooding has been reported to reduce asphaltene precipitation [31,32].

2.2. *Direct Thickening of CO_2*

The density and viscosity of CO_2 gas can be increased by thickening it with polymers, which reduces gas channeling issues brought on by gravity overlap and viscous fingering phenomena [33,34]. Brien believes that the increase in density and viscosity can be achieved by controlling the concentration of the added polymer [35]. Polymers enhance the density and viscosity of CO_2 at different levels by dissolving them in CO_2. In general, the higher the molecular weight of a polymer, the greater its viscosity, and the better its thickening effect on CO_2. However, the higher the molecular weight of the polymer, the lower its solubility in CO_2, which is unfavorable for CO_2 thickening. Therefore, using low molecular weight polymers to thicken CO_2 is also a feasible option. Siloxane polymers have been proven to be an effective CO_2 thickener [36]. Bac tested the thickening ability of polydimethylsiloxane (PDMS) on $scCO_2$ at 2500 psi and 130 °F. It was found that the viscosity of CO_2 increased from 0.04 cp to 1.2 cp after thickening. In addition, the use of toluene results in the higher solubility of polymers under the same pressure conditions. Bac conducted CO_2 core displacement experiments and found that, after adding polymers, the oil recovery rate increased, the gas breakthrough was delayed, and the oil recovery rate increased by 3.4–9% from its original value [37].

There is also a special polymer, which can form a three-dimensional grid structure through a cross-linking reaction between molecular chains, and the network can swell in water. This kind of polymer is called a gel [38–40]. Because gels have good plugging properties, they often block high-permeability channels in CO_2-EOR, thus inhibiting gas channeling caused by formation heterogeneity [41]. There are two solutions for polymer gels used for CO_2 consistency control. The first solution is in situ gel plugging, which injects the solution composed of a polymer monomer, cross-linking agent, and auxiliary agent into the formation to form a gel in the formation and block the migration of CO_2 in the high-permeability channel. The second scheme is pre-crosslinked gel plugging, which can be directly injected into the formation after the gel has been completely formed. Alternatively, it can be processed into particles and prepared with a solution, and then injected into the formation [42]. Durucan et al. carried out the core displacement experiment of supercritical CO_2 oil displacement, injected polyacrylamide-based polymer gel into the core, and then conducted the CO_2 displacement experiment again, and found that the permeability of CO_2 decreased by 99% [43].

2.3. *Foam Injection*

Foam is a gas dispersion system surrounded by liquid film prepared and stabilized by a surfactant [44,45]. Surfactants are amphiphilic compounds, which means they are composed of hydrophilic heads and hydrophobic tails. They are generally divided into four types (according to the charge of the head group): non-ionic, anionic, cationic, and zwitterionic surfactants. Yan et al. found that foam has greater effective viscosity, which can alleviate gravity overlap and the viscous fingering phenomenon during CO_2 flooding, and improve sweep efficiency during CO_2 flooding. Foam can also control the local flow resistance of CO_2, forcing it into the low-permeability area and displacing the crude oil [46]. Through core displacement studies, Boud and Holbrook demonstrated for the first time that foam may be used to improve oil recovery by gas flooding. Additionally, foam can be produced in reservoir rocks under both miscible and immiscible circumstances using this water-soluble foaming agent. Ren et al. tested the effects of three different types of surfactants on CO_2 flooding. The first two surfactants were 2 EH-PO_5-EO_{15} and 2 EH-PO_5-EO_9, both of which were nonionic surfactants. The third type was the water-soluble anionic surfactant CD-1045. The phase behavior experiments conducted showed that none of these three surfactants could significantly reduce the interfacial tension between water and crude oil. However, all of them can significantly improve crude oil recovery. Compared with pure CO_2 flooding, the three surfactants can increase oil recovery by 71%, 92%, and 54%, respectively. Moreover, the effect of improving the oil recovery is closely related to the injection scheme [47]. Zhang et al. used UC_{11}AMPM, SDS, and their mixture as foaming agents, respectively, to prepare CO_2 foams, and tested the effect of temperature on the stability of these three foams. With the increase in temperature, the stability of the foams decreased, and the foam produced by the UC_{11}AMPM and SDS mixture had the best temperature resistance [48]. Combining multiple surfactants can achieve better oil recovery effects, but the proportion of different surfactants will have a significant impact on the oil displacement effect. Attarhamed and zoveidavianpoor found that the foaming performance of the mixture of AOS and TX-100 in aqueous CO_2 foam was improved compared with that of AOS and TX-100 alone [49]. Memon et al. used AOS, TX-100, and a third surfactant, rose amidopropylamine oxide (LMDO), to control the fluidity of CO_2 and improve oil recovery. After the water flooding of Berea sandstone using different combinations of CO_2 and surfactant solutions at 1400 psi and 96 °C, surfactant alternating gas (SAG) injection was performed. According to core oil displacement experiments, CO_2-SAG based on (0.6 wt% AOS + 0.6 wt% TX-100) achieved the highest recovery rate [50].

There are three main options for introducing surfactants into oil recovery processes. First, CO_2 foam is generated from the outside and then injected into the porous medium. Secondly, the surfactant solution and CO_2 can be injected together at the same time to form foam in porous media. Thirdly, carbon dioxide and surfactant solutions can be alternately injected, known as SAG injection. The advantages of surfactants mainly lie in reducing viscosity fingering, gravity segregation, and early CO_2 breakthrough by changing the magnitude of viscosity and gravity. In addition to fixing CO_2, surfactants also tend to reduce the IFT between reservoir fluids, reduce capillary forces, and thus improve crude oil recovery. The synergistic effect of multiple surfactants may produce a better profile control effect than a single surfactant, and this profile control effect is closely related to the proportion of different types of surfactants, which will also an important development direction of foam profile control and flooding in the future.

2.4. Nanoparticle Injection

An NP (nanoparticle) is defined as a material composed of particles with sizes between 1 nm and 100 nm [51,52]. In terms of CO_2-EOR, nanoparticles enhance oil recovery through two pathways: improving the mobility ratio of CO_2 to crude oil and reducing asphaltene precipitation during CO_2 flooding [53,54]. Lu et al. designed a CO_2 core displacement experiment and injected Al_2O_3 nanoparticles into the core. They found that they adsorbed asphaltene in a solution prepared from toluene and dissolved asphaltene, which means that these NPs can be used to suppress the deposition of asphaltene during CO_2 injection in porous media. A concentration of 0.5 wt% nanoparticles and a volume ratio of 0.1 nanofluid slugs to CO_2 slugs are considered the best conditions for inhibiting asphaltene damage during CO_2 flooding. Compared to the cyclic injection mode, continuous CO_2 and nanofluid injection may be more effective. The higher the mass fraction of Al_2O_3 nanoparticles, the lower the strength of asphaltene precipitation and the greater the decrease in interfacial tension [55]. Other studies have also reached the same conclusion that nanoparticles can reduce the interfacial tension between crude oil and CO_2 and reduce asphaltene precipitation in crude oil [56,57]. Ehsan et al. simulated the viscosity increasing effect of Al_2O_3 nanoparticles with particle diameters of 1 nm, 2 nm, and 3 nm on $scCO_2$ in an environment of 380 K and 20 MPa. Particles with a diameter of 1 nm have the weakest effect on CO_2 viscosity, resulting in a 3.67-fold increase in CO_2 viscosity. The author also compared the viscosity increasing effect of spherical Al_2O_3 nanoparticles and columnar CuO nanoparticles on $scCO_2$, and found that the viscosity increasing effect of CuO was 3.4 times lower than that of Al_2O_3 [58].

Because the surfactant is easy to be adsorbed on the rock surface and decomposes itself, the stability of foam in the formation is poor, and it is not suitable for large-scale application. Nanoparticles can effectively improve the stability of foam in the formation, which has attracted the attention of researchers. At present, there are several views on the mechanism of nanoparticles improving the strength of foam (see Figure 2): (1) nanoparticles will gather at the node intersection of the foam liquid film, hinder the liquid flow between liquid films, reduce the water loss rate of the foam liquid film, and thus improve the stability of the foam liquid film; (2) nanoparticles will form a single layer, double layer, and network of bridging particles between foam liquid films to hinder the coalescence and water loss of the foam, thus improving the stability of the foam [59,60]. Among them, the network aggregation of nanoparticles has the strongest stabilizing effect on foam. AttarHamed et al. investigated the effect of the diameter and concentration of SiO_2 nanoparticles on the anionic surfactant effect of α-AOS-CO_2 foam stability. The concentrations of the nanoparticles were 0.1 wt%, 0.3 wt%, 0.5 wt%, and 1 wt%, respectively. The diameters of the nanoparticles were 15 nm, 70 nm, and 250 nm. The final experimental results are shown in the figure. When the particle concentration was low, the larger the particle diameter, the better the stability effect of the foam [61]. Bayat et al. compared the stabilization effect of TiO_2, CuO, Al_2O_3, and SiO_2 nanoparticles on CO_2 foam. When the concentration of nanoparticles was 0.008 wt%, the stabilization effect was the best. When using SiO_2 nanoparticles under the same conditions, the maximum increase in crude oil recovery was 17.4%. The main reason for the poor stability of nanoparticles in foam is that nanoparticles are easily adsorbed on the rock surface and agglomerated. Therefore, the better the dispersity of particles in the system, the better the stability of the foam [62–64].

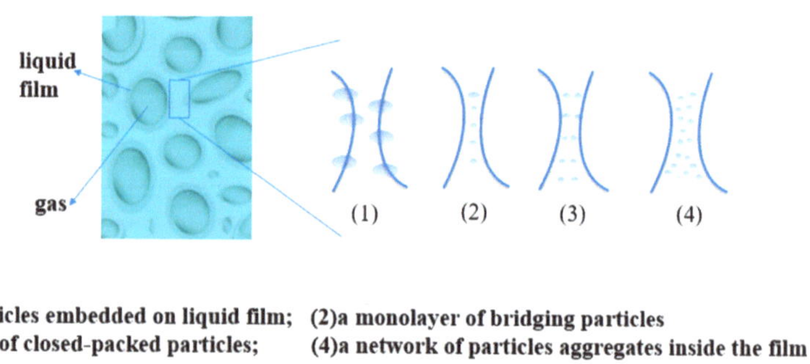

Figure 2. Mechanism of nanoparticles enhancing the stability of foam.

The above four methods for preventing and controlling oil and gas migration in CO_2 flooding have their own advantages and disadvantages, as summarized in Table 1. In order to further solve the problem of gas migration during CO_2 flooding, in addition to making up for the shortcomings of existing technologies, efforts should also be made to develop new CO_2-enhanced oil recovery technologies.

Table 1. Comparison between the different methods of CO_2 mobility control.

Method	Advantage	Disadvantage	Influence Factor
WAG	Reduces CO_2 loss and generate economic benefits; Increases the sweep coefficient; Relieves sticky fingering; Delays CO_2 breakthrough; Reduces the mobility ratio; Maintains CO_2 mixing with crude oil.	Corrodes pipelines; Unable to alleviate the phenomenon of gravitational differentiation; Initiates asphaltene precipitation; Initiates inorganic salt precipitation; Causes stress damage to the tubing; Causes water lock effect; There are many parameters involved.	Reservoir factors (temperature, pressure, thickness, porosity, permeability, saturation, heterogeneity); CO_2 rheological properties and density; Composition and viscosity of crude oil; Water composition and mineralization; Injection parameters (injection rate, injection pressure, slug ratio); Injection scheme; Spacing between injection and production wells.
Polymers for direct thickening of CO_2	It can be mixed with CO_2 to form a thermodynamically stable solution; Increases CO_2 density and viscosity; Relieves early CO_2 breakthrough and viscous fingering.	The solubility of polymers is limited by pressure, molecular weight, and molecular chain structure, making it difficult to meet the requirements in many cases.	Reservoir factors (temperature, pressure, thickness, porosity, permeability, saturation, heterogeneity); CO_2 rheological properties and density; Composition and viscosity of crude oil; Polymer type, molecular weight, and molecular chain structure; Injection scheme.
In situ polymer gels	Good injectability; Reduces formation heterogeneity.	Sensitive to reservoir conditions.	Reservoir factors (temperature, pressure, thickness, porosity, permeability, saturation, heterogeneity); CO_2 rheological properties and density; Composition and viscosity of crude oil; Injection scheme; Injection parameters (injection pressure, injection rate, injection fluid concentration).
Preformed polymer gels	Reduces formation heterogeneity; Low sensitivity to reservoir.	Difficulty in injection and inability to act on deep formations; Only applicable to formations with strong heterogeneity or developed fractures.	Reservoir factors (temperature, pressure, thickness, porosity, permeability, saturation, heterogeneity); CO_2 rheological properties and density; Composition and viscosity of crude oil; Injection scheme; Injection parameters (injection pressure, injection rate, injection fluid concentration).

Table 1. *Cont.*

Method	Advantage	Disadvantage	Influence Factor
Foam	Relieves sticky fingering; Relieves gravity differentiation; Relieves early breakthroughs; Reduces interfacial tension; Changes wettability; Easy to inject; Prevents and controls sedimentation.	Sensitive to temperature and pressure, prone to cracking; Material exchange with crude oil results in a decrease in stability; Adsorbs on the surface of rocks, resulting in ineffective sealing; Short life cycle; Changes the properties of crude oil.	Reservoir factors (temperature, pressure, thickness, porosity, permeability, saturation, heterogeneity); CO_2 rheological properties and density; Composition and viscosity of crude oil; Injection scheme; Type and concentration of surfactants, molecular structure; Injection parameters (injection rate, injection pressure).
Nanoparticle	Reduces the mobility ratio; Prevents and controls asphaltene precipitation; Changes the wettability of rocks; Improves the stability and viscosity of foam; Reduces interfacial tension; Improves CO_2 rheological properties.	Nanoparticles are prone to coalescence, blocking the roar channel, and failing; Large particle sizes can pollute the environment.	Reservoir factors (temperature, pressure, thickness, porosity, permeability, saturation, heterogeneity); CO_2 rheological properties and density; Composition and viscosity of crude oil; Injection scheme; Nanoparticle type, particle size, hydrophilicity, concentration.

3. Pathways of CO_2 Leakage

CO_2 geological storage is the process of trapping CO_2 emitted during the burning of fossil fuels before it enters the atmosphere, moving it via pipelines to the burial location, and then sealing it in a supercritical state in the formation (which typically includes deep salt water layers, abandoned gas layers, and abandoned oil and gas reservoirs). CO_2 leakage is the term used to describe the phenomenon in which stored CO_2 migrates to the ground along faults, wellbore fractures, and other formation fractures, or where it re-enters the atmosphere due to geological events (such as earthquakes or volcanic eruptions) or human activity, ultimately leading to the failure of storage (see Figure 3).

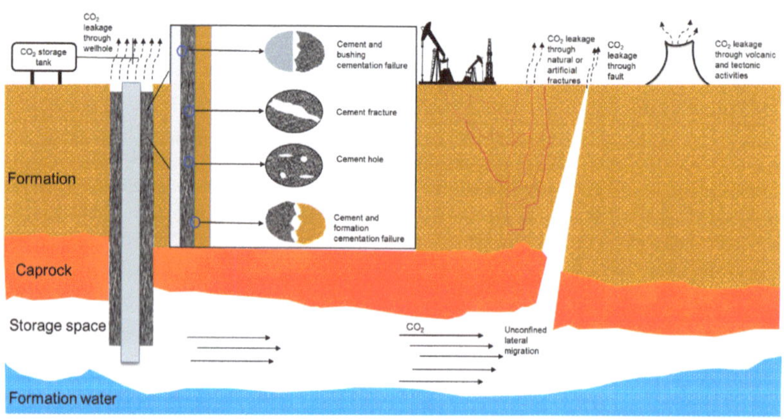

Figure 3. The pathways of CO_2 leakage.

3.1. Potential Inflence of CO_2 Leakage

The hazards caused by CO_2 leakage can be roughly divided into two categories: global risks and regional risks. Global risk refers to the change in the global climate caused by CO_2 leakage, while local risk refers to the damage to the local ecological environment caused by CO_2 leakage [65].

After CO_2 leakage, it will enter the atmosphere again and cause a secondary greenhouse effect [66]. Patil et al. believe that after CO_2 leakage, the oxygen content in the soil will be significantly reduced, leading to a decrease in crop yields such as grass and soybeans. In the experiment they designed, when the CO_2 flow rate reached 1 L/min, the yield of soybeans decreased by half [67]. The United States Department of Energy designed the "Frio Brine Pioneer Experiment". In the experiment, the researchers injected 1600 t carbon dioxide into the sandstone layer 1550 m deep underground in an oil field northeast of Houston, Texas. Carbon dioxide caused the pH value of the brine in the storage formation to drop from nearly neutral 6.5 to 3.0, and dissolved a large number of carbonate rocks, leading to CO_2 leakage and the pollution of drinking water. When the concentration of CO_2 is too high, it can cause suffocation. When CO_2 is excessively injected into the formation, it can cause excessive injection pressure to cause fractures in the formation, and, additionally, it may trigger earthquakes.

3.2. The Pathways of CO_2 Leakage

There are two ways in which CO_2 might leak. The first is known as engineering leakage and is brought on by either artificially created reservoir damage or a poorly wellbore integrity. When CO_2 enters the formation and leaks due to natural geological activities or geological characteristics during the migration process, this leakage pathway is called natural leakage [68].

3.2.1. Engineering Leakage Pathways

When the integrity of an oil well is poor, such as when there are some small cracks on the wellbore wall or near the wellbore, CO_2 will leak along these cracks [69]. When the depth of the well is low, the injected CO_2 cannot reach the designated storage location. The low density of CO_2 will continuously migrate above the formation, ultimately leading to leakage. CO_2 leakage can be avoided by restoring wellbore integrity, but this is only economically feasible when the number of wellbore repairs is small. It is not economically feasible to prevent CO_2 leakage by repairing poorly sealed wells. Excessive CO_2 injection pressure can lead to formation fractures and become a potential pathway for CO_2 leakage. In addition, other human activities such as oil and gas engineering operations such as crude oil extraction and exploration can also lead to the generation of formation fractures and CO_2 leakage [70].

3.2.2. Natural Leakage Pathways

Natural leakage is a CO_2 leakage caused by geological characteristics or the geological activities of reservoirs, and is not related to human activities [71]. The mechanisms of natural CO_2 leakage typically include the following. Firstly, the poor sealing of the cover layer, such as the presence of cracks, may lead to CO_2 migration through the cover layer and leakage [72–74]. Secondly, the presence of high-permeability channels such as fractures or faults in the formation can lead to CO_2 migration to the ground and leakage along the fractures or faults. Bentz et al. believed that these leakage channels are usually caused by geological activities such as earthquakes and tectonic movements [75]. Therefore, geological activities are also a potential factor leading to CO_2 leakage. The deep salt water layer serves as a good CO_2 storage site, and salt water can provide a sealing effect on CO_2. Due to the low density of CO_2, it will continue to migrate laterally until it bypasses the saline layer and leaks along areas with poor sealing of the formation.

3.3. CO_2-Sealing Agents

Manceau et al. suggested that existing CO_2 plugging materials can prevent CO_2 leakage using the following three techniques: (1) mitigating CO_2 leakage by restoring wellbore integrity; (2) obstructing the high-speed migration of CO_2 in cracks or faults; (3) improving the sealing of the cover layer to CO_2 [68]. Wu et al. classified six types of sealants used to alleviate cement-related wellbore leakage. They are cementitious materials, temperature-activated resins, nano-strengthening sealants, gels, geopolymer cements, and low melting point alloys [66]. Based on the existing literature, this article summarizes the types, characteristics, functional mechanisms, and scope of applications of seven CO_2-sealing materials, and points out the advantages and disadvantages of each material systems. The seven materials are Portland cement, geopolymer cement, resin, gel, biofilm barrier, nanoparticles, and foam, as summarized in Table 2.

Table 2. Formation CO_2 leakage prevention and sealing agent.

Plugging Agent	Composition	Working Region	Sealing Mechanism	Advantages	Disadvantages	References
Portland cement	21–67% alite; 6–18% ferrite; 0–49% belite; 1–17% aluminate.	Wellbore and near wellbore area.	Direct sealing.	Having long-term stability; low price.	Poor acid resistance; high CO_2 emissions during production.	[76,77]
Geopolymer cement	48–56% SiO_2; 23–35% Al_2O_3; 3–15% Fe_2O_3; 0–8% CaO; 0.2–1.4% MgO; 0.7–0.85% K_2O; 0.2–0.5% SO_3; 0.3–0.5% alkaline liquid/fly ash.	Wellbore and near wellbore area.	Direct sealing.	High temperature resistance, still usable at 1000 °C; strong acid resistance; the shrinkage rate is small, usually around 0.05%; the CO_2 emission during the production process is about 20% of that of Portland cement, and the energy consumption is about 25% of that of the former.	The production steps are cumbersome; involves some highly corrosive materials, posing a threat to human life and safety; needs to go through the curing process.	[78,79]
Resin sealing system	Phenolic resin, epoxy resin, furan resin, and hardener.	Wellbore and near wellbore area.	Direct sealing.	High adhesive strength and easy to adjust viscosity; has good thermal stability and long-term stability; good acid resistance.	The preparation process is complex; high cost; difficult to control crosslinking reaction.	[80]
Gel	It is divided into organic gel systems and inorganic gel systems, represented by silicate gel.	Casing, microcracks, oil storage media.	Injection of polymer solution or gel particles to form blocky gel and block CO_2 flow channel.	Good injection performance; controllable crosslinking time; can solve complex CO_2 leakage problems.	The sealing effect is generally poor; poor thermal stability and acid resistance; cracks with a width exceeding 2 mm cannot be sealed.	[81,82]

Table 2. *Cont.*

Plugging Agent	Composition	Working Region	Sealing Mechanism	Advantages	Disadvantages	References
Biofilm	A type of bacteria that can produce urease.	Enhances the integrity of wellbore cement.	Induction of urea hydrolysis to form calcium carbonate precipitation, blocking CO_2.	Able to penetrate the pores of the formation and play a role in the near wellbore area; environmentally friendly; biomineralization process is controllable; good acid resistance.	High production costs; need to continuously transport nutrients; uneven distribution of precipitation; unable to block large-scale leaks.	[83,84]
Nanoparticles	Metal oxide particles or polymer particles with a diameter between 1 and 100 nm.	Reduces CO_2 mobility; as a strengthening agent for cement, gel, and foam.	Deep reservoirs and deep salt water layers.	It can greatly improve the strength of gel, foam, and cement; does not pollute the environment; has good stability and acid resistance.	High cost; complex preparation process.	[85,86]
Foam	Aqueous solutions and foaming agents of surfactants.	Reduces the mobility of CO_2 in high permeability channels.	Near wellbore area and deep reservoir.	The injection performance is good and can reach the deep part of the formation.	Low sealing strength and lack of long-term stability.	[87,88]

4. CO_2 Responsive Gel

CO_2-responsive intelligent gel refers to a kind of gel the structure and properties of which can undergo particular transformation when contacting CO_2. The stimulus source of this type of gel is CO_2, which can avoid the accumulation of a large number of impurities in the response process using traditional stimulus sources. By utilizing the intelligent CO_2 response characteristics, it can be applied to sealing CO_2 during CO_2 flooding and geological storage [13].

4.1. CO_2 Responsive Functional Group

The process of synthesizing CO_2 responsive gels involves adding functional groups that are responsive to CO_2 to the conventional gel's molecular chain. As summarized in Figure 4, there are several known CO_2 responsive functional groups: primary amine groups, amidine groups, guanidine groups, tertiary amine groups, azole-containing heterocyclic groups, carboxylic acid [89].

Primary amine groups: CO_2 can react with primary amines at room temperature to generate negatively charged carbamate salts, and some of them are positively charged through protonation, forming bicarbonate between them, thus achieving the salt bridge bonding of primary amine groups, which is the basis for primary amine groups having CO_2 responsiveness.

Amidine groups: Slightly stronger in alkalinity, they will be hydrolyzed to a certain extent in aqueous solution, resulting in deviations when calculating the degree of protonation of the amidine group from the amount of CO_2 injected. When amidine group encounters CO_2, it will generate protonated amidine salt compounds. After CO_2 is discharged, amidine salt will appear deprotonated and return to the original state again.

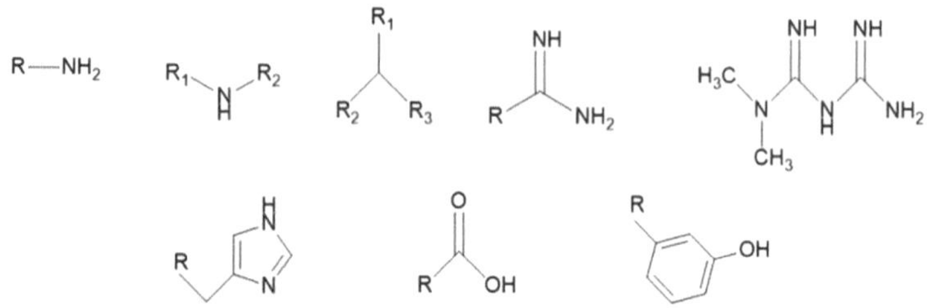

Primary amine; Secondary amine; tertiary amine; Amidine; guanidino

Heterocyclic azoles; carboxyl; phenol

Figure 4. CO_2 responsive groups.

Guanidine: Guanidine is the most basic CO_2-responsive functional group, and the addition of aromatic substituents slows down alkalinity. Guanidine containing N-H bonds can form carbamates or forms of salts other than bicarbonate. The alkalinity of the guanidine group is the highest among several functional groups at present, so it has the strongest CO_2 responsiveness, which also leads to more energy consumption during its deprotonation process. At the same time, the higher temperature required to convert bicarbonate or guanidine formate into neutral guanidine is an advantage for applications that require high temperatures.

Tertiary amino groups: They are common weakly alkaline molecular groups that are found in many polymers. Tertiary amino groups can undergo changes with CO_2 to form cationic tertiary amine salts. When the CO_2 concentration decreases, a reversible deprotonation reaction will occur, and the goal of a CO_2 reversible response will be achieved finally.

Azazole-containing heterocycles: Numerous organisms include heterocyclic compounds including azole, and the molecular fragments of these compounds are present in peptides, genetic elements, and amino acids. In addition, azazole heterocyclic compounds can act as functional groups in response to CO_2 and exhibit weak alkalinity. This polymer can produce protonated histamine salts in response to CO_2 in solution. Imidazole functional groups can be thought of as a novel class of CO_2-adsorbent materials since they are more stable following protonation than amidine and tertiary amine groups.

Carboxylic acid and phenols: These groups are present in the solution as anions when CO_2 is not present. The pH of the solution steadily drops as the CO_2 level rises until it reaches a critical value (corresponding to the solution pKa), and the functional groups progressively become neutral, which indicates a steady decline in the polymer's solubility. Typically, these groups are added to the solution as anions or acids. In cases when the neutral form's solubility in aqueous solution is restricted, the critical point could be noticeably higher than the pKa.

The advantages and disadvantages of each responsive group are summarized in Table 3. Based on the characteristics of each responsive group, its corresponding polymer or itself usually has the following advantages and disadvantages for CO_2 flooding profile control and CO_2 geological storage.

Table 3. Characteristics of each functional group.

CO_2-Responsible Group	Advantage	Disadvantage
Amino groups (primary, secondary, tertiary)	The synthesis route is simple and there are many mature industrial monomers available for use; The group in aqueous solution has low alkalinity and is not easily hydrolyzed.	Slow reaction rate with CO_2; Unable to respond to low concentration of CO_2.
Amine and guanidine groups	Has rapid and sensitive CO_2 responsiveness.	The group in aqueous solution has strong alkalinity and is easy to hydrolyze; Complex synthesis route.
Azazole-containing heterocycles	The salt generated after responding to CO_2 is relatively stable; Widely present in organisms.	The working conditions are harsh and not suitable for geological environments.
Carboxyl and phenolic hydroxyl groups	Has good acid resistance; Easy to synthesize, with many mature industrial monomers available for use.	Sealing of wellbore cracks not applicable.

4.2. CO_2 Responsivee Gels

The studies of CO_2 responsive gels are summarized in Table 4. As early as 2003, Carretti et al. synthesized a class of polyacrylamide-based polymers (PAAm), which contained fatty alcohols and 1-methyl-2-pyrrolidone in organic solvents with a large number of primary amine groups on their side chains at room temperature. After the continuous introduction of CO_2 into the system, the system gradually changes from the initial colloidal state to a gel, and gels can absorb a high concentration of CO_2. The primary amino group has obvious advantages in the synthesis of CO_2-responsive gels, but the primary amino gel is relatively slow in its response to CO_2 [90]. In 2005, Darabi et al. found that amidine groups are responsive to CO_2, which inspired a series of amidine gel to be designed and synthesized [91]. In 2011, Yan et al. successfully prepared CO_2-responsive and breathable polymer vesicles using amphiphilic block copolymers containing amidine groups for the first time. They first synthesized the functional monomer N-guanidododecylacrylamide containing an amidine group, and polymerized it with polyethylene glycol macromolecules through atom transfer radical polymerization to synthesize polyethylene glycolb poly (N-guanidododecylacrylamide) (PEO-b-PAD). After introducing CO_2 into the system, the amidine group reacts with CO_2 to form a protonation amidine salt structure, and the volume of the vesicles will increase significantly. When argon is continuously injected into the system, the amidine salt structure will undergo deprotonation, and then return to the solid state [92].

The synthesis of amidine groups is relatively complex, and the degree of their protonation is reduced due to their easy hydrolysis in water. In 2013, Zou et al. polymerized dimethylaminoethyl methacrylate with a tertiary amino group, which was CO_2-responsive, and with the increase in CO_2 concentration, the degree of protonation of the polymer also increased [93]. In 2012, Hoshino et al. used 3-dimethylaminopropylacrylamide (DMAPM) containing a tertiary amine as a functional monomer to copolymerize with the thermosensitive monomer NIPAM and a crosslinking agent. A micro gel was formed in the solution. The molecular side chain of the gel contains a large number of tertiary amine groups, which can react with CO_2, and fix CO_2 in the gel to achieve the purpose of fixing CO_2. This is the first successful example of using microscopic materials in solution for controllable CO_2 absorption and release [94]. Nitrazole-containing heterocyclic compounds (usually diazoles) are widely present in organisms. Azazole heterocyclic molecular fragments can be found in amino acids, peptides, and genetic materials. At the same time, the azazole-containing heterocyclic molecules are also weakly alkaline and can respond to CO_2 to generate protonation histamines. The protonated functional group is more stable than the previous protonated functional groups, and can be used to prepare an intelligent responsive gel for CO_2 adsorption.

Table 4. CO_2 responsive gel.

Author, Year	Responsive Functional Group	Synthesis Process	Responsive Performance of Gel	Reference
Carretti et al., 2003	Primary amine	Polymerization of allylamine to form polyacrymide.	The polymer is dissolved in the organic solvent containing fatty alcohol and 1-methyl-2-pyrrolidone, and CO_2 gas is continuously injected, which makes the polymer change from the initial colloidal state to chemical gel.	[90]
Wang et al., 2012	Primary amine	Synthesis of a class of crosslinked porous polymer microspheres from N-methyl-N-vinylformamide (MVF) and Din-vinylformamide (DVFE).	By adjusting the ratio of MVF and DVFE monomers to control the size of polymer microspheres, the absorbable surface area inside the microspheres can be maximized. BET experiments have shown that the maximum adsorption area can reach 246 mg/g, and the total adsorption amount of CO_2 can reach 97 mg/g.	[95]
Nagai et al., 2011	Primary amine	Regulating conventional polyaniline increases its relative molecular weight.	CO_2 is continuously injected into the 1% polymer solution, and the solution undergoes a sol–gel–sol transition.	[96]
Suzuki et al., 2011	Primary amine	In the presence of 1,8-diazabicyclo and undecen-7-ene (DBU), a novel hydrogel was prepared by the crosslinking reaction of PAAm with CO_2.	The CO_2 absorption capacity of the hydrogel at room temperature is 2.8 times that of the traditional PAAm, and the absorption capacity and absorption efficiency of the polymer hydrogel are basically unchanged after multiple absorption of CO_2.	[97]
Xu et al., 2004	Primary amine	Firstly, four urea-substituted calixarene host units were designed, and small molecule monomers with primary amine groups were obtained through selective modification.	When CO_2 is introduced into the solution, the dimer can form linear non-covalent bond polymer through ammonium carbamate bridge bond. When competitive polar solvents are added to the solution, the hydrogen bond between the host molecule calix tetraarene is destroyed, leading to the disintegration of the supermolecule polymer.	[98]
Xiong et al., 2017	Primary amine	Mixing HMPAM and TMPDA.	After CO_2 was introduced, the solution turned into gel, and the zero-shearing viscosity and elastic modulus increased by 360 and 400 times respectively.	[99]
Yan et al., 2023	Secondary amine	0.5 wt% NaSal and 3 wt% ENPD were dissolved in deionized water to form the initial system.	After injecting CO_2, the viscosity of the system continues to increase and does not change after reaching a certain stage.	[100]

Table 4. *Cont.*

Author, Year	Responsive Functional Group	Synthesis Process	Responsive Performance of Gel	Reference
Yan et al., 2011	Amidine	Polyethylene glycol b-poly (N-amido dodecyl acrylamide) (PEO-b-PAD) was synthesized by aAtom transfer radical polymerization (ATRP) of N-amido dodecyl acrylamide and polyethylene glycol macromolecules. The PEO segment in this block copolymer is hydrophilic while the PAD segment is hydrophobic and can self-assemble into a vesicle structure in solution.	Alternating CO_2/Ar injection can drive the vesicle to move towards a compound expansion/contraction movement.	[93]
Zhou et al., 2009	Amidine	The (Diphenylphosphine) ethylene terephthalate derivative containing amidine group was connected to the side group of partially azide functionalized polystyrene by Staudinger coupling method.	After dissolving the polymer in a DMF/water mixture and alternately introducing CO_2 and N_2, the conductivity of the system undergoes a rapid impulse change. It is proven that the polymer responds to CO_2 gas through the protonation/deprotonation process of amidine group.	[101]
Guo et al., 2011	Amidine	A class of homopolymers containing amidine groups on its side groups were prepared by RAFT method using 4-chlorostyrene as a monomer.	Alternating the stimulation of CO_2/N_2 can achieve the transfer of substances in the organic/aqueous phase.	[102]
Yan et al., 2013	Amidine	A complex triblock copolymer of polyethylene glycol b poly (N-aminododecylacrylamide) b polystyrene (EAS) has been developed. Due to its amphiphilicity, the copolymer can self-assemble into micrometer-sized hollow tubular structures in aqueous solution.	After introducing CO_2, the polymer gradually transforms from a hollow tubular body to a spherical vesicle, then into a columnar body, and finally evolves into a spherical micelle form.	[103]
Zhang et al., 2012	Amidine	Lotion polymerization of N-amido dodecyl acrylamide with styrene monomer in CO_2 gas environment.	At 60 °C, continuous introduction of N_2 neutral lotion molecules into the product will cause condensation; when CO_2 gas is introduced, ionic lotion repel each other statically, leading to reversible dispersion.	[104]
Su et al., 2012	Amidine	Direct lotion polymerization of styrene monomer in 2,2-azabis (2-neneneba imidazoline) di Bicarbonate initiator.	After CO_2 is introduced, colloidal particles between lotion repel and cause lotion dispersion.	[105]
Han et al., 2012	Tertiary amine	Copolymerization of DMAEMA with N-isopropylacrylamide (NIPAM) monomer using RAFT method to synthesize P (DMAEMA co NIPAM) random copolymer.	After CO_2 is introduced, the solution protonation degree increases, and LCST increases from 35 °C to 60 °C.	[106]

Table 4. *Cont.*

Author, Year	Responsive Functional Group	Synthesis Process	Responsive Performance of Gel	Reference
Yan et al., 2013	Tertiary amine	Polyacrylamide and Diethylamine ethyl polymethacrylate (PDMA-b-PDEAEMA) were chimed. Copolymerize into a certain proportion of monomers containing Coumarin side groups and carry out photo crosslinking.	After introducing CO_2, the vesicles expand and the polymer volume increases.	[107]
Han et al., 2012	Tertiary amine	Polymethacrylic acid oligopoly polyethylene glycol ester co polymethacrylic acid Dimethylamine ethyl ester b polyethylene glycol b polymethacrylic acid oligopoly polyethylene glycol ester co polymethacrylic acid Dimethylamine ethylene ester.	After CO_2 is introduced, the gel dissociates into sol, and after Ar is introduced, the sol gradually transforms into gel.	[108]
Zhao et al., 2013	Tertiary amine	Aggregating DMAEMA.	After introducing CO_2, the particle size increases and the solution transitions from a suspended state to a glassy state.	[109]
Lei et al., 2015	Tertiary amine	Using MBA as crosslinking agent and V-50 as initiator, copolymerize FS and DMAEMA.	After CO_2 is introduced, gel particles expand and gel particles shrink when CO_2 is removed.	[110]
Zhang et al., 2015	Tertiary amine	Copolymerization of pyridine and NIPAM.	After CO_2 is introduced, gel particles expand and gel particles shrink when CO_2 is removed.	[111]
Zhang et al., 2015	Tertiary amine	PDMAEMA-PEO-PDMAEMA triblock copolymer was synthesized using ATRF method.	After CO_2 is introduced, the sol changes into gel, and the process is reversible.	[112]
Chen at al, 2015	Tertiary amine	DEAEMA stabilized by PEGMA was obtained through lotion polymerization.	These PDEAEMA-PEGMA micro gel collapse within 5 s after CO_2 bubbling into solution.	[113]
Zhao at al, 2019	Tertiary amine	Mixing sodium oleate (NaOA) and the small organic counterion 2, 6, 10-trimethyl-2, 6, 10-triazaundecane (TMTAD) in a 3:1 M ratio.	After introducing CO_2, the system transitions from an aqueous solution to a viscoelastic fluid, followed by a milky white solution.	[114]
Wu et al., 2023	Tertiary amine	Mix AM and DMAPMA, add AIPI to form a responsive polymer, and finally add PEI to form a responsive gel.	When the pH of the solution is 4.5 and the immersion time is 48 h, the mass of responsive gel can be increased by 7–18 times.	[115]
Quek et al., 2013	Heterocyclic azoles	Synthesized a class of polymers with histamine side groups using RAFT polymerization method.	The polymer can respond to CO_2 in solution to generate protonation histamines. Because the imidazole functional group is more stable than amidine group and tertiary amine group after protonation.	[116]

5. Application of CO_2-Responsive Gel in CO_2 Gas Channeling and Leakage Mitigation

For the application of a CO_2 responsive gel in CO_2-enhanced oil recovery and geological storage, in addition to the performance of the gel itself, such as rheology, swelling, and acid resistance, etc. The environmental conditions(pressur, tempreture, etc.) in the action area of gel also greatly affect the action effect of gel Figure 5.

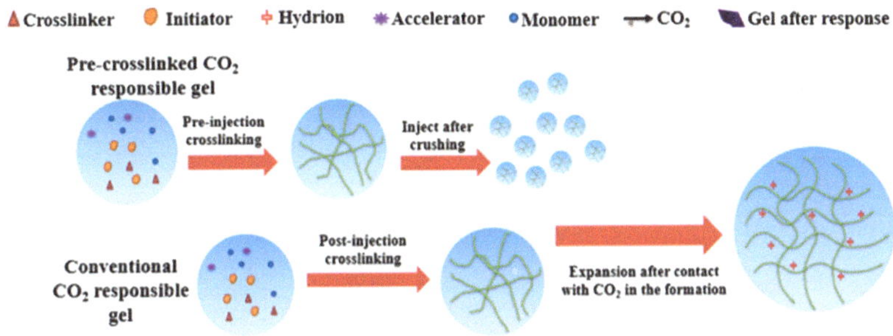

Figure 5. The responsive mechanism of CO_2 responsive gel.

The key point is to observe some characteristics of gel in the actual application process, such as evaluating the plugging strength of gel by the breakthrough pressure of CO_2 after injection of gel, or evaluating gel's ability to adjust CO_2 mobility by resistance coefficient and residual resistance coefficient. This paper summarizes some work on the evaluation of CO_2 responsive gel for CO_2 plugging ability and gas channeling inhibition ability in recent years (as shown in Table 5).

Table 5. CO_2 responsive gel plugging CO_2 experiment.

Author, Year	Synthetic Ingredient	Rheology	Experimental Design	Main Findings	Ref.
Mingwei Zhao, 2023	Sodium salicylate, acetone, sodium fluoride, erucic acid N,N-dimethyl-1,3-propanediamine (99%).	CO_2 sensitivity: the relationship between the viscosity of gel at 25 °C and the amount of CO_2 injected; rheological properties: viscosity modulus, elastic modulus, relaxation time.	CO_2 plugging experiment and parallel core floods (80 °C, backpressure 4 MPa, 1038 mD).	The CO_2 sealing rate reaches 97.45%, and as the permeability increases, the sealing rate decreases. When the maximum permeability is 2000 mD, the sealing rate is 95%. The sealing rate increases with injection volume. Increased the recovery factor of low permeability cores by 18.7%.	[114]
Dexiang Li, 2019	Modified polyacrylamide, methylamine, resorcinol.	Not mentioned.	CO_2 plugging test with gel (80 °C, backpressure 10.28 MPa, 30 mD).	Water phase infiltration reduces by 85%, and the resistance coefficient during CO_2 flow is greater than 29. The gel is easy to absorb near the injection points. The closer it is to the leakage point, the better the plugging effect will be. Gel stability declines under high temperatures.	[117]
JF Ho, 2016	Polyacrylic acid	Liquidity conforms to the power law equation.	Gel plugging CO_2 experiment: remove calcium ions with $Na_5P_3O_{10}$.	Gel can significantly increase the pressure gradient of gel's CO_2 retention, use $Na_5P_3O_{10}$ to reduce the weakening of calcium ions on gel strength, and extend the action time of gel.	[118]

Table 5. *Cont.*

Author, Year	Synthetic Ingredient	Rheology	Experimental Design	Main Findings	Ref.
Shayan, 2017	Polyacrylic acid	It has shear dilution properties and conforms to the Herschel–Bulkley model. Under high calcium concentration, stability is weakened.	Gel plugging strength test (supercritical CO_2, backpressure 1100 psi, 70 °C).	Using sodium phosphate as a chelating agent, increase the pressure gradient of supercritical CO_2 interception to 70 psi/ft. The formation hypothesis of gel deposition layer is verified, and the protective effect of gel on cement integrity is confirmed.	[119]
Du et al., 2022	AM, $AFAPE_20$.	Shrinks under high temperature conditions, with a weight loss of 84.6% at 700 °C and a shear rate below 250 s^{-1}. A pseudoplastic fluid, and an expansive fluid above a critical shear rate.	Plugging experiment (90 °C, backpressure 21 MPa).	The closer the particle diameter to the average pore size, the better the sealing effect. The sealing rate >90%. A significant power law relationship between resistance factor and residual resistance factor and concentration and rate of injected gel particles.	[120]
Pu W, 2021	AM, $AFAPE_20$.	Shrinks under high temperature conditions, with a weight loss of 84.6% at 700 °C and a shear rate below 250 s^{-1}. A pseudoplastic fluid, and an expansive fluid above a critical shear rate.	Plugging experiment (68 °C, backpressure 3 MPa).	After the formation of gel plugging, CO_2 injection pressure increased 68 times; the sealing rate reached 99%. After the formation of gel plugging, the CO_2 flooding recovery rate increased by 23%.	[121]
Luo et al., 2022	N-erucamidopropyl-N,N-dimethylamine.	$UC_{22}AMPM$ solution will become power-law fluid after reacting with CO_2.	Core flooding test (45 °C, backpressure 17 MPa).	The recovery rate of UC_22AMPM solution WAG increased by 8% compared to WAG.	[122]
Du et al., 2022	CO_2-responsive gel and CO_2-responsive wormlike micelles.	The flow state follows a power law equation.	Core flooding experiments (45 °C, backpressure 21 MPa).	The higher the matrix permeability, the poorer the system's ability to improve oil recovery; the larger the crack width, the worse the gel plugging effect.	[123]
Du et al., 2022	CO_2-responsive gel and CO_2-responsive wormlike micelles.	The flow state follows a power law equation.	Core slab models (45 °C, backpressure 0.5 MPa).	Effect of alternate injection of gel system and CO_2 slug on oil recovery single injection of gel system.	[123]
Ye et al., 2022	N. N-Dimethyloctylamide, propyl tertiary amine, Sodium p-toluenesulfonate.	Not mentioned.	Plugging and core flooding experiment (25 °C, backpressure 3 MPa).	The sealing rate for CO_2 is 99.2%. Improved CO_2 flooding oil recovery by 20%.	[124]
Li et al., 2016	Acrylamide, methamphetamine, Resorcinol.	Strength code after reaction with CO_2 is H.	Sand pack experiments (90 °C, backpressure 10.28 MPa).	The CO_2 gas drive permeability of sandstone has decreased by 93.8%. Oil recovery factor of low permeability core increased by 46.5%.	[125]

Table 5. *Cont.*

Author, Year	Synthetic Ingredient	Rheology	Experimental Design	Main Findings	Ref.
Welch et al., 2020	Methacrylate	Not mentioned.	Stability tests and Plugging experiments (confining pressure 0.5 MPa, maximum temperature 73 °C).	High salinity brine will reduce the stability of gel particles; when the ratio of monomer to surfactant is lower than 22.1, gel particles are not easy to gather and have good stability. The fluid loss rate during the initial sealing stage decreased by 93%. The seal fails after the pressure exceeds 5 MPa.	[126]
Ji et al., 2023	Sodium silicate, modified polyphenol, methena-Mine.	Gel strength code: G. After CO_2 is injected, the viscosity decreases from 28.3 mPa·s to 20.5 mPa·s.	Plugging experiment (90 °C).	After the injection of gel, the injection pressure increased from 0.18 MPa to 0.64 MPa, the plugging ratio reached 70.3%, and the oil recovery increased by 10.4%.	[127]

6. Further Development of CO_2 Responsive Gels and Concluding Remarks

The CO_2-responsive gel has good profile control and plugging performance, as demonstrated by numerous core-scale plugging studies. This suggests that the CO_2 responsive gel has great application value in the field of CO_2-EOR and CO_2 geological storage. At present, the application of CO_2 responsive gels on the reservoir scale is still lacking precedent, and its popularization and application in the oil and gas industry still need to solve the following problems. (1) All of the current CO_2 responsive groups depend on H^+ generation. In hotter conditions, CO_2 becomes less soluble in formation water, which inhibits the generation of H^+. The use of CO_2 responsive gels in deep, high-temperature formation is restricted, and its sensitivity is reduced. It is necessary to explore new CO_2 responsive materials. (2) The existing CO_2 responsive gels are mainly used to qualitatively evaluate their CO_2 responsiveness according to the change in their macro phase state before and after CO_2 response. There is a lack of a systematic quantitative characterization method of gel CO_2 responsiveness in order to screen out the ideal CO_2 responsive gel. Moreover, CO_2-responsive groups can be used not only in the preparation of CO_2-responsive gels, but also in the preparation of CO_2-responsive ionic solutions, which have broad application prospects in the field of CO_2 capture [128,129]. (3) The current CO_2-responsive gels are based on the mechanism of protonation of the CO_2 responsive groups, which causes the volume expansion of the gels, and thus seal the CO_2 migration channel of the formation. However, the protonation process of the responsive group is reversible. When the responding gel encounters a large amount of non-acid gas in the formation or the formation temperature increases, the CO_2-responsive gel will return to the state before the response, thus losing the sealing effect on the formation. Further efforts are required to make the gels more stable and controllable. (4) In order to improve the mechanical strength, salt resistance, and temperature resistance, the CO_2 responsive gel has been suggested to be combined with other technologies to meet the requirements of actual production in oil fields. For example, foam compound systems, organic inorganic gel compound systems, and nanoparticle compound systems.

Author Contributions: Literature collection and organization, Y.D., Y.Z. and X.W.; thesis writing, Y.D.; paper revision, Y.Z., Y.L., M.F. and Z.R. All authors have read and agreed to the published version of the manuscript.

Funding: This research was funded by National Key Research and Development Program of China under grant (2022YFE0206700), the Science Foundation of China University of Petroleum, Beijing

(Nos. 2462022BJRC001, 2462021QNXZ012, and 2462021YJRC012), and the funding from the State Key Laboratory of Petroleum Resources and Prospecting (No. PRP/indep-1-2103).

Data Availability Statement: The article is a review paper and no new data have been established yet.

Conflicts of Interest: Authors Ming Feng was employed by the company CNPC Engineering Technology R&D Company Limited, Beijing, China. The remaining authors declare that the research was conducted in the absence of any commercial or financial relationships that could be construed as a potential conflict of interest.

References

1. Li, C.; Gu, G. Green House Effect and Control Technology on Carbon Dioxide. *Environ. Prot. Sci.* **2000**, *2*, 13–15.
2. Zhang, F.; Hu, D. Research on Greenhouse Gas Monitoring and Emission Reduction Under the Backgroung of Carbon Peak and Carbon Neutrality. *China Resour. Compr. Util.* **2021**, *11*, 186–188.
3. IEA. *Energy Technology Perspectives 2020: Special Report on Carbon Capture, Utilization and Storage*; IEA: Paris, France, 2020.
4. Yao, M. Prospects of Carbon Capture, Utilization and Storage Technology and CO_2 Flooding Enhanced Oil Recovery Technology under the Goal of "Double Carbon". *Yunnan Chem. Technol.* **2023**, *50*, 9–11.
5. Yao, J.; Han, H.; Yang, Y.; Song, Y.; Li, J. A review of recent progress of carbon capture, utilization, and storage (CCUS) in China. *Appl. Sci.* **2023**, *13*, 1169. [CrossRef]
6. Rui, Z.; Li, Y. Development Suggestions for Key Technologies of CO_2-Enhanced Oil and Gas Recovery and Geological Sequestration. *Sci. Technol. Foresight* **2023**, *2*, 145–160.
7. Zhang, N. Research on Carbon Neutrality Realization Path of Petrochemical Industry in China. *Guangdong Chem. Ind.* **2021**, *48*, 88–89.
8. Jiang, S.; Zhang, K.; Du, F.; Cui, G. Progress and Prospects of CO_2 Storage and Enhanced Oil, Gas and Geothermal Recovery. *Earth Sci.* **2023**, *48*, 2733–2749.
9. Hu, T.; Wang, Y.; Xu, T.; Tian, H.; Rui, Z.; Zhao, Y.; Liu, T. A novel technique for quantifying the fate of CO_2 injected in oil reservoir for geological utilization and storage. *Energy Rep.* **2023**, *9*, 5350–5361. [CrossRef]
10. Daggash, H.A.; Heuberger, C.F.; Mac Dowell, N. The role and value of negative emissions technologies in decarbonising the UK energy system. *Int. J. Greenh. Gas Control* **2019**, *81*, 181–198. [CrossRef]
11. Lou, W.; Wang, Z.; Zhang, J.; Liu, H.; Sun, B.; Zhang, F. Flow regime evolution mechanism and falling flux prediction model for bypass injection of viscous liquid in vertical T-junction. *Chem. Eng. J.* **2023**, *476*, 146601. [CrossRef]
12. Liu, Y.; Liu, Q. Review of gel systems for CO_2 geological storage leakage and conformance control for enhanced oil recovery: Mechanisms, recent advances, and future perspectives. *J. Pet. Sci. Eng.* **2022**, *219*, 111110. [CrossRef]
13. Zheng, C.; Zhang, L.; Xu, J. Construction and performance evaluation of nano-microemulsion oil displacement system. *Xinjiang Oil Gas* **2023**, *19*, 89–94.
14. Kumar, S.; Mandal, A. A comprehensive review on chemically enhanced water alternating gas/CO_2 (CEWAG) injection for enhanced oil recovery. *J. Pet. Sci. Eng.* **2017**, *157*, 696–715. [CrossRef]
15. Farajzadeh, R.; Andrianov, A.; Spe, R.; Krastev, R.; Hirasaki, G.J.; Rossen, W.R. Foam-oil interaction in porous media: Implications for foam assisted enhanced oil recovery. *Adv. Colloid Interface Sci.* **2012**, *183–184*, 1–13. [CrossRef] [PubMed]
16. Gao, S.; Hu, Z.; Hou, J. Experimental Study on Anti-channeling during CO Flooding for Low Permeability Reservoirs. *Spec. Oil Gas Reserv.* **2013**, *20*, 105–108+147.
17. Li, X.; Hou, Y.; Gu, Y.; Zhang, C.; Mou, C.; Qi, N.; Pan, L. Experimental study on conductivity of acid etched fractures in dolomite reservoirs. *Editor. Dep. Pet. Geol. Recovery Effic.* **2021**, *28*, 88–94.
18. Liu, Y.; Liu, Q. A review of channeling blocking gel systems for CO_2 flooding. *Pet. Geol. Recovery Effic.* **2022**, *30*, 122–134.
19. Wang, H. Study on the Gas Channeling Mechanism and Sealing Technology of CO_2 Flooding. Master's Thesis, Northeast Petroleum University, Daqing, China, 2017.
20. Bai, B.; Sun, X. Development of Swelling-Rate Controllable Particle Gels to Control the Conformance of CO_2 Flooding. In Proceedings of the SPE Improved Oil Recovery Conference, Virtual, 31 August–4 September 2020.
21. Panjalizadeh, H.; Alizadeh, A.; Ghazanfari, M.; Alizadeh, N. Optimization of the WAG Injection Process. *Pet. Sci. Technol.* **2015**, *33*, 294–301. [CrossRef]
22. Gbadamosi, A.O.; Kiwalabye, J.; Junin, R.; Augustine, A. A review of gas enhanced oil recovery schemes used in the North Sea. *J. Pet. Explor. Prod. Technol.* **2018**, *8*, 1373–1387. [CrossRef]
23. Awan, A.R.; Teigland, R.; Kleppe, J. A survey of North Sea enhanced-oil-recovery projects initiated during the years 1975 to 2005. *SPE Reserv. Eval. Eng.* **2008**, *11*, 497–512. [CrossRef]
24. Leeuwenburgh, O.; Meekes, S.; Chitu, A. Optimizing CO_2—EOR operations by improved conditioning of reservoir models to time-Lapse seismic data. In Proceedings of the Offshore Technology Conference, Rio de Janeiro, Brazil, 29 October 2015.
25. Kamali, F.; Hussain, F.; Cinar, Y. An experimental and numerical analysis of water-alternating-gas and simultane-ous-water-and-gas displacements for carbon dioxide enhanced oil recovery and storage. *SPE J.* **2017**, *22*, 521–538. [CrossRef]
26. Han, L.; Gu, Y. Optimization of miscible CO_2 water-alternating-gas injection in the bakken formation. *Energy Fuels* **2014**, *28*, 6811–6819. [CrossRef]

27. Mansoori, G.A. Modeling of asphaltene and other heavy organic depositions. *J. Pet. Sci. Eng.* **1997**, *17*, 101–111. [CrossRef]
28. Khormali, A. Effect of water cut on the performance of an asphaltene inhibitor package: Experimental and modeling analysis. *Pet. Sci. Technol.* **2022**, *40*, 2890–2906. [CrossRef]
29. Wang, X.; Gu, Y. Oil recovery and permeability reduction of a tight sandstone reservoir in immiscible and miscible CO_2 flooding processes. *Ind. Eng. Chem. Res.* **2011**, *50*, 2388–2399. [CrossRef]
30. Lei, H.; Yang, S.; Zu, L.; Wang, Z.; Li, Y. Oil Recovery Performance and CO_2 Storage Potential of CO_2 Water-Alternating-Gas Injection after Continuous CO_2 Injection in a Multilayer Formation. *Energy Fuels* **2016**, *30*, 8922–8931. [CrossRef]
31. Wang, Q.; Yang, S.; Lorinczi, P.; Glover, P. Experimental investigation of oil recovery performance and permeability damage in multilayer reservoirs after CO_2 and water–alternating-CO_2 (CO_2–WAG) flooding at miscible pressures. *Energy Fuels* **2019**, *34*, 624–636. [CrossRef]
32. Zhang, S.; She, Y.; Gu, Y. Evaluation of polymers as direct thickeners for CO_2 enhanced oil recovery. *J. Chem. Eng. Data* **2011**, *56*, 1069–1079. [CrossRef]
33. Zhang, Y.; Huang, S.; Luo, P. Coupling immiscible CO_2 technology and polymer injection to maximize EOR performance for heavy oils. *J. Can. Pet. Technol.* **2010**, *49*, 27–33. [CrossRef]
34. Lee, J.J.; Cummings, S.; Dhuwe, A.; Enick, R.; Beckman, E.J.; Perry, R.M.; Doherty, M.; O'Brien, M. Development of small molecule CO_2 thickeners for EOR and fracturing. In Proceedings of the SPE Improved Oil Recovery Symposium, Tulsa, OK, USA, 12–16 April 2014; OnePetro: Richardson, TX, USA, 2014.
35. O'Brien, M.J.; Perry, R.J.; Doherty, M.D.; Jason, J.L.; Aman, D.; Beckman, E.J.; Robert, M.E. Anthraquinone siloxanes as thickening agents for supercritical CO_2. *Energy Fuels* **2016**, *30*, 5990–5998. [CrossRef]
36. Lee, J.J.; Cummings, S.D.; Beckman, E.J.; Robert, M.E.; Ward, A.B.; Mark, D.D.; Michael, J.O.; Robert, J.P. The solubility of low molecular weight Poly (Dimethyl siloxane) in dense CO_2 and its use as a CO_2-philic segment. *J. Supercrit. Fluids* **2017**, *119*, 17–25. [CrossRef]
37. Bac, J.H.; Irani, C.A. Laboratory investigation of viscosified CO_2 process. *SPE Adv. Technol. Ser.* **1993**, *1*, 166–169.
38. Sakai, T.; Katashima, T. Relationship between physical properties of tetra-PEG gels and polymer chains. *Kobunshi* **2014**, *63*, 860–861.
39. Sugimura, A.; Asai, M.; Matsunaga, T.; Akagi, Y.; Sakai, T.; Noguchi, H.; Shibayama, M. Mechanical properties of a polymer network of Tetra-PEG gel. *Polym. J.* **2013**, *45*, 300–306. [CrossRef]
40. Al-Ali, A.H.; Schechter, D.S.; Lane, R.H. Application of polymer gels as conformance control agents for carbon dioxide EOR WAG floods. In Proceedings of the SPE International Symposium on Oilfield Chemistry, Houston, TX, USA, 13–16 February 2001; OnePetro: Richardson, TX, USA, 2013.
41. Bai, B.; Zhou, J.; Yin, M. A comprehensive review of polyacrylamide polymer gels for conformance control. *Pet. Explor. Dev.* **2015**, *42*, 525–532. [CrossRef]
42. Durucan, S.; Korre, A.; Shi, J.Q.; Govindan, R.; Mosleh, M.H. The use of polymer-gel solutions for CO_2 flow diversion and mobility control within storage sites. *Energy Procedia* **2016**, *86*, 450–459. [CrossRef]
43. Binks, B.P.; Campbell, S.; Mashinchi, S.; Piatko, M.P. Dispersion behavior and aqueous foams in mixtures of a vesicle-forming surfactant and edible nanoparticles. *Langmuir* **2015**, *31*, 2967–2978. [CrossRef]
44. Hosseini-Nasab, S.M.; Zitha, P.L.J. Investigation of certain physicalchemical features of oil recovery by an optimized alkalisurfactantfoam (ASF) system. *Colloid. Polym. Sci.* **2017**, *295*, 1873–1886. [CrossRef]
45. Massarweh, O.; Abushaikha, A.S. The use of surfactants in enhanced oil recovery: A review of recent advances. *Energy Rep.* **2020**, *6*, 3150–3178. [CrossRef]
46. Yan, W.; Miller, C.A.; Hirasaki, G.J. Foam sweep in fractures for enhanced oil recovery. *Colloids Surf. A Physicochem. Eng. Asp.* **2006**, *282–283*, 348–359. [CrossRef]
47. Ren, G.; Nguyen, Q.P. Understanding aqueous foam with novel CO_2-soluble surfactants for controlling CO_2 vertical sweep in sandstone reservoirs. *Pet. Sci.* **2017**, *14*, 330–361. [CrossRef]
48. Zhang, P.; Diao, Y.; Shan, Y.; Pei, S.; Ren, S.; Zhang, L.; Yang, H. Experimental investigation of amine-surfactant CO_2 foam for smart mobility control during CO_2 flooding. *J. Pet. Sci. Eng.* **2020**, *184*, 106511. [CrossRef]
49. AttarHamed, F.; Zoveidavianpoor, M. The foaming behavior and synergistic effect in aqueous CO_2 foam by in situ physisorption of alpha olefin sulfonate and Triton X-100 surfactants and their mixture. *Pet. Sci. Technol.* **2014**, *32*, 2376–2386. [CrossRef]
50. Memon, M.K.; Elraies, K.A.; Al-Mossawy, M.I. Impact of new foam surfactant blend with water alternating gas injection on residual oil recovery. *J. Pet. Explor. Prod. Technol.* **2017**, *7*, 843–851. [CrossRef]
51. Auffan, M.; Rose, J.; Bottero, J.Y.; Lowry, J.P.; Jolivet, J.P.; Wiesner, M.R. Towards a definition of inorganic nanoparticles from an environmental, health and safety perspective. *Nat. Nanotechnol.* **2009**, *4*, 634–641. [CrossRef]
52. Jeevanandam, J.; Barhoum, A.; Chan, Y.S.; Dufresne, A.; Danquah, M.K. Review on nanoparticles and nanostructured materials: History, sources, toxicity and regulations. *Beilstein J. Nanotechnol.* **2018**, *9*, 1050–1074. [CrossRef] [PubMed]
53. Jafari, S.; Khezrnejad, A.; Shahrokhi, O.; Ghazanfari, M.H.; Vossoughi, M. Experimental investigation of heavy oil recovery by continuous/WAG injection of CO_2 saturated with silica nanoparticles. *Int. J. Oil Gas Coal Technol.* **2015**, *9*, 169–179. [CrossRef]
54. Kazemzadeh, Y.; Malayeri, M.R.; Riazi, M.; Parsaei, R. Impact of Fe_3O_4 nanoparticles on asphaltene precipitation during CO_2 injection. *J. Nat. Gas Sci. Eng.* **2015**, *22*, 227–234. [CrossRef]
55. Lu, T.; Li, Z.; Fan, W.; Zhang, X.; Lv, Q. Nanoparticles for inhibition of asphaltenes deposition during CO_2 flooding. *Ind. Eng. Chem. Res.* **2016**, *55*, 6723–6733. [CrossRef]

56. Hassanpour, S.; Malayeri, M.R.; Riazi, M. Utilization of Co_3O_4 nanoparticles for reducing precipitation of asphaltene during CO_2 injection. *J. Nat. Gas Sci. Eng.* **2016**, *31*, 39–47. [CrossRef]
57. Hassanpour, S.; Malayeri, M.R.; Riazi, M. Asphaltene Precipitation during Injection of CO_2 Gas into a Synthetic Oil in the Presence of Fe_3O_4 and TiO_2 Nanoparticles. *J. Chem. Eng. Data* **2018**, *63*, 1266–1274. [CrossRef]
58. Mahdavi, E.; Khaledialidusti, R.; Barnoush, A. Rheological properties of super critical CO_2 with Al_2O_3: Material type, size and temperature effect. *J. Mol. Liq.* **2019**, *289*, 111037. [CrossRef]
59. Fehr, A.; Telmadarreie, A.; Berton, P.; Bryant, S. Synergy Between Commodity Molecules and Nanoparticles as Steam Mobility Control Additives for Thermal Oil Recovery. In *SPE Annual Technical Conference and Exhibition?* SPE: Houston, TX, USA, 2020. [CrossRef]
60. Rognmo, A.U.; Al-Khayyat, N.; Heldal, S.; Vikingstad, I.; Eide, O.; Fredriksen, S.B.; Alcorn, Z.P.; Graue, A.; Bryant, S.L.; Kovscek, A.R.; et al. Performance of silica nanoparticles in CO_2 foam for EOR and CCUS at tough reservoir conditions. *SPE J.* **2020**, *25*, 406–415. [CrossRef]
61. Attarhamed, F.; Zoveidavianpoor, M.; Jalilavi, M. The incorporation of silica nanoparticle and alpha olefin sulphonate in aqueous CO_2 foam: Investigation of foaming behavior and synergistic effect. *Pet. Sci. Technol.* **2014**, *32*, 2549–2558. [CrossRef]
62. Bayat, A.E.; Rajaei, K.; Junin, R. Assessing the effects of nanoparticle type and concentration on the stability of CO_2 foams and the performance in enhanced oil recovery. *Colloids Surf. A Physicochem. Eng. Asp.* **2016**, *511*, 222–231. [CrossRef]
63. Horozov, T.S. Foams and foam films stabilised by solid particles. *Curr. Opin. Colloid. Interface Sci.* **2008**, *13*, 134–140. [CrossRef]
64. Daryasafar, A.; Shahbazi, K. Using nanotechnology for CO_2-foams stabilization for application in enhanced oil recovery. *Int. J. Energy A Clean Environ.* **2018**, *19*, 217–235. [CrossRef]
65. Bai, Y.E.A. Research Progress of CO_2, Geological Storage Leakage Path and Monitoring Method under Dual Carbon Targets. *J. Yulin Univ.* **2021**, *31*, 43–46.
66. Wu, J.; Ma, J. A Discussion about Potential Risks of Geological Storage of CO_2. *Environ. Sci. Surv.* **2012**, *31*, 89–93.
67. Patil, R.H.; Colls, J.J.; Steven, M.D. Effects of CO_2 gas as leaks from geological storage sites on agro-ecosystems. *Energy* **2010**, *35*, 4587–4591. [CrossRef]
68. Manceau, J.C.; Hatzignatiou, D.G.; De Lary, L.; Jensen, N.B.; Reveillrer, A. Mitigation and remediation technologies and practices in case of undesired migration of CO_2 from a geological storage unit—Current status. *Int. J. Greenh. Gas Control* **2014**, *22*, 272–290. [CrossRef]
69. Nordbotten, J.M.; Celia, M.A. *Geological Storage of CO_2: Modeling Approaches for Large-Scale Simulation*; John Wiley and Sons: Hoboken, NJ, USA, 2011.
70. Esposito, A.; Benson, S.M. *Remediation of Possible Leakage from Geologic CO_2 Storage Reservoirs into Groundwater Aquifers*; Elsevier Ltd.: Amsterdam, The Netherlands, 2011.
71. Yang, Y.; Su, X.; Fan, Z.; Wang, D.; Li, J. Numerical investigation of naphthalene deposition dynamics during CO_2 leakage. *Int. J. Greenh. Gas Control* **2020**, *92*, 102860. [CrossRef]
72. Li, Z.; Dong, M.; Li, S.; Huang, S. CO_2 sequestration in depleted oil and gas reservoirs—Caprock characterization and storage capacity. *Energy Convers. Manag.* **2006**, *47*, 1372–1382. [CrossRef]
73. Miocic, J.M.; Johnson, G.; Bond, C.E. Uncertainty in fault seal parameters: Implications for CO_2 column height retention and storage capacity in geological CO_2 storage projects. *Solid. Earth* **2019**, *10*, 951–967. [CrossRef]
74. Wu, B.; Arjomand, E.; Tian, W.; Dan, B.; Yan, S. Sealant Technologies for Remediating Cement-Related Oil and Gas Well Leakage. CSIRO. 2020. Available online: https://gisera.csiro.au/wp-content/uploads/2020/06/GISERA_G6_Task-1_Literature-Review__.pdf (accessed on 10 October 2023).
75. Bentz, D.P. Three-dimensional computer simulation of Portland cement hydration and microstructure development. *J. Am. Ceram. Soc.* **1997**, *80*, 3–21. [CrossRef]
76. Chen, W.; Brouwers, H.J.H.; Shui, Z. Reactivity prediction of ground granulated blastfurnace slag and simulation of hy-dration. *Wuhan Ligong Daxue Xuebao/J. Wuhan Univ. Technol.* **2008**, *30*, 12–15. [CrossRef]
77. Bishop, M.; Bott, S.G.; Barron, A.R. A new mechanism for cement hydration inhibition: Solid-state chemistry of calcium nitrilotris (methylene) triphosphonate. *Chem. Mater.* **2003**, *15*, 3074–3088. [CrossRef]
78. Amran, M.; Debbarma, S.; Ozbakkaloglu, T. Fly ash-based eco-friendly geopolymer concrete: A critical review of the long-term durability properties. *Constr. Build. Mater.* **2021**, *270*, 121857. [CrossRef]
79. Khalifeh, M.; Hodne, H.; Saasen, A.; Integrity, O.; Eduok, E.I. Usability of geopolymers for oil well cementing applications: Reaction mechanisms, pumpability, and properties. In Proceedings of the SPE Asia Pacific Oil & Gas Conference and Exhibition, Perth, Australia, 25–27 October 2016; OnePetro: Richardson, TX, USA, 2016.
80. Sanabria, A.E.; Knudsen, K.; Leon, G.A. Thermal activated resin to repair casing leaks in the middle east. In Proceedings of the Abu Dhabi International Petroleum Exhibition & Conference, Abu Dhabi, United Arab Emirates, 7–10 November 2016; Society of Petroleum Engineers: Abu Dhabi, United Arab Emirate, 2016.
81. Seright, R.; Brattekas, B. Water shutoff and conformance improvement: An introduction. *Pet. Sci.* **2021**, *18*, 450–478. [CrossRef]
82. Zhu, D.; Xu, Z.; Sun, R.; Fang, X.; Gao, D.; Jia, X.; Hu, J.; Weng, J. Laboratory evaluation on temporary plugging performance of degradable preformed particle gels (DPPGs). *Fuel* **2021**, *289*, 119743. [CrossRef]

83. Kirkland, C.M.; Thane, A.; Hiebert, R.; Hyatt, R.; Kirksey, J.; Cunningham, A.B.; Gerlach, R.; Spangler, L.; Phillips, A. Addressing wellbore integrity and thief zone permeability using microbially-induced calcium carbonate precipitation (MICP): A field demonstration. *J. Pet. Sci. Eng.* **2020**, *190*, 107060. [CrossRef]
84. Cunningham, A.B.; Lauchnor, E.; Eldring, J.; Esposito, R.; Mitchell, A.C.; Gerlach, R.; Phillips, A.J.; Ebigbo, A.; Spangler, L.H. Abandoned well CO_2 leakage mitigation using biologically induced mineralization: Current progress and future directions. *Greenh. Gases Sci. Technol.* **2013**, *3*, 40–49. [CrossRef]
85. Rathnaweera, T.D.; Ranjith, P.G. Nano-modified CO_2 for enhanced deep saline CO_2 sequestration: A review and perspective study. *Earth-Sci. Rev.* **2020**, *200*, 103035. [CrossRef]
86. Chen, S.; Wang, R.; Yi, Y.; Zhang, X.; Xu, J.; Shang, J. Experimental study on film permeability and stability of nanofiber enhanced foam. *Xinjiang Pet. Nat. Gas* **2023**, *19*, 81–88.
87. Hopmann, C.; Latz, S. Foaming technology using gas counter pressure to improve the flexibility of foams by using high amounts of CO_2 as a blowing agent. *Polymer* **2015**, *56*, 29–36. [CrossRef]
88. Karakas, M.; Alcorn, Z.P.; Graue, A. CO_2 foam field pilot monitoring using transient pressure measurements. In Proceedings of the Society of Petroleum Engineers (SPE), Virtual, Online, 26–29 October 2020.
89. Yan, Q.; Zhao, Y. Carbon dioxide responsive polymers and their application progress. *Polym. Mater. Sci. Eng.* **2014**, *30*, 170–178.
90. Carretti, E.; Dei, L.; Baglioni, P.; Weiss, R.G. Synthesis and characterization of gels from polyallylamine and carbon dioxide as gellant. *J. Am. Chem. Soc.* **2003**, *125*, 5121–5129. [CrossRef]
91. Darabi, A.; Jessop, P.G.; Cunningham, M.F. CO_2-responsive polymeric materials: Synthesis, self-assembly, and functional applications. *Chem. Soc. Rev.* **2016**, *45*, 4391–4436. [CrossRef]
92. Yan, Q.; Zhou, R.; Fu, C.; Zhang, H.; Yin, Y.; Yuan, J. CO_2-responsive polymeric vesicles that breathe. *Angew. Chem. Int. Ed.* **2011**, *50*, 4923–4927. [CrossRef]
93. Zou, Y.; Brooks, D.E.; Kizhakkedathu, J.N. A novel functional polymer with tunable LCST. *Macromolecules* **2008**, *41*, 5393–5405. [CrossRef]
94. Hoshino, Y.; Imamura, K.; Yue, M.; Inoue, G.; Miura, Y. Reversible absorption of CO_2 triggered by phase transition of amine-containing micro-and nanogel particles. *J. Am. Chem. Soc.* **2012**, *134*, 18177–18180. [CrossRef] [PubMed]
95. Wang, H.; Jessop, P.G.; Liu, G. Support-Free Porous Polyamine Particles for CO_2 Capture. *ACS Macro Lett.* **2012**, *1*, 944–948. [CrossRef] [PubMed]
96. Nagai, D.; Suzuki, A.; Kuribayashi, T. Synthesis of Hydrogels from Polyallylamine with Carbon Dioxide as Gellant: Development of Reversible CO_2 Absorbent. *Macromol. Rapid Commun.* **2011**, *32*, 404–410. [CrossRef] [PubMed]
97. Nagai, D.; Suzuki, A.; Maki, Y.; Takeno, H. Reversible chain association/dissociation via a CO_2 responsive crosslinking/decrosslinking system. *Chem. Commun.* **2011**, *47*, 8856–8858. [CrossRef] [PubMed]
98. Xu, H.; Rudkevich, D.M. CO_2 in Supramolecular Chemistry: Preparation of Switchable Supramolecular Polymers. *Chem. Eur. J.* **2004**, *10*, 5432–5442. [CrossRef] [PubMed]
99. Xiong, C.; Peng, K.; Tang, X.; Ye, Z.; Shi, Y.; Yang, H. CO_2-responsive self-healable hydrogels based on hydrophobically-modified polymers bridged by wormlike micelles. *RSC Adv.* **2017**, *7*, 34669–34675. [CrossRef]
100. Yan, X.; Zhao, M.; Yan, R.; Wang, X.; Dai, C. Statistical analysis of gelation mechanism of high-temperature CO_2-responsive smart gel system. *J. Mol. Liq.* **2023**, *377*, 121521. [CrossRef]
101. Zhou, K.; Li, J.; Lu, Y.; Zhang, G.; Xie, Z.; Wu, C. Re-examination of Dynamics of Polyeletrolytes in Salt-Free Dilute Solutions by Designing and Using a Novel Neutral—Charged—Neutral Reversible Polymer. *Macromolecules* **2009**, *42*, 7146–7154. [CrossRef]
102. Guo, Z.; Feng, Y.; Wang, Y.; Wang, J.; Wu, Y.; Zhang, Y. A novel smart polymer responsive to CO_2. *Chem. Commun.* **2011**, *47*, 9348–9350. [CrossRef]
103. Yan, Q.; Zhao, Y. Polymeric Microtubules That Breathe: CO_2-Driven Polymer Controlled-Self-Assembly and Shape Transformation. *Angew. Chem. Int. Ed.* **2013**, *52*, 9948–9951. [CrossRef]
104. Zhang, Q.; Yu, G.; Wang, W.J.; Yuan, H.; Li, B.; Zhu, S. Preparation of N_2/CO_2 triggered reversibly coagulatable and redispersible latexes by emulsion polymerization of styrene with a reactive switchable surfactant. *Langmuir* **2012**, *28*, 5940–5946. [CrossRef] [PubMed]
105. Su, X.; Jessop, P.G.; Cunningham, M.F. Surfactant-Free Polymerization Forming Switchable Latexes That Can Be Aggregated and Redispersed by CO_2 Removal and Then Readdition. *Macromolecules* **2012**, *45*, 666–670. [CrossRef]
106. Han, D.; Tong, X.; Boissiere, O.; Zhao, Y. General strategy for making CO_2-switchable polymers. *ACS Macro Lett.* **2012**, *1*, 57–61. [CrossRef]
107. Yan, B.; Han, D.; Boissière, O.; Ayotte, P.; Zhao, Y. Manipulation of block copolymer vesicles using CO_2: Dissociation or "breathing". *Soft Matter* **2013**, *9*, 2011–2016. [CrossRef]
108. Han, D.; Boissiere, O.; Kumar, S.; Tong, X.; Tremblay, L.; Zhao, Y. Two-way CO_2-switchable triblock copolymer hydrogels. *Macromolecules* **2012**, *45*, 7440–7445. [CrossRef]
109. Zhao, C.; Zhang, J.; Yuan, G.; Han, C.C. CO_2-triggered liquid–solid switching through a jamming mechanism. *RSC Adv.* **2013**, *3*, 9645–9648. [CrossRef]
110. Lei, L.; Zhang, Q.; Shi, S.; Zhu, S. Oxygen and carbon dioxide dual gas-responsive and switchable microgels prepared from emulsion copolymerization of fluoro-and amino-containing monomers. *Langmuir* **2015**, *31*, 2196–2201. [CrossRef] [PubMed]

111. Zhang, Q.M.; Ahiabu, A.; Gao, Y.; Serpe, M.J. CO_2-switchable poly (N-isopropylacrylamide) microgel-based etalons. *J. Mater. Chem. C* **2015**, *3*, 495–498. [CrossRef]
112. Zhang, L.; Qian, J.; Fan, Y.; Feng, W.; Tao, Z.; Yang, H. A facile CO_2 switchable nanocomposite with reversible transition from sol to self-healable hydrogel. *RSC Adv.* **2015**, *5*, 62229–62234. [CrossRef]
113. Chen, Y.; Zhao, T.; Wang, B.; Qiu, D.; Ma, N. Highly sensitive CO_2-responsive polymeric microgels that respond within seconds. *Langmuir* **2015**, *31*, 8138–8145. [CrossRef]
114. Zhao, M.; Yan, X.; Wang, X.; Yan, R.; Dai, C. The development of a smart gel for CO_2 mobility control in heterogeneity reservoir. *Fuel* **2023**, *342*, 127844. [CrossRef]
115. Wu, Y.; Liu, Q.; Liu, D.; Cao, X.; Yuan, B.; Zhao, M. CO_2 responsive expansion hydrogels with programmable swelling for in-depth CO_2 conformance control in porous media. *Fuel* **2023**, *332*, 126047. [CrossRef]
116. Quek, J.Y.; Roth, P.J.; Evans, R.A.; Davis, T.P.; Lowe, A.B. Reversible addition–fragmentation chain transfer synthesis of amidine-based, CO_2-responsive homo and AB diblock (Co) polymers comprised of histamine and their gas-triggered self-assembly in water. *J. Polym. Sci. Part A Polym. Chem.* **2013**, *51*, 394–404. [CrossRef]
117. Li, D.; Zhang, L.; Ren, S.; Rui, H. Leakage mitigation during CO_2 geological storage process using CO_2 triggered gelation. *Ind. Eng. Chem. Res.* **2019**, *58*, 3395–3406. [CrossRef]
118. Ho, J.F.; Patterson, J.W.; Tavassoli, S.; Patterson, J.W.; Shafiei, M.; Huh, C.; Bommer, P.M.; Bryant, S.L.; Balhoff, M.T. The use of a pH-triggered polymer gelant to seal cement fractures in wells. In Proceedings of the SPE Annual Technical Conference and Exhibition, Houston, TX, USA, 28–30 September 2015; OnePetro: Richardson, TX, USA, 2015.
119. Tavassoli, S.; Ho, J.F.; Shafiei, M.; Huh, C.; Bommer, P.; Bryant, S.; Balhoff, M.T. An experimental and numerical study of wellbore leakage mitigation using pH-triggered polymer gelant. *Fuel* **2018**, *217*, 444–457. [CrossRef]
120. Du, D.; Zou, B.; Pu, W.; Wei, X.; Liu, R. Injectivity and plugging characteristics of CO_2-responsive gel particles for enhanced oil recovery in fractured ultra-low permeability reservoirs. *J. Pet. Sci. Eng.* **2022**, *214*, 110591. [CrossRef]
121. Pu, W.; Du, D.; Fan, H.; Chen, B.; Yuan, C.; Varfolomeev, M.A. CO_2-responsive preformed gel particles with interpenetrating networks for controlling CO_2 breakthrough in tight reservoirs. *Colloids Surf. A Physicochem. Eng. Asp.* **2021**, *613*, 126065. [CrossRef]
122. Luo, X.J.; Wei, B.; Gao, K.; Jing, B.; Huang, B.; Guo, P.; Yin, H.; Feng, Y.; Zhang, X. Gas channeling control with an in-situ smart surfactant gel during water-alternating-CO_2 enhanced oil recovery. *Pet. Sci.* **2023**, *20*, 2835–2851. [CrossRef]
123. Du, D.; Chen, B.; Pu, W.; Zhou, X.; Liu, R.; Jin, F. CO_2-responsive gel particles and wormlike micelles coupling system for controlling CO_2 breakthrough in ultra-low permeability reservoirs. *Colloids Surf. A Physicochem. Eng. Asp.* **2022**, *650*, 129546. [CrossRef]
124. Ye, S.; Zhang, W.; Zhai, Z.; Song, B.; Shang, X.; Song, Z. Fully bio-based CO_2-responsive pickering gel emulsions stabilized by cellulose nanocrystals combined with a rosin-based surfactant. *J. Mol. Liq.* **2022**, *368*, 120794. [CrossRef]
125. Li, D.X.; Zhang, L.; Liu, Y.M. CO_2-Triggered Gelation for Mobility Control and Channeling Blocking during CO_2 Flooding Processes. *Pet. Sci.* **2016**, *13*, 247–258. [CrossRef]
126. Welch, N.; Gilbertson, R.; Boukhalfa, H.; Roback, R.; Carey, J.W.; Goodman, H.; Rösli, U.; Gisiger, J.; Espie, T.; Christian, M. Effectiveness of a Smart Hydrogel in Well Leakage Remediation. In Proceedings of the Paper presented at the SPE Annual Technical Conference and Exhibition, Virtual, 27 October 2020; SPE: Houston, TX, USA, 2020. [CrossRef]
127. Ji, W.; Dai, C.; Cao, Y.; Yang, N.; Wang, T.; Zhao, G. A novel CO_2-resistant dispersed particle gel for gas channeling control in low-permeability reservoirs. *J. Mol. Liq.* **2023**, *374*, 121251. [CrossRef]
128. Suo, X.; Fu, Y.; Do-Thanh, C.L.; Qiu, L.; Jiang, D.; Mahurin, S.M.; Yang, Z.; Dai, S. CO_2 chemisorption behavior in conjugated carbanion-derived ionic liquids via carboxylic acid formation. *J. Am. Chem. Soc.* **2022**, *144*, 21658–21663. [CrossRef] [PubMed]
129. Mahdavi, E.; Zebarjad, F.S.; Ayatollahi, S.; Taghikhani, V. Experimental investigation on the effect of Asphaltene types on the interfacial tension of CO_2–hydrocarbon systems. *Energy Fuels* **2015**, *29*, 7941–7947. [CrossRef]

gels

MDPI

Article

Field and Experimental Investigations on the Effect of Reservoir Drill-In Fluids on Penetration Rate and Drilling Cost in Horizontal Wells

Neamat Jameel [1] and Jagar A. Ali [2,3,*]

[1] Department of Natural Resources Engineering and Management, School of Science and Engineering, University of Kurdistan Hawler, Erbil 44001, Iraq; nimet.jamil@gmail.com

[2] Department of Petroleum Engineering, Faculty of Engineering, Soran University, Soran P.O. Box 624, Iraq

[3] Department of Geology, Palacký University, 17. Listopadu 12, 77146 Olomouc, Czech Republic

* Correspondence: jagar.pet@gmail.com; Tel.: +964-7-5187-83616

Abstract: In this study, the reservoir drill-in fluid (RDF) was modified and optimized to improve the rheological properties and reduce the filtration properties of the drilling fluid used for drilling the oil-bearing zone horizontally. In polymer science, degradation generally refers to a complex process, by which a polymeric material exposed to the environment and workload loses its original properties. Degradation is usually an unwanted process. In certain cases, however, controlled polymer degradation is useful. For instance, it can improve the processability of the polymer or can be used in recycling or natural decomposition of waste polymer. Thus, the drilling fluid and parameter data of 30 horizontal wells that were drilled in the south of Iraq were collected using several reservoir drill-in fluids (RDFs), including FLOPRO, salt polymer mud (SPM), non-damaged fluid (NDF), and FLOPRO_PTS-200 (including the polymer thermal stabilizer). The obtained results showed that the polymer temperature stabilizer (PTS-200) enabled reducing the filtration rate by 44.33% and improved the rheological properties by 19.31% as compared with FLOPRO. Additionally, the average cost of NDF and SPM drilling fluids for drilling the horizontal section of the selected wells is around USD 96,000 and USD 91,000, respectively. However, FLOPRO-based drilling fluid showed less cost for drilling the horizontal section, which is USD 45,000.

Keywords: reservoir drill-in fluid; FLOPRO; non-damaging fluid; salt polymer mud; fluid loss; rate of penetration; drilling cost

check for updates

Citation: Jameel, N.; Ali, J.A. Field and Experimental Investigations on the Effect of Reservoir Drill-In Fluids on Penetration Rate and Drilling Cost in Horizontal Wells. *Gels* **2023**, *9*, 510. https://doi.org/10.3390/gels9070510

Academic Editors: Qing You, Guang Zhao and Xindi Sun

Received: 30 May 2023
Revised: 9 June 2023
Accepted: 21 June 2023
Published: 24 June 2023

1. Introduction

Reservoir drill-in fluids (RDFs) made from biopolymers and synthetic polymers have a variety of applications, such as fluid loss reduction, viscosity improvement, and suspension stability [1–3]. Bio- and synthetic polymers deteriorate at high-temperature ranges within the harsh conditions of the wellbore [4–7]. On the other hand, while drilling with water-based mud (WBM), certain polymers are frequently employed to reduce filtration loss in permeable intervals. Although they frequently operate well, they are unable to completely limit the fluid loss and mud invasion into the reservoir, and they may seriously damage the formation, including negative changes in surface wettability [8,9]. Biopolymers are used to make up reservoir drilling fluid, such as non-damaging fluid (NDF) and salt polymer mud (SPM). Therefore, during drilling under High-Pressure High-Temperature (HPHT) conditions, they are not suitable, as the polymer breaks down at the high bottom hole temperature of the well. Hence, a continuous treatment is needed, for the polymer breaks down, which increases the cost of the drilling [10–13]. However, FLOPRO can be used to overcome the high-temperature problem of drilling inside the borehole [14].

Developments in science and technology, especially over the last two decades, have led to the production of several synthetic polymers worldwide that have resistance to higher temperatures, such as FLOTROL polymers that are used to make up FLOPRO drilling fluid. FLOTROL polymers are more stable at higher temperatures, with less chemical concentration, non-formation damage, zero non-productive time for drilling fluid, and the ability to drill more than 2000 m horizontally at a cost that is significantly lower than commercial polymers [15,16].

The FLOPRO fluid is a proven water-based drill-in system with non-damaging characteristics, a compatibility breaker, an extremely low friction factor for low pump pressures, high ROP, and extremely high capacity for cuttings transport, and a global environmental compliance application revealed that this fluid had consistent performance and rheology. It was suited for drilling horizontal wells because of its good hole cleaning, preventing the development of cutting beds [17,18]. The main components of the FLOPRO solution are fresh (or sea) water, calcium carbonate, the polymer preparations FLO-VIZ and FLO-TROL, sodium or potassium salts, and LUB lubricating additives. FLOPRO is used to provide the lowest skin damage of productive horizons with deviated and horizontal wells. Predicting formation damage in cased-hole and open-hole completion wells is important since the primary objective of reservoir drill-in fluids (RDFs) is to minimize formation damage and provide a thin filter cake that can be removed by differential pressure. This is especially important when the damage is well-bore induced and is brought on by reservoir drill-in fluids. Cake filter removal has been shown to be an effective method for determining induced damage and gauging the effectiveness of drill-in fluids [19,20].

Table 1 illustrates the summary of the performed studies on the application of different types of polymers in the drilling fluid circulation. Samavati and Abdullah (2015) discovered that using gilsonite at concentration of 17.5 gm/cm^3 reduced the polymer breakdown by 72%, improving the viscosity and decreasing the fluid loss compared with the starch under HPHT conditions [17]. In 2019, Aruther and co-workers stated that adding a novel high-temperature polymer within WBM at a concentration of 7 ppb can increase the thermal stability of WBM to be able to withstand 400 °F and maintain its original properties [21]. In addition, two biopolymers, A and B, were reported to positively influence the fluid loss and formation damage reduction by 60% compared with the clay-free starch-containing drilling mud [8]. Moreover, Akpan et al. (2018) investigated the effect of polyglycol on the drilling fluid rheological properties at the concentration of 0.7 wt.%, and their results showed that the polyethylene glycol additive maintained the suspension capability of the drilling fluid formulations. These additives can be used to stabilize the water-based drilling fluids containing biopolymers at 150–232 °C without using expensive and formation-damaging synthetic polymers [14]. Recently, Wallace and co-workers (2020) studied the effect of the polymeric rheology modifier at the concentration of 5 mg in 350 mL with polymer thermal stabilizer at 2 vol.%. Their outcome shows that synthetic polymer holds more excellent rheology suspension than xanthan gum by 48% [22]. The primary objective of this study is to optimize the types and concentrations of chemicals/additives used to develop reservoir-friendly drilling fluids with high tolerance to temperature to horizontally drill specific sections of the Mishrif and Saadi formations that are in the Basra province in the south of Iraq. For this purpose, the field data and laboratory measurements of several RDFs used in the horizontal interval of 30 wells in the south of Iraq were analyzed. Several drilling parameters, including the lost circulation, buildup volume, drilling cost, rate of penetration, and drilling fluid rheological and filtrations properties, were considered.

Table 1. Summary of the published research studies investigated on polymer temperature stabilizer, synthetic polymer, and biopolymer within the reservoir drilling fluid.

Reference	Polymer Temperature Stabilizer	Concentration	Mechanism	Study Outcomes
Samavati and Abdullah [17]	Gilsonite	17.5 cm^3/350 mL	Rheological improvement	Combined with thermal stabilizers, improved the viscosity of WBM by 22% and lowered fluid losses by 25%.
Galindo et al. [18]	Novel high-temperature polymer	7 ppb	Rheological improvement	WBM can withstand temperatures of 400 °F while maintaining its viscosity, excellent suspension, low shear strengths, shale stability, and filtration control by 18%.
Akpan et al. [15]	Polyglycol	0.7 wt.%	Rheological and filtration improvement	Synthetic polymers to stabilize water-based drilling fluids containing biopolymers improved viscosity by 6%, and the filtration rate was reduced by 14%.
Tehrani et al. [23]	PAC	6 gm of PAC in 350 mL	Rheological improvement	The efficiency with 46% fluid loss reduction and the highest value of plastic viscosity, yield point, and gel strength was around 34%.
Al-Otaibi et al. [24]	Xanthan gum and glycol	3 mg in 350 mL	Rheological improvement	Increased the plastic viscosity by 38% at 170–280 °F.
Huang et al. [25]	Laponite	25 and 50 wt.%	Resistance to temperature improvement	Laponite could increase the onset deposition temperature of solid-state AAD terpolymer and substantially increase the high-temperature viscosity of 2 wt.% AAD terpolymer water solution.
Zaboli et al. [19]	Hydrophobic silica NPs	2 wt.%	Resistance to temperature improvement	Hydrophilic or hydrophobic silica NPs phase separation occurred after only a few minutes. By contrast, the emulsions containing the modified silica NPs with contact angles around 92 and 115 were stable for months and days, respectively.
Chen H, et al. [26]	Novel hyper-cross-linked polymer (ACP)	3 gm in 400 mL	Filtration improvement	The preferred one (ACP-5) can reduce the filtrate volume of oil-based drilling fluid by over 90% with a small dosage (3 g in 400 mL drilling fluid) after hot rolling for 16 h at 840 °F.
Zhang et al. [27]	Viscosity stabilizer (PB-854)	2-tert-Butylphenol, paraformaldehyde, phloroglucinol 2:1:2.5	Resistance to temperature improvement	The results show that PB-854 has good high-temperature stability and could effectively protect the polymer at the high temperature.

2. Results and Discussion

2.1. Rheological Properties

In this section, the rheological measurements—including the plastic viscosity (µp), apparent viscosity (µa), yield point, and gel strength of the FLOPRO; non-damaging fluid (NDF); salt polymer mud; and FLOPRO with PTS-200 drilling fluids—are presented and discussed. The measurements included plastic viscosity, yield point, and gel strengths at 10 and 10 min under different temperature conditions, which are shown in Table 2. As can be seen, the plastic and apparent viscosities of the base sample are 13 and 7 cP, respectively, while the yield point is 25 and 14 lb/100 ft^2, and the 10 s and 10 min gel strengths were 8 and 12 lb/100 ft^2, respectively.

The rheological properties of 6 prepared samples of FLOPRO drilling fluid with 1.4 lb/bbl of FLO-VIS powder, 6 samples of salt polymer mud drilling fluid with 1.4 lb/bbl of DUO-VIS powder, 6 samples of non-damaging fluid prepared with 1.6 lb/bbl of DUO-VIS powder, and 7 samples of FLOPRO with different concentrations of PTS-200 are shown in Table 2 and Figures 1 and 2. In the sample of FLOPRO drilling fluid, the plastic viscosity (PV) started from 13 cP and then reduced to 8 cP. However, the gel strengths at 10 s and 10 min were also found to be 8 lb/100 ft^2, but after the temperature increased, the gel strength was reduced to 4 lb/100 ft^2, and the gel strength at 10 min started from 10 lb/100 ft^2 reduced to 5 lb/100 ft^2 due to polymer breaking down. Furthermore, the yield point was also reduced in the sample of FLOPRO drilling fluid starting from 25 lb/100 ft^2 reduced to 19 lb/100 ft^2 (see Figures 1a and 2a). In the sample of non-damaging fluid (RDF), the temperature effect on rheological properties, including the plastic viscosity (PV), started from 11 cP and then reduced to 6 cP, which means losses of 45.45% on their specific with temperature formed are fragile for this mud. Nevertheless, the gel strength of 10 s started at 6 lb/100 ft^2 and then reduced to 3 lb/100 ft^2, and the gel at 10 min started at 8 lb/100 ft^2 and then reduced to 5 lb/100 ft^2.

Table 2. Measured values of the plastic viscosity (μp), apparent viscosity (μa), yield point, and gel strength (10 s and 10 min) of the FLOPRO, non-damaging fluid, salt polymer mud, and FLOPRO with PTS-200 drilling fluid.

Drilling Fluid	Sample	Temperature	YP	μ_p	Gel Strength (lb/100 ft^2)	
		°F	(Ib/100 ft^2)	(cP)	Gel_{inital}	Gel_{final}
FLOPRO	FLOPRO_1	220	25	13	8	10
	FLOPRO_2	280	25	11	7	9
	FLOPRO_3	320	23	10	7	9
	FLOPRO_4	360	23	9	6	8
	FLOPRO_5	400	20	8	6	7
	FLOPRO_6	500	19	8	4	5
NDF	NDF_1	220	24	11	6	8
	NDF_2	280	21	11	6	7
	NDF_3	320	19	10	5	7
	NDF_4	360	17	10	4	6
	NDF_5	400	15	8	4	6
	NDF_6	500	14	6	3	5
SPM	SPM_1	220	25	12	8	10
	SPM_2	280	24	12	8	9
	SPM_3	320	23	10	8	9
	SPM_4	360	22	9	7	9
	SPM_5	400	20	8	5	7
	SPM_6	500	18	7	4	5
FLOPRO with PTS-200	FLOPRO_PTS-200_1	220	23	13	8	10
	FLOPRO_PTS-200_2	280	25	12	8	12
	FLOPRO_PTS-200_3	320	26	12	9	12
	FLOPRO_PTS-200_4	360	24	11	7	11
	FLOPRO_PTS-200_5	400	25	9	8	11
	FLOPRO_PTS-200_6	500	21	9	6	8
	FLOPRO_PTS-200_7	500	23	11	7	10

In addition, the yield point also reduced from 24 to 14 lb/100 ft^2, which means losses of 41.66% on their rheological properties with a high-temperature shape are frail for this mud (see Figures 1b and 2b). In the sample of salt polymer mud RDF, the plastic viscosity was found from 12 lb/100 ft^2, then reduced to 7 lb/100 ft^2. However, the gel strength was affected by the temperature: gel 10 s from 8 lb/100 ft^2 reduced to 4 lb/100 ft^2 and gel 10 min from 10 to 5 lb/100 ft^2. Furthermore, the yield point was reduced from 25 to 18 lb/100 ft^2 (see Figures 1c and 2c). Polymer temperature stabilizer (PTS-200) was employed to protect the polymer from breaking down and added to the FLOPRO sample; for plastic viscosity, the result shows that it was precisely 20.08% more stable than FLOPRO without PTS-200, and salt polymer and NDF, about 26.28 and 30.07%, respectively. Furthermore, the gel strength of both 10 s and 10 min FLOPRO with PTS-200 is more stable when compared with FLOPRO without PTS-200, NDF, and salt polymer mud about 49.7, 44, and 37.5% separately. Moreover, the yield point of FLOPRO with PTS-200 also showed better performance than

FLOPRO without PTS-200, NDF, and salt polymer mud about 24, 41.66, and 28, respectively (see Figures 1d and 2d).

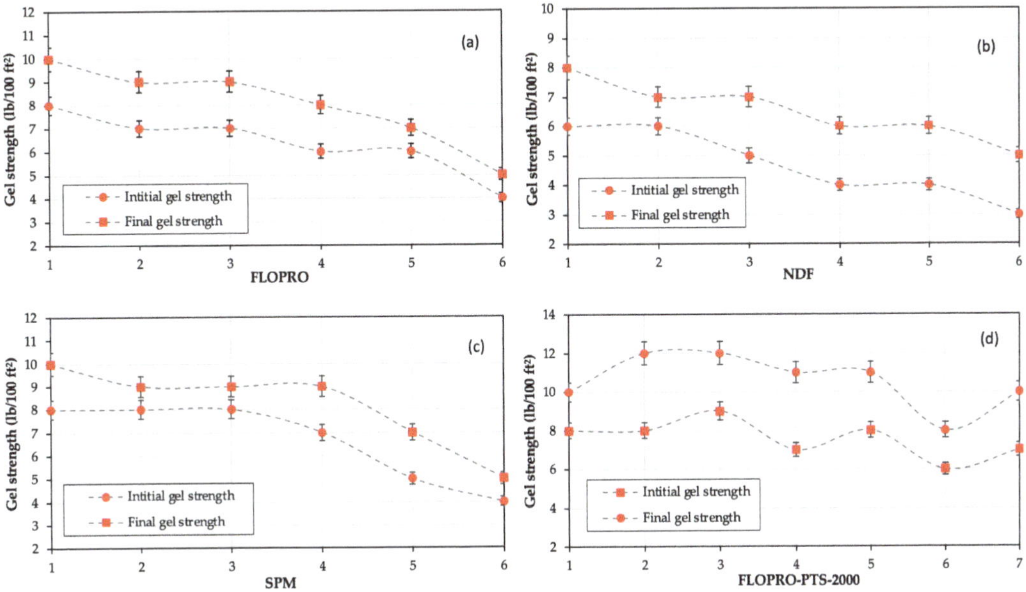

Figure 1. The results of the gel strength at 10 s and 10 min of the reservoir drilling fluid: (**a**) FLOPRO, (**b**) non-damaging fluid, (**c**) salt polymer mud, and (**d**) FLOPRO with PTS-200.

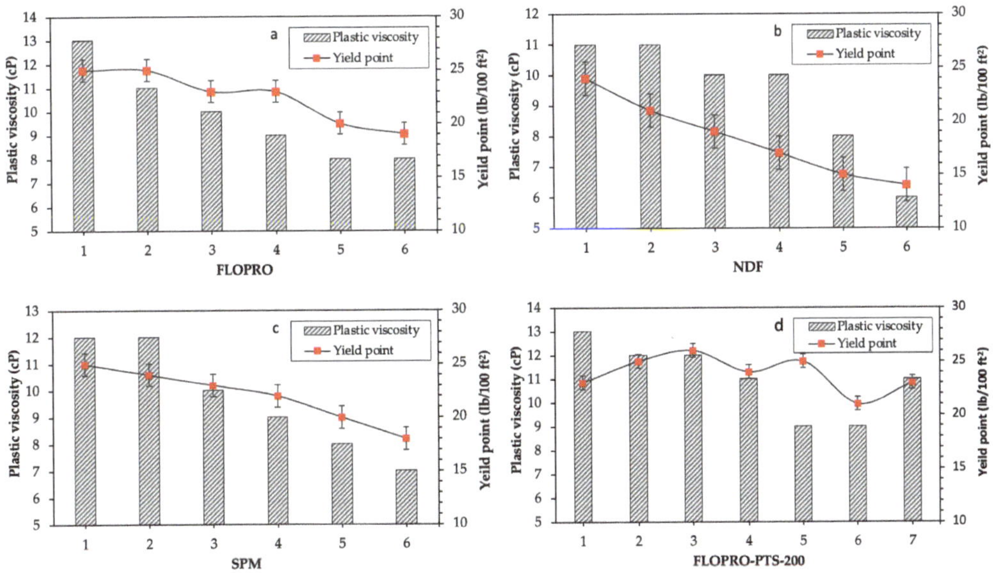

Figure 2. The results of the Yield Point of the reservoir drilling fluid (RDF): (**a**) FLOPRO, (**b**) non-damaging fluid, (**c**) salt polymer mud, and (**d**) FLOPRO with PTS-200.

2.2. Filtration Properties and Filter Cake Thickness

For the filtration characteristics to obtain better filtration control, 2 types of polymers added to each RDF of the FLOPRO drilling fluid (RDF) were prepared with 6.3 lb/bbl of FLOTROL and 3.5 lb/bbl of M-I PAC UL powder. In total, 6 samples were prepared: salt polymer mud drilling fluid (RDF) prepared with 5 lb/bbl of M-I PAC UL and 4.2 lb/bbl of Polysal powder was used for six samples; Non-Damaging Fluid (RDF) was prepared with 6.8 lb/bbl of PAC LV and 5 lb/bbl of Starch powder and also used for 6 samples. The materials were studied and determined under different temperature ranges from 220 to 500 °F. Table 3 displays the HP/HT fluid losses of three reservoir drilling fluids. As is obvious, the fluid loss is increasing with the increasing of the temperate from 220 to 500 °F.

Table 3. Measured values of the HPHT fluid losses with temperature applied, including initial and final thickness of filter paper, and filter cake thickness of the FLOPRO (RDF), non-damaging fluid, salt polymer mud, and FLOPRO with PTS-200 drilling fluid.

Drilling Fluid	Sample	Temp.	HPHT Fluid Loss	Filter Cake Thickness		
				Initial	Final	Average
		°F	cm^3/30 min	1/32″	1/32″	1/32″
FLOPRO	FLOPRO_1	220	9.1	0.004	0.0045	0.0005
	FLOPRO_2	280	9.3	0.004	0.0047	0.0007
	FLOPRO_3	320	10.1	0.004	0.005	0.001
	FLOPRO_4	360	10.5	0.004	0.0055	0.0015
	FLOPRO_5	400	10.9	0.004	0.0059	0.0019
	FLOPRO_6	500	11.8	0.004	0.007	0.003
NDF	NDF_1	220	10.1	0.004	0.005	0.001
	NDF_2	280	10.6	0.004	0.0055	0.0015
	NDF_3	320	10.9	0.004	0.0059	0.0019
	NDF_4	360	12.8	0.004	0.0081	0.0041
	NDF_5	400	13.5	0.004	0.0089	0.0049
	NDF_6	500	15.6	0.004	0.009	0.005
Salt Polymer mud	SPM_1	220	9.7	0.004	0.0048	0.0008
	SPM_2	280	10.3	0.004	0.0051	0.0011
	SPM_3	320	10.7	0.004	0.0057	0.0017
	SPM_4	360	12.3	0.004	0.0079	0.0039
	SPM_5	400	13.1	0.004	0.0085	0.0045
	SPM_6	500	14.5	0.004	0.011	0.007
FLOPRO with PTS-200	FLOPRO_PTS-200_1	220	9.1	0.004	0.0045	0.0005
	FLOPRO_PTS-200_2	280	9.3	0.004	0.0047	0.0007
	FLOPRO_PTS-200_3	320	9.6	0.004	0.0048	0.0008
	FLOPRO_PTS-200_4	360	9.9	0.004	0.0049	0.0009
	FLOPRO_PTS-200_5	400	10.2	0.004	0.0051	0.0011
	FLOPRO_PTS-200_6	500	10.6	0.004	0.0055	0.0015
	FLOPRO_PTS-200_7	500	10.2	0.004	0.0051	0.0011

Figure 3 illustrates the filtration rate of the FLOPRO, non-damaging fluid, salt polymer mud, and FLOPRO with PTS-200 drilling fluid for 30 min. As can be seen, the NDF fluid had the highest filtration rate, which increases with temperature to 15.6 mL at 30 min. Generally, the salt polymer mud drilling fluids prepared from M-I PAC UL and Polysal powders showed better performance in reducing the filtration rate due to creating sufficient filter cakes, which are impermeable at about 14.5 m. While the FLOPRO drilling fluids were prepared with FLOTROL and M-I PAC, fluid losses of 11.8 mL were recorded. In addition, PTS-200 after being added to FLOPRO showed the best performance; the minimum filtration rate of 10.2 mL at 30 min was obtained. During the investigation of the drilling fluid with the optimum temperature for polymer before breaking down, it was found that FLOTROL started to break down at 320 °F; meanwhile, drilling was required to treat the active system. However, PAC LV and M-I PAC UL polymer started to break down at 280 °F. Furthermore, Polysal polymer started to break down at 170 °F. Moreover, adding 2 wt.% of PTS-200 to FLOPRO (RDF) at 500 °F obtained the best performance to reduce filtration and protect the polymer from breaking down when the previous sample of FLOPRO at 500 °F with 1.5% PTS-200 was about 10.78% (see Figure 3).

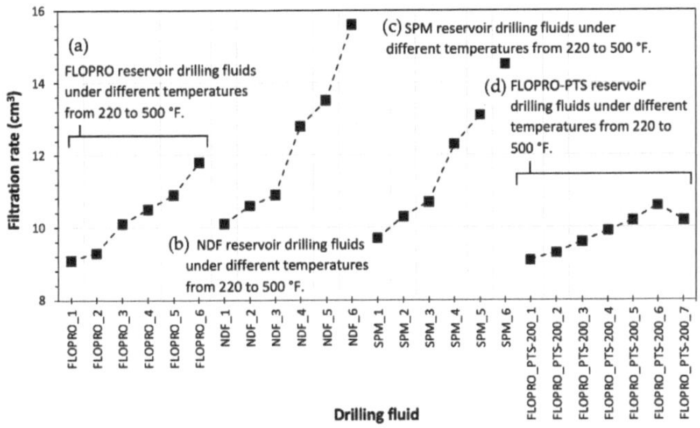

Figure 3. Measured filtration rates at 30 min of the developed reservoir drilling fluids (RDFs) under different temperatures from 220 to 500 °F: (**a**) FLOPRO, (**b**) non-damaging fluid, (**c**) salt polymer mud, and (**d**) FLOPRO with PTS-200.

Furthermore, Figure 4 illustrates the filter cake thickness of three types of the reservoir drilling fluids; FLOTROL filtration polymer shows better resistance to temperature, and the thickness increased by 83%. However, in the sample of non-damaging fluid, polysal showed poor results, with the lowest resistance to the temperature and a higher filtration rate at 500 °F. Moreover, the thickness of the filter cake increased by 94.44%. Furthermore, the sample of salt polymer mud showed slightly better resistance to the temperature, and the filter cake thickness was thinner than non-damaging fluid, but thicker than FLOPRO; the result showed 92.85%. Moreover, adding a polymer temperature stabilizer (PTS-200) to RDF can protect the polymer from breaking down by 66.66%, but the filter cake thickness increases by 40%.

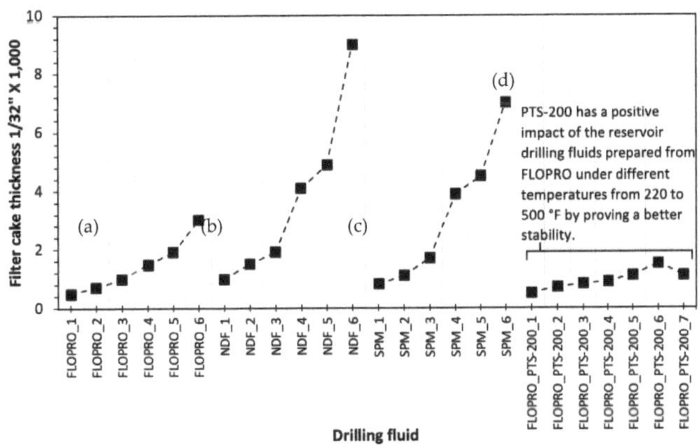

Figure 4. Measured filter cake thicknesses of 3 types of reservoir drilling fluids (RDFs) under different temperatures from 220 to 500 °F: (**a**) FLOPRO, (**b**) non-damaging fluid, (**c**) salt polymer mud, and (**d**) FLOPRO with PTS-200.

2.3. *Drilling Parameters*

2.3.1. Rate of Penetration

Overall, FLOPRO drilling fluid was used to drill 22 wells, whereas salt polymer mud (SPM) was used to drill 4 wells and non-damaging drilling fluid to drill 4 wells. The well was drilled using FLOPRO with a maximum ROP of 37 m/h for Well X-9. The minimum ROP was 16 m per hour in Well X-16. The well's highest ROP while using NDF was 14 m per hour in Well X-14, and the minimum ROP was 11 m/h in Well X-1. Furthermore, the greatest ROP for the salt polymer-drilled well was 26 m/h. The minimum ROP was 15 m per hour, as shown in Figure 5. Figure 5 displays the ROP for all wells that were drilled utilizing the three different types of RDF. Among all the used drilling fluids, the maximum ROP of 37 m/h was obtained when FLOPRO was used in drilling Wells X-8 and X-25. However, NDF illustrated the minimum ROP of 11 m/h in Well X-11. Overall, NDF reduced the ROP in all four wells that were used for drilling the horizontal section of the reservoir.

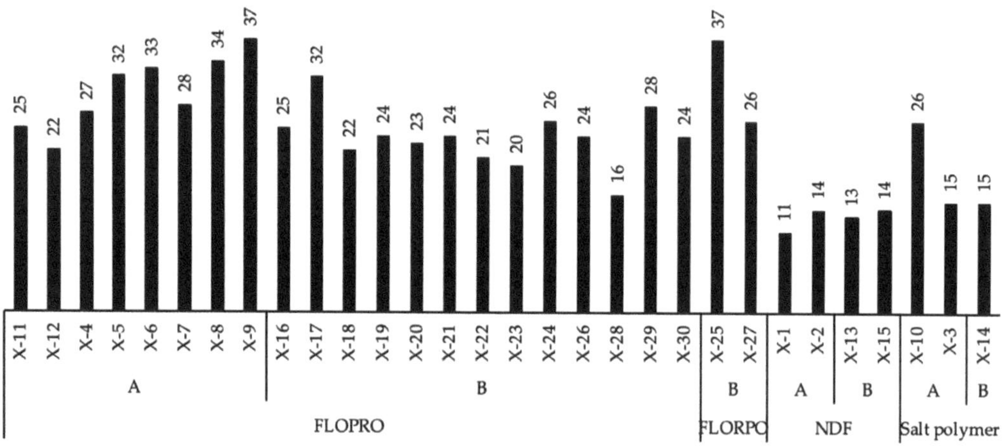

Figure 5. Average rate of penetration (ROP) of drilling the horizontal section of the reservoir of Wells X-1 to 30 using FLOPRO, non-damaging fluid, and salt polymer reservoir drilling fluids.

2.3.2. Cost per Meter (USD/m)

The cost per meter calculation of the drilling fluid used for drilling the horizontal section of thirty wells, excluding the drilling rig and other services in drilling, is shown in Figure 6. FLOPRO's average cost is roughly USD 29.64, compared with salt polymer mud's average cost of USD 38.28 and NDF's average cost of USD 48.67. As a result, FLOPRO is more affordable and performs better compared with the other types of used reservoir drilling fluids. The overall cost of the FLOPRO drilling fluid supplied for drilling about 2000 is 41.52% less than the SPM drilling fluid supplied for 4 wells and the NDF drilling fluid used in 4 wells. The highest cost of USD 83.3 for drilling a meter of the reservoir section horizontally was recorded for Well X-14 using SPM drilling fluid when only 811 m were drilled. However, the minimum cost of USD 14.8 for drilling a meter of the reservoir horizontally was obtained when drilling Well X-8 using FLOPRO reservoir drilling fluid when drilling 798 m (see Figure 6). In addition, the number of meters drilled of the reservoir section horizontally are shown in Figure 6. The horizontal section in almost all wells is about 2000 m. The minimum number of meters were drilled in Well X-9, which is 736 m, and the maximum was 2114.5 m in Well X-22.

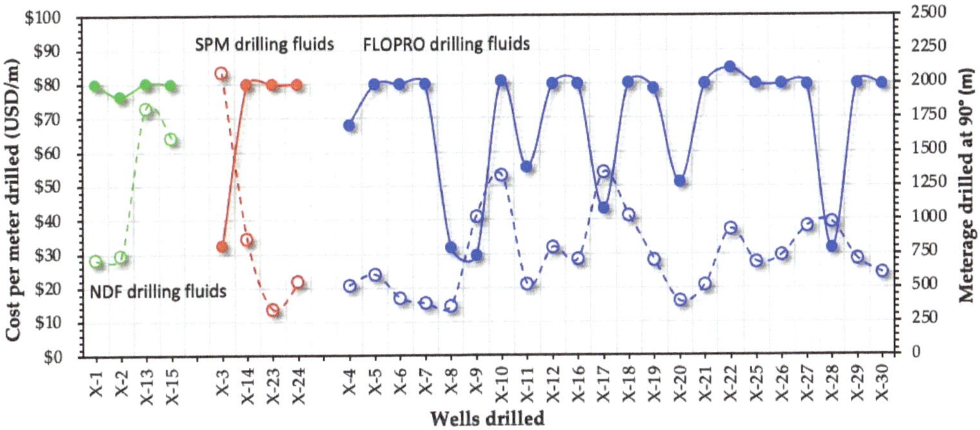

Figure 6. Measured depth and the cost per meter (USD/m) of drilling the horizontal section of the reservoir of Wells X-1 to 30 using FLOPRO, non-damaging fluid, and salt polymer reservoir drilling fluids.

2.3.3. Formation Losses

Overall, 30 wells were drilled with 3 different types of reservoir drilling fluids. Drilling with salt polymer mud resulted in no downhole losses, whereas drilling with NDF resulted in 2 wells with average losses of 82.22 m^3. In addition, 14 FLOPRO-drilled wells experienced downhole losses, with an average losses rate of roughly 60.85 m^3, as shown in Figure 7. As is clear, the highest losses of 152.2 m^3 happened within Well X-19, which was drilled using FLOPRO drilling fluid, while Well X-25 showed a minimum loss of the FLOPRO drilling fluid of 17 m^3. The loss of NDF drilling fluid is also high in both Wells X-13 and X-14, which was 99 and 65.45, respectively.

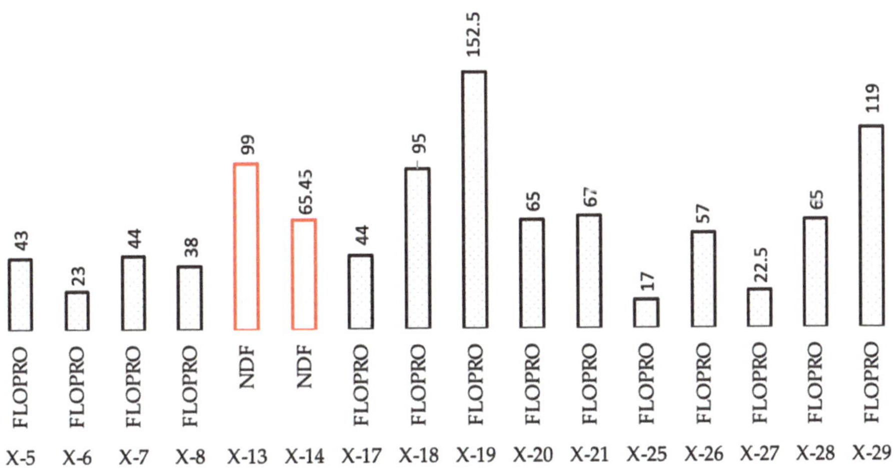

Figure 7. Total losses of the used drilling fluid in the reservoir section of studied wells.

2.4. Cost Analysis

2.4.1. Cost of Reservoir Section, Volume Build-Up, and Cost Per Barrel (USD/bbl)

NDF drilling fluid was used in Wells X-1, 2, 13, and 15, and SPM RDF was used in Wells X-3, 14, 23, and 24, as shown in Figure 8a, while FLOPRO drilling fluid was used in 22 wells (see Figure 8b). The average costs of NDF and SPM drilling fluids were 342 and

267 USD/m^3, respectively, and USD 165.11 was the average cost of each cubic meter of FLOPRO drilling fluid used in drilling the reservoir section horizontally. Hence, FLOPRO's end-of-well cost is significantly lower than that of NDF and salt polymer. The volume buildup of NDF and SPM drilling fluids was starting from 146.5 m^3 used in Well X-23 to the highest amount of 880 m^3 in Well X-15, as shown in Figure 8a. Furthermore, the costs of each cubic meter of SPM and NDF RDFs used in Well X-14 and Well X-1 are 133.57 and 653 USD/m^3, respectively. The FLOPRO buildup volume varied from 143 to 445.5 m^3, along with its cost per m3, which started from 58 to 314 USD/m^3 (see Figure 8b).

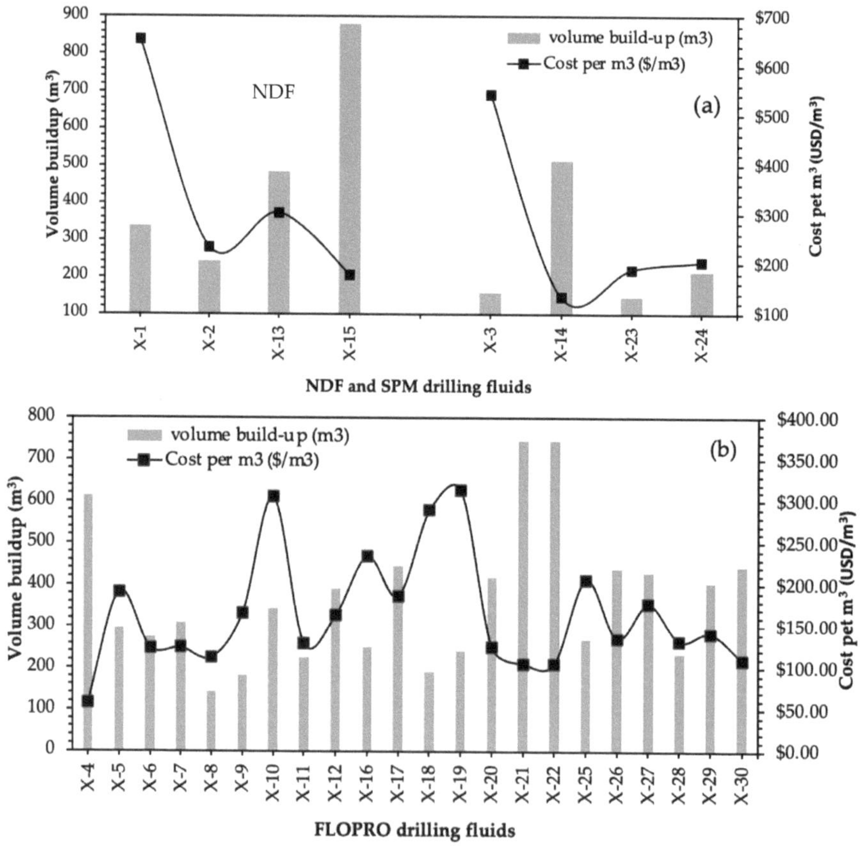

Figure 8. Volume buildup and its cost of the used reservoir drilling fluids in 30 wells: (**a**) SPM and NDF drilling fluids and (**b**) FLOPRO drilling fluid.

2.4.2. Completion Cost (USD/bbl)

Wells that were drilled with FLOPRO fluid had a lower overall completion cost compared with those drilled with NDF and SPM fluids because the NDF needed a D-Destroyer system to remove the filter cake from the borehole wall, and SPM needed a $CaCl_2$ filtration unit to complete the Saadi Formation. The NDF fluid was used in the completion of Wells X-1, 2, 13, 14, 28, and 29, and the maximum cost was recorded for Well X-1, which is USD 76,844. The cost of other wells by NDF was much less, that is, around USD 10,000. In addition, the completion cost of almost all wells by FLOPRO is around USD 10,000, expect for Wells X-22 and X-23. The average completion cost with FLOPRO is USD 8034 for all 20 wells used, from the minimum of USD 9945 to the maximum rate of USD 75,072.

Hence, the average cost of the completion with FLOPRO is lower compared with the NDF and SPM fluids, which are USD 21,818 and USD 16,947, respectively (see Figure 9).

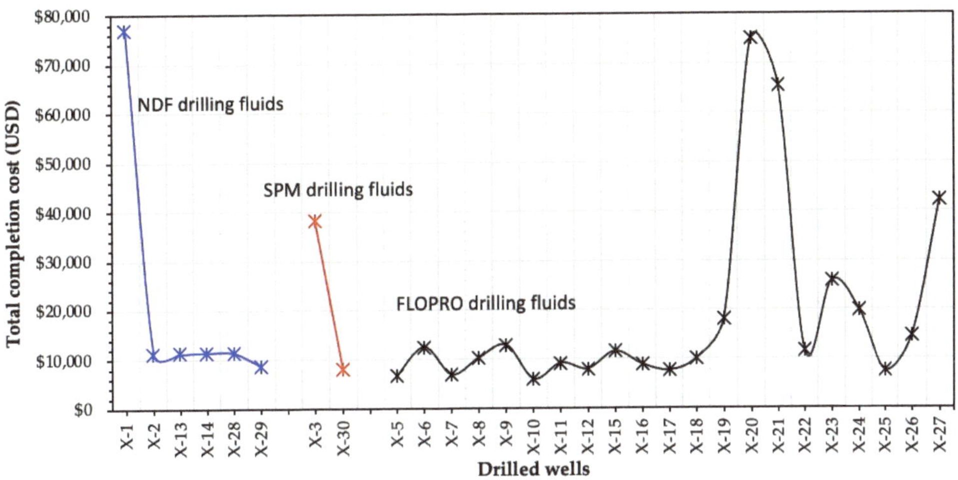

Figure 9. Measured values of the completion cost of 30 wells with types of reservoir drilling fluid.

2.5. Challenges and Prospective the Reservoir Drill-in Fluid

The breakdown temperature of polymers depends on the type of polymer and the drilling conditions. Some polymers, such as starch, can break down at temperatures as low as 150 degrees Fahrenheit. Others, such as FLOTROL, can withstand temperatures up to 300 degrees Fahrenheit when a polymer temperature stabilizer (PTS-200) is added. PTS-200 is a solution that can benefit modern industry challenges and promote the strategic opportunities of oil and gas companies [28–32].

It is important to take steps to protect polymers from breaking down. One of the main challenges in the oil industry is the lifecycle of polymers in long-term and short-term projects. As long as polymers break down, continuous treatment is required to keep drilling fluid parameters in the acceptable range. At the end, the well concentration and cost will be much higher than the planned cost of drilling the well [33–38].

After the end of a recent project, data were collected and compared from 30 turnkey project wells that used 3 different types of drilling fluid (RDF). The results indicated that FLOPRO had a positive economic impact and improved the KPI performance for drilling parameters. However, the study was limited by the fact that no core samples were taken during the drilling operation. This means that there was no core sample that could be used in the lab to test the filter cake removal performance of FLOTROL. Additionally, PTS-200 was not used in the NDF and salt polymer mud.

Overall, the results of this study suggest that FLOPRO is a promising new RDF that can help to improve drilling efficiency and reduce its costs. However, further research is needed to confirm the filter cake removal performance of FLOTROL and to assess its impact on skin damage.

3. Conclusions

The main goal of this study was to formulate a drilling fluid using an optimum concentration of polymer temperature stabilizer (PTS-200) to protect the polymers from breaking down at a temperature greater than 220 degrees Fahrenheit in order to achieve better filtration and rheological properties and lower cost. During the investigation period, starch polymer, an anti-filter loss polymer, began to degrade at 170 °F, whereas FLOTROL (modified starch) degraded at 320 °F. The optimum concentrations of polymer temperature stabilizer (PTS-200) were discovered at a temperature above the 320 degrees Fahrenheit required for 1% of PTS-200, based on achieving the lowest filtration rate. However, wells with greater temperatures, such as 500 °F, required using 2 wt.% of PTS-200, based on the archiving of better filtration and rheological properties. The results showed that polymers used to make FLOPRO, including FLOTROL and M-I PAC UL, are more stable than those used to make NDF and salt polymer mud, such as starch and PAC LV, which will break down at temperatures over 320 °F and result in viscosity that is 13.66% less stable and 31.14% less suitable for filtering.

4. Materials and Methods

4.1. Materials

Various additives were used in this study, such as modified natural (Duo-Vis) and synthetic viscosifier (FLO-VIS), sodium carbonate (Na_2CO_3), starch, biopolymer filtration (Polysal), poly anionic cellulose ultra-low viscosity (M-I PAC UL), polyanionic cellulose low-viscosity (PAC LV), synthetic polyanionic cellulose low-viscosity (FLOTROL) polymer, caustic soda (NaOH), NaCl salt, chloromethyl-isothiazolinone (M-I Cide), low-cost surfactant, drilling torque reducer (drillzone), ester lubricant (Lube XLS) and water-soluble brine lubricant (Safe Lube), phosphate-base corrosion inhibitor (Qonqor 404), and polymer temperature stabilizer (PTS-200). Each of the above chemicals and additives was added for a specific purpose, such as fluid loss control, viscosity improvement, corrosion inhibition, lubricity improvement, and temperature stability. All the mentioned chemicals were provided by the M-I SWACO—Schlumberger company (Basra, Iraq) with a purity of 99.8%.

4.2. Field Data and Study Area

Field data of drilling horizontal sections of 30 wells (X-1 to X-30) in the south of Iraq were collected. The drilled reservoir sections were Mishrif Formation from the middle cretaceous and Saadi Formation from the late cretaceous, which are limestone white chalky. The thicknesses of the Saadi and Mishrif formations are 136 m (from 2017 to 2153 m MD) and 110 m (from 2242 to 2352 m MD), respectively. Both formations are drilled horizontally at a 90° angle with an open-hole section length of about 2000 m. Table 4 shows the collected data from wells X-1-30. As can be seen, two drilling rigs, A and B, were used for drilling these wells. In addition, the meterage drilled, true of vertical depth (TVD), times taken to drill the sections, and horizontally drilled meterage are shown.

Table 4. Field data were collected from 30 wells in South Iraq using RDFs through the reservoir horizontally.

No.	Rig	Well	TVD	Meterage Drilled (MD)	Days to Finish Well	Meterage Drilled at 90°
1	A	X-1	2137	4867	56	1997
2	A	X-2	2405.88	4590	60	1910
3	A	X-3	2134	3520	37	811
4	A	X-4	2369.15	4686	38	1700
5	A	X-5	2397.75	4778	26	2000
6	A	X-6	2383.62	4770	27	2000
7	A	X-7	2416.96	5114	27	2000
8	A	X-8	2413.85	3729	22	798
9	A	X-9	2431.47	4176	22	736
10	A	X-10	2387	4826	47	2019
11	A	X-11	2408.23	4114	41	1386
12	A	X-12	2366	4946	46	2000
13	B	X-13	2379	5015	76	2005
14	B	X-14	2364.13	5456	52	1995
15	B	X-15	2477.01	5120	54	2000

Table 4. *Cont.*

No.	Rig	Well	TVD	Meterage Drilled (MD)	Days to Finish Well	Meterage Drilled at 90°
16	B	X-16	2371	4690	46	2003
17	B	X-17	2367	3767	24	1081
18	B	X-18	2373.31	4855	25	2005
19	B	X-19	2368.1	4917	23	1960
20	B	X-20	2379.77	4434	35	1276
21	B	X-21	2427.79	4839	30	1997
22	B	X-22	2442.46	5571	45	2114.5
23	B	X-23	2371	4684	39	1997
24	B	X-24	2106.39	4523	31	2000
25	B	X-25	2368	4681	26	2000
26	B	X-26	2370	5026	42	1997
27	B	X-27	2363.5	4695	35	1995
28	B	X-28	2135	3411	37.8	789
29	B	X-29	2371	4690	46	2003
30	B	X-30	2413.54	5201	31	1995

4.3. Preparation of the Drilling Fluids

Excluding the base sample, four types of RDFs were prepared. The developed RDFs are categorized into FLOPRO (six samples), salt polymer mud, SPM (six samples), non-damaging fluids, NDF (six samples), and an additional seven samples of FLOPRO with polymer temperature stabilizer (PTS-200). All drilling fluids were prepared using a hot plate stirrer and Hamilton Beach mixer for 30 min. The composition of the formulated drilling fluids is shown in Table 5, which includes the concentration of each chemical/additive used within each different drilling fluid sample.

Table 5. Composition of the formulated reservoir drill-in fluids used in this study.

Drilling Fluid Component	Reservoir Drilling Fluid (RDF)									
	FLOPRO	SPM	NDF	FLOPRO_PTS-200						
				1	2	3	4	5	6	7
Water, mL	350	350	350	350	350	350	350	350	350	350
Soda ash, gm	0.7	0.7	0.7	0.7	0.7	0.7	0.7	0.7	0.7	0.7
Caustic Soda, gm	0.5	0.5	0.7	0.5	0.5	0.5	0.5	0.5	0.5	0.5
FLO-Vis, gm	1.4	-		1.4	1.4	1.4	1.4	1.4	1.4	1.4
FLOTROL, gm	6.3	-	-	6.3	6.3	6.3	6.3	6.3	6.3	6.3
Safe Carb-20, gm	14	-	-	14.0	14.0	14	14	14	14	14
Safe Lube, cm^3	2.0	1.5	-	2.0	2.0	2.0	2	2	2	2
M-I Cide, cm^3	0.2	0.2	-	0.2	0.2	0.2	0.2	0.2	0.2	0.2
M-I PAC UL, gm	3.5	-	5.0	3.5	3.5	3.5	3.5	3.5	3.5	3.5
Qonqor 404, cm^3	1.0	-	-	1.0	1.0	1.0	1.0	1.0	1.0	1.0
DUO-Vis, mg	-	1.6	1.4	-	-	-	-	-	-	-
PAC LV, mg	-	6.8	-	-	-	-	-	-	-	-
$CaCO_3$ F, mg	-	14.0	14.0	-	-	-	-	-	-	-
Starch, mg	-	5.0	-	-	-	-	-	-	-	-
NaCl, mg	-	-	1.11	-	-	-	-	-	-	-
Lube XLS	-	-	2.0	-	-	-	-	-	-	-
DrillZone, cm^3	-	-	1.0	-	-	-	-	-	-	-
Polysal, mg	-	-	4.2	-	-	-	-	-	-	-
$ZnCO_3$, mg	-	-	0.7	-	-	-	-	-	-	-
PTS-200	-	-	-	-	0.5	1.0	1.0	1.5	1.5	2.0

4.4. Rheological Measurements of the Drilling Fluids

An API standard viscometer (FANN 35) was used to measure the rheological parameters of all prepared drilling fluid samples, including apparent viscosity, plastic viscosity, yield point, and gel strength under different temperature conditions of 220 to 500 °F. The apparent viscosity, plastic viscosity, and yield point are calculated using readings of 300 and 600 RPM of the viscometer rotor using Equations (1)–(3). The device is modified to measure gel strengths at 10 s and 10 min by observing the greatest, or maximum, deflection of the dial prior to the gel breaking. The recorded deflections of the rotating viscometer at various speeds allowed for the determination of shear loads, shear rates, and drilling fluid's 10 s and 10 min gel strengths.

$$\text{Plastic Viscosity } (\mu p)\ (cP) = 600 \text{ rpm reading} - 300 \text{ rpm reading} \quad (1)$$

$$\text{Apparent Viscosity } (\mu a)\ (cP) = 600 \text{ rpm reading}/2 \quad (2)$$

$$\text{Yield Point } (\tau y)\ (Ib/100\ ft2) = 300 \text{ rpm reading} - \mu p \quad (3)$$

4.5. Filtration Measurements of the Drilling Fluids

The filtration properties of the prepared drilling fluids at different concentrations of FLOTROL, M-I PAC UL, polysal, starch, and PAC LV were studied using a Series 300 HPHT Filter Press (M-I SWACO—Schlumberger company, Basra, Iraq) at 600 psi and 220–500 °F. The drilling fluid sample was put into the filter cell for each of the samples depicted in Table 5. The filter press is set up for the tests, and a graduated cylinder is positioned beneath the filtrate tube. As soon as the equipment is ready, timing is started, and the test begins. The test may continue for 30 min, and the amount of filtrate in the graduated cylinder was recorded.

4.6. Analysis of Drilling Field Data

Field data collected from 30 horizontal wells are for all the different types of RDFs mentioned in Table 5; NDF was used in drilling Wells X-1, 2, 13, and 15; SPM was used within Wells X-3, 14, 2,3, and 24; and FLOPRO was used in other wells. The profile of Well X-1 is shown in Figure 10, which includes a horizontal section of the reservoir that the field data are taken from. However, the rest of the wells almost have the same well profile. The mentioned drilling fluids were used under different temperature conditions through the reservoir interval of 2000 m horizontally. Field data will be focusing on the rate of penetration (ROP), cost of drilling in the reservoir section, cost per meter (USD/m), completion cost, volume build-up, cost per barrel (USD/bbl), and downhole losses using the following formulas:

$$\text{Cost per meter (USD/m)} = \text{Cost of reservoir section (USD)/meterage drilled (m)} \quad (4)$$

$$\text{Rate of Penetration (m/h)} = \text{Meterage drilled in reservoir section/day to complete section} \quad (5)$$

$$\text{Cost per barrel (USD/bbl)} = \text{Cost of reservoir section/volume build-up} \quad (6)$$

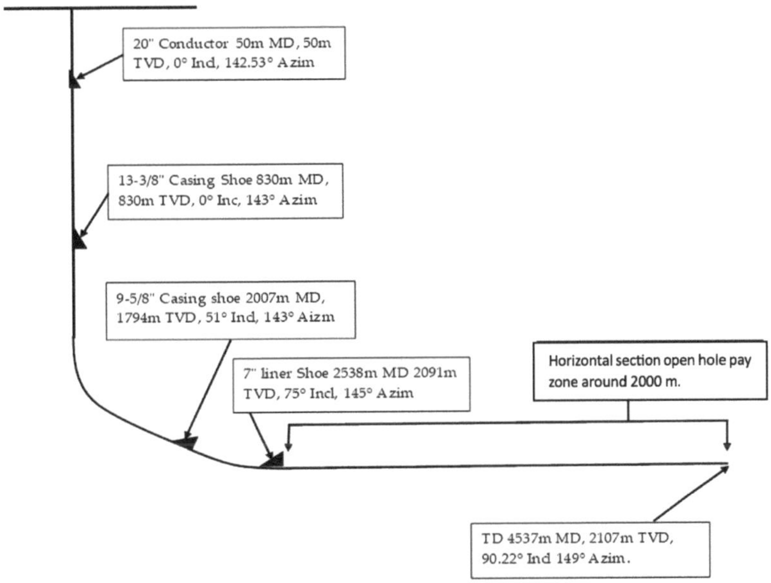

Figure 10. Well profile includes size and depth of casing, and a horizontal section of the reservoir that field data are taken from.

Author Contributions: Conceptualization, N.J. and J.A.A.; methodology, N.J. and J.A.A.; investigation, N.J.; data curation, N.J.; writing—original draft preparation, N.J.; writing—review and editing, J.A.A.; visualization, J.A.A. All authors have read and agreed to the published version of the manuscript.

Funding: This research received no external funding.

Institutional Review Board Statement: Not applicable.

Informed Consent Statement: Not applicable.

Data Availability Statement: The authors can provide any necessary information on the study upon request.

Acknowledgments: The authors would like to acknowledge Mohammed Jamal Al-Hayani M-I Swaco PSD Manager, a Schlumberger company, for his support of this study.

Conflicts of Interest: The authors declare no conflict of interest.

References

1. Annis, M.R.; Smith, M.V. *Drilling Fluids Technology*; EXXON Company: Spring, TX, USA, 1996; p. 366.
2. Caenn, R.; Darley, H.C.; Gray, G.R. *Composition and Properties of Drilling and Completion Fluids*; Gulf Professional Publishing: Houston, TX, USA, 2011.
3. Cargnel, R.D.; Luzardo, J.P. Particle Size Distribution Selection of CaCO3 in Drill-In Fluids Theory and Applications. In Proceedings of the Latin American and Caribbean Petroleum Engineering Conference, Caracas, Venezuela, 21–23 April 1999; SPE (53937), M-I Drilling Fluids de Venezuela; OnePetro: Richardson, TX, USA, 1999.
4. Tahr, Z.; Ali, J.A.; Mohammed, A.S. Sustainable aspects behind nano-biodegradable drilling fluids: A critical review. *Geoenergy Sci. Eng.* **2023**, *222*, 211443. [CrossRef]
5. Tahr, Z.; Ahmed, M.; Jagar, A.A. Surrogate models to predict initial shear stress of clay bentonite drilling fluids incorporated with polymer under various temperature conditions. *Arab. J. Geosci.* **2022**, *15*, 1449. [CrossRef]
6. Ali, J.A.; Ahmed, R.N.; Abdullah, A.D.; Ali, N.H.; Kalhury, A.M.; Sabir, A.N.; Manshad, A.K.; Keshavarz, A.; Mohammadi, A.H. Development of a Nanobiodegradable Drilling Fluid Using *Prosopis farcta* Plant and Pomegranate Peel Powders with Metal Oxide Nanoparticles. *ACS Omega* **2022**, *7*, 31327–31337. [CrossRef] [PubMed]

7. Ali, J.A.; Hamadamin, A.B.; Ahmed, S.M.; Mahmood, B.S.; Sajadi, S.M.; Manshad, A.K. Synergistic Effect of Nanoinhibitive Drilling Fluid on the Shale Swelling Performance at High Temperature and High Pressure. *Energy Fuels* **2022**, *36*, 1996–2006. [CrossRef]
8. Chean, L.; Raoof, G.; Mehdi, S.; Arshad, R.; Minou, R.; Nikoo, F.; Vamegh, R.; Jose, V. A new mud design to reduce formation damage in sandstone reservoirs. *J. Pet. Sci. Eng.* **2019**, *181*, 106221.
9. Offenbacher. *Reservoir Drill-In Fluid System Design Manual Handbook*, 3rd ed.; Offenbacher: Houston, TX, USA, 2016; pp. 130–135.
10. Wang, Z.; Wu, Y.; Luo, P.; Tian, Y.; Lin, Y.; Guo, Q. Poly (sodium P-styrene sulfonate) modified Fe3O4 nanoparticles as effective additives in water-based drilling fluids. *J. Pet. Sci. Eng.* **2018**, *165*, 786–797. [CrossRef]
11. Howard, S.; Kaminski, L.; Downs, J. *Xanthan Stability in Format Brines-Formulating Non-Damaging Fluids for High-Temperature Applications*; SPE (174228-MS); Format Brine Ltd.: Tealby, UK, 2015.
12. Marinescu, P.; Young, S.; Rameses, G.; Swaco, G.-I. A Schlumberger Company. New Generation Ultra-High Temperature Synthetic-Based Drilling Fluid; Development and Best Practices for Extreme Condition and ECD Management. In Proceedings of the Abu Dhabi International Petroleum Exhibition and Conference, Abu Dhabi, UAE, 10–13 November 2014; SPE (172056-MS). OnePetro: Richardson, TX, USA, 2014.
13. Zhuang, G.; Zhang, Z.; Peng, S.; Gao, J.; Jaber, M. Enhancing oil-based drilling fluids' rheological properties and thermal stability by synergetic use of organo-montmorillonite and organo-sepiolite. *Appl. Clay Sci.* **2018**, *161*, 505–512. [CrossRef]
14. Akpan, E.U.; Enyi, G.C.; Nasr, G.G.; Yahaya, A.A. Stabilizing biopolymers in water-based drilling fluids at high temperatures using antioxidants, a formatted salt, and polyglycol. *J. Eng. Technol.* **2018**, *7*, 469–486.
15. Lipatov, E.; Krivova, N. Influence of drilling fluids on the filtration properties of reservoirs. *IOP Conf. Ser. Mater. Sci. Eng.* **2019**, *663*, 012066. [CrossRef]
16. Elmgerbi, A.; Thonhauser, G.; Fine, A.; Hincapie, E.R.; Borovina, A. Experimental approach for assessing filter-cake removability derived from reservoir drill-in fluids. *J. Pet. Explor. Prod. Technol.* **2021**, *11*, 4029–4045. [CrossRef]
17. Samavati, R.; Abdullah, N. The Thermal Stability and Fluid Loss Control Enhancement of Fufu-WBM by Addition of Carbon Black and Gilsonite, in HTHP Circumstances. *Int. J. Sci. Res. Chem. Eng.* **2015**, *2*, 23–30. [CrossRef]
18. Galindo, K.A.; Zha, W.; Zhou, H.; Deville, J. High Temperature, High-Performance Water-Based Drilling Fluid for Extreme High-Temperature Wells. In Proceedings of the SPE International Symposium on Oilfield Chemistry, The Woodlands, TX, USA, 13–15 April 2015; SPE (173773-MS); OnePetro: Richardson, TX, USA, 2015.
19. Zaboli, M.D.; Sahraei, E.; Pourabbas, B. Hydrophobic silica nanoparticle-stabilized invert emulsion as drilling fluid for deep drilling. *Pet. Sci.* **2017**, *14*, 105–115. [CrossRef]
20. Wallace, S.; Beeson, J.; Dalton, F.; David, E.; Henry, I. Engineered Reservoir Drilling Fluid Help Maximize Production in Gulf of Mexico Deepwater Horizontal. In Proceedings of the SPE International Conference and Exhibition on Formation Damage Control, Lafayette, LA, USA, 19–21 February 2020; SPE (199281-MS). OnePetro: Richardson, TX, USA, 2020.
21. Aruther, K.G.; Peden, J.M. The Evaluation of Drilling Fluid Filter Cake Properties and their Influence on Fluid Loss. In Proceedings of the International Meeting on Petroleum Engineering, Tianjin, China, 1–4 November 1988; SPE (17617); OnePetro: Richardson, TX, USA, 1988; pp. 705–717.
22. Al-Mehailani, M.; Abdullah, N.; Al-Anzi, G.A.; Hussain, I.; Haider, B. Carefully Designed Water-Based Drill-In Fluid and Filter Cake Removal system Improve Reservoir Productivity and Reduce Well Completion Costs. In Proceedings of the 8th European Formation Damage Conference, Scheveningen, The Netherlands, 27–29 May 2009; SPE (122327); OnePetro: Richardson, TX, USA, 2009.
23. Tehrani, M.; Poppleton, A.; Guarneri, A.; Carminati, S. Water-Based Drilling Fluid For HP/HT Application. International Symposium on Oilfield Chemistry. In Proceedings of the International Symposium on Oilfield Chemistry, Houston, TX, USA, 28 February–2 March 2007; SPE (105485); OnePetro: Richardson, TX, USA, 2007.
24. Al-Otaibi, M.B.; Nasr-El-Din, H.A.; Al-Tameimi, Y.M. Wellbore Clean up by Water Jetting and Enzyme Treatments in MRC Wells Case Histories. In Proceedings of the SPE/IADC Middle East Drilling Technology Conference and Exhibition, Dubai, United Arab Emirates, 12–14 September 2005.
25. Huang, X.; LV, K.; Sun, J.; Lu, J.; Bai, Y.; Shen, H.; Wang, J. Enhancement of thermal stability of drilling fluid using laponite nanoparticles under extreme temperature conditions. *J. Mater. Lett.* **2019**, *248*, 146–149. [CrossRef]
26. Chen, Y.; Wu, R.; Zhou, J.; Chen, H.; Tan, Y. A novel hyper-cross-linked polymer for high-efficient fluid-loss control in oil-based drilling fluids. *Colloids Surf. A Physicochem. Eng. Asp.* **2021**, *626*, 127004. [CrossRef]
27. Zhang, Y.; Luo, H.; Shi, L.; Hongjun Huang, H. Synthesis of high-temperature viscosity stabilizer used in the drilling fluid. *IOP Conf. Ser. Earth Environ. Sci.* **2018**, *121*, 052029. [CrossRef]
28. Li, J.; Sun, J.; Lv, K.; Ji, Y.; Ji, J.; Liu, J. Nano-Modified Polymer Gels as Temperature- and Salt-Resistant Fluid-Loss Additive for Water-Based Drilling Fluids. *Gels* **2022**, *8*, 547. [CrossRef]
29. Qiyan Zhang, Q.; Chen, X.; Zhang, B.; Zhang, T.; Lu, W.; Chen, Z.; Liu, Z.; Kim, H.S.; Donovan, B.; Warzoha, R.J.; et al. High-temperature polymers with record-high breakdown strength enabled by rationally designed chain-packing behavior in blends. *Matter* **2021**, *4*, 2448–2459. [CrossRef]
30. Kato, K.L.; Cameron, R.E. A Review of the Relationship Between Thermally-Accelerated Ageing of Paper and Horrification. *Cellulose* **1999**, *6*, 23–40. [CrossRef]

31. Gałko, G.; Sajdak, M. Trends for the Thermal Degradation of Polymeric Materials: Analysis of Available Techniques, Issues, and Opportunities. *Appl. Sci.* **2022**, *12*, 9138. [CrossRef]
32. Bo, C.; Yun, X.; Xin, W.; Chunming, H.; Yuebin, G.; Guifu, D.; Wei, J.; Xuemei, Y.; Zhihe, X. A New High-Temperature Polymer Fracturing Fluid. *IOP Conf. Ser. Earth Environ. Sci.* **2018**, *186*, 012028. [CrossRef]
33. Hassan, R. Thermal Degradation of Paper: The Structural Changes of Fibers. *Egypt. J. Archaeol. Restor. Stud. EJARS* **2016**, *6*, 71–84.
34. Madorsky, S.L.; Sidney Straus, S. Thermal Degradation of Polymers at High Temperatures. *J. Res. Natl. Bur. Stand A Phys. Chem.* **1959**, *63*, 261–268. [CrossRef] [PubMed]
35. Thomas, D.C. Thermal Stability of Starch- and Carboxymethyl Cellulose-Based Polymers Used in Drilling Fluids. *SPE J.* **1982**, *22*, 171–180. [CrossRef]
36. Karakosta, K.; Mitropoulos, A.C.; Kyzas, G.Z. A review in nano-polymers for drilling fluids applications. *J. Mol. Struct.* **2021**, *1227*, 129702. [CrossRef]
37. Gautam, S.; Guria, C. Optimal Synthesis, Characterization, and Performance Evaluation of High-Pressure High-Temperature Polymer-Based Drilling Fluid: The Effect of Viscoelasticity on Cutting Transport, Filtration Loss, and Lubricity. *SPE J.* **2021**, *25*, 1333–1350. [CrossRef]
38. Tsiglianu, P.; Romasheva, N.; Nenko, A. Conceptual Management Framework for Oil and Gas Engineering Project Implementation. *Resources* **2023**, *12*, 64. [CrossRef]

gels

Article

Study on Water-Soluble Phenolic Resin Gels for High-Temperature and High-Salinity Oil Reservoir

Yunling Ran, Guicai Zhang *, Ping Jiang and Haihua Pei

School of Petroleum Engineering, China University of Petroleum (East China), Qingdao 266580, China; upcrhelm20@163.com (Y.R.); jiangping@upc.edu.cn (P.J.); peihaihua@upc.edu.cn (H.P.)
* Correspondence: zhanggc@upc.edu.cn

Abstract: High water cut of produced fluid is one of the most common problems in reservoir development. At present, injecting plugging agents and other profile control and water plugging technologies are the most widely used solutions. With the development of deep oil and gas resources, high-temperature and high-salinity (HTHS) reservoirs are becoming increasingly common. Conventional polymers are prone to hydrolysis and thermal degradation under HTHS conditions, making polymer flooding or polymer-based gels less effective. Phenol–aldehyde crosslinking agent gels can be applied to different reservoirs with a wide range of salinity, but there exist the disadvantage of high cost of gelants. The cost of water-soluble phenolic resin gels is low. Based on the research of former scientists, copolymers consisting of acrylamide (AM) and 2-Acrylamido-2-Methylpropanesulfonic acid (AMPS) and modified water-soluble phenolic resin were used to prepare gels in the paper. The experimental results show that the gelant with 1.0 wt% AM-AMPS copolymer (AMPS content is 47%), 1.0 wt% modified water-soluble phenolic resin and 0.4 wt% thiourea has gelation time of 7.5 h, storage modulus of 18 Pa and no syneresis after aging for 90 days at 105 °C in simulated Tahe water of 22×10^4 mg/L salinity. By comprehensively comparing the effectiveness of the gels prepared by a kind of phenolic aldehyde composite crosslinking agent and modified water-soluble phenolic resin, it is found that the gel constructed by the modified water-soluble phenolic resin not only reduces costs, but also has shorter gelation time and higher gel strength. The oil displacement experiment with a visual glass plate model proves that the forming gel has good plugging ability and thus improves the sweep efficiency. The research expands the application range of water-soluble phenolic resin gels, which has an important implication for profile control and water plugging in the HTHS reservoirs.

Keywords: low-cost gel; AM-AMPS copolymer; high-temperature and high-salinity resistant gel; oil displacement experiment

Citation: Ran, Y.; Zhang, G.; Jiang, P.; Pei, H. Study on Water-Soluble Phenolic Resin Gels for High-Temperature and High-Salinity Oil Reservoir. *Gels* **2023**, *9*, 489. https://doi.org/10.3390/gels9060489

Academic Editor: Michael Gradzielski

Received: 22 April 2023
Revised: 21 May 2023
Accepted: 2 June 2023
Published: 14 June 2023

1. Introduction

The problem of high water cut in oilfield development needs to be solved urgently, as it has the potential to reduce the utilization efficiency of injected water. A large amount of injected water flows through the high permeability layer, resulting in premature water emergence in oilfield development and a low sweep efficiency of the injected water [1]. To solve the above problems, common method is to inject selective plugging agents into high-permeability channels such as fractures [2,3]. Phenolic gel is one of the most commonly used plugging agents, the use of which is mainly divided into two methods: one is to make gelants by adding phenol and aldehyde solutions to the polymer solutions [4–14]; the other is to directly add a water-soluble phenolic resin solution into the polymer solutions [2,3,15–23].

The first type of gels has been widely studied and applied in reservoirs with different temperatures and salinities [4–14]. This paper focuses on the second kind of gels. Compared with the phenol–aldehyde crosslinking system gels, the latter can be prepared at a low cost. At present, some scientists have studied some gelling behaviors using water-soluble phenolic resin to replace the phenol–aldehyde crosslinking agent system [15–23]. According to the microstructure of chrome gel and phenolic resin gel prepared by Zhang et al. [16], the network structure of the phenolic resin gel is more prone to deformation under external forces. At high magnification rates, the phenolic gel network is more developed than the dendritic cluster structure of the chrome gel. Under the conditions of low temperature and salinity, the properties of the gels formed by cross-linking water-soluble phenolic resin with different polymers were also studied. Shang et al. [17] prepared a gel system formed by partially hydrolyzed polyacrylamide (HPAM) crosslinked with organic chromium and water-soluble phenolic resin at a temperature lower than 80 °C, a salt concentration lower than 16,000 mg/L, and a pH value between 6 and 8. The blocking rate of the system in the simulated core was reported to be 96%. Cui et al. [18] determined that gels with good thermal stability and shear resistance could be obtained when the concentration of HPAM was greater than 600 mg/L and the concentration of water-soluble phenolic resin was more than 1400 mg/L. Zhao et al. [19] selected phenolic resin cross-linked non-ionic PAM to prepare gels for deep profile control. The salinity in their experiment was 492.08 mg/L, and the experiment temperature was 80 °C. Gu et al. [20] used phenolic resin to crosslink non-ionic polyacrylamide to prepare gels and observed that when the concentration of the polymer and the crosslinking agent increased, so did the gels' viscosity and strength, but the viscosity and strength decreased when the concentration of salt or pH value increased. The phenolic resin gel system could maintain good stability at 95 °C under alkaline conditions with a salt concentration lower than 30,000 mg/L or a pH value greater than 7.

In the past five years, water-soluble phenolic resin has been further applied in oil fields. In 2019, Ge et al. [2] and Wu et al. [3] prepared a water-soluble phenolic resin gel at 70 °C for the low porosity and ultra-low permeability fractured reservoir in the Honghe Oilfield. The gel had good long-term stability after 180 days of heat treatment, and the strength was more than 10 Pa. In 2022, the impact of shear rate on the gelling behavior of phenolic resin gel in porous media was investigated by Yu et al. [21]. The findings revealed that the dynamic gel time was minimally influenced by the injection speed, whereas the gel strength was significantly affected. Guo et al. [22] used partially hydrolyzed polyacrylamide (HPAM) as the main agent, water-soluble phenolic resin (WSPR) as the crosslinking agent, and nano-SiO_2 as the stabilizer. The gelation time could be varied between 3 h and 23 h at 110 °C and 12,124 mg/L of salinity, and within 180 days, the gel's stability was still excellent. Xu et al. [23,24] chose AM-AMPS copolymer as the gel-forming agent and studied the plugging performance of a water-soluble phenolic resin gel system under the conditions of a salt concentration of 41,110 mg/L and a temperature range of 80–90 °C. Qu et al. [25] presented a gel with gelation time of 26–34 h at 55 °C using polyacrylamide, chromium acetate and water-soluble phenolic resin for the purpose of controlling water coning. In 2023, Zhi et al. [26] prepared a weak gel that demonstrated excellent resistance to temperature and salt by utilizing a crosslinking agent in conjunction with the phenolic resin when the salinity was 40,300.86 mg/L at 120 °C. From the above studies, it can be found that during the preparation of gels, most of the water-soluble phenolic resins are used at temperatures between 70 °C and 90 °C, and the reservoir salinity is usually lower than 4×10^4 mg/L. It is because water-soluble phenolic resins would precipitate and stick to walls in high salinity water, resulting in poor stability and unsatisfactory gelling properties.

In view of this, copolymers consisting of acrylamide (AM) and 2-acrylamido-2-methylpropanesulfonic acid (AMPS) and a modified water-soluble phenolic resin are used to prepare gels in this paper. The experimental results show that the gelant with 1.0 wt% AM-AMPS copolymer (AMPS content is 47%), 1.0 wt% modified water-soluble phenolic resin, and 0.4 wt% thiourea has a gelation time of 7.5 h, a storage modulus of 18 Pa, and no syneresis after aging for 90 days at 105 °C in simulated Tahe water with a salinity of 22×10^4 mg/L. That is to say, compared to the application conditions of the gels prepared by other scientists using water-soluble phenolic resin, this gel can be used in a higher salinity condition. The state changes of the modified water-soluble phenolic resin gel added to simulated Tahe water were observed, and the plugging performance of the modified water-soluble phenolic resin gel in porous media was evaluated in the experiment which proved that the forming gel had good stability and plugging ability. This research expands the application range of water-soluble phenolic resin gels, which has an important implication for profile control and water plugging in the HTHS reservoirs.

2. Results and Discussion

2.1. Evaluation of Gelation Performances of the Modified Soluble Phenolic Resin Gels

Phenol–aldehyde crosslinking agent gels can be applied to different reservoirs with a wide range of salinities, but they have the disadvantage of high cost of gelants. The cost of water-soluble phenolic resin is relatively low, so we are trying to use it to prepare a cheaper gel to meet the needs of the HTHS reservoirs.

The water-soluble phenolic resin is prone to self-polycondensation at room temperature and has poor solubility under high salinity. Clearly, this is not a good situation for making high-temperature and high-salinity resistant gels. Liu et al. [27] prepared a modified water-soluble phenolic resin by adding p-hydroxybenzoic acid. Their research shows that when the dosage of p-hydroxybenzoic acid is 10%, the phenolic resin can stably exist in Tahe water for more than 4 weeks at 50 °C. In other words, the modified water-soluble phenolic resin has excellent solubility in high-salinity water. Additionally, the research papers by Zhang et al. [28] and Guo et al. [29] showed that AM-AMPS copolymer could enhance the long-term stability of the prepared gel and reduce its syneresis rate in the HTHS environment.

Referring to the preparation method of conventional water-soluble phenolic resin, adding AM-AMPS copolymer solutions to the water-soluble phenolic resin modified by adding p-hydroxybenzoic acid may generate a uniformly high-strength gel to achieve a plugging effect. AM-AMPS copolymers with AMPS content of 0%, 25%, and 47% are referred to as PAM, AM-AMPS 30 and AM-AMPS 50, respectively. They were selected as gelling agents. The modified water-soluble phenolic resin was used as a crosslinking agent, and thiourea with a mass fraction of 0.4 wt% was added as a stabilizer to investigate the gelling performance of the prepared gels in simultated Tahe water at a temperature of 105 °C.

2.1.1. Gelation Time of the Modified Soluble Phenolic Resin Gels

At a temperature of 105 °C and a salinity of 22×10^4 mg/L, copolymers with different AMPS content were crosslinked with the modified water-soluble phenolic resin. The gelation time of the prepared gels is shown in Figure 1.

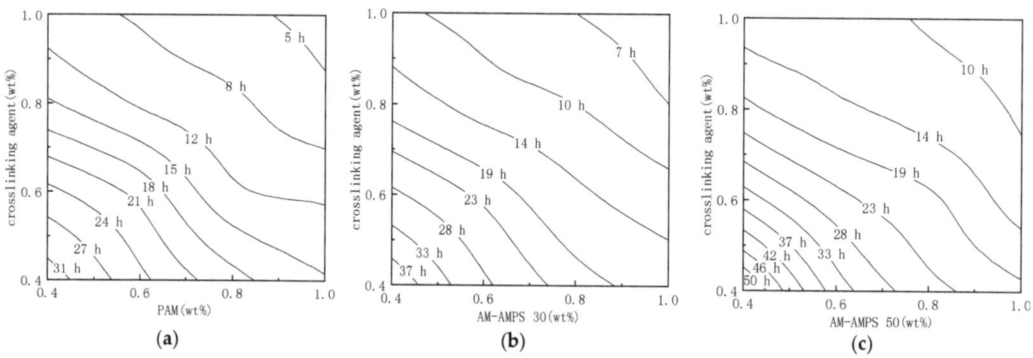

Figure 1. Gel formation time of different component gels aged at 105 °C: (**a**) Gel formation time of PAM and modified soluble phenolic resin; (**b**) Gel formation time of AM-AMPS 30 and modified soluble phenolic resin; (**c**) Gel formation time of AM-AMPS 50 and modified soluble phenolic resin.

It is evident that the gelation time and the mass fraction of the polymer and crosslinking agent have a negative relationship when the content of AMPS is constant. When the mass fractions of the polymer and crosslinking agent are both 0.4–1.0 wt%, the gelation time with PAM as the main agent is 4–32 h; the gelation time of an AM-AMPS copolymer with AMPS content of 25% as the main agent is 5–40 h; and the gelation time of an AM-AMPS copolymer with AMPS content of 47% as the main agent is 7.5–55 h. That is to say, when the AMPS content increases, the gelation time becomes significantly prolonged, which is similar to the results of Xu et al. [23,24].

2.1.2. Storage Moduli of the Modified Soluble Phenolic Resin Gels

As seen in Figure 2, the storage moduli of the gels prepared by crosslinking polymers with different AMPS content and the modified water-soluble phenolic resin at 105 °C after 5 days were measured. It can be seen that the strength has a good positive association with the mass fraction of the polymer and crosslinking agent. When the mass fractions of the polymer and crosslinking agent are both 0.4–1.0 wt%, the strength of AM-AMPS gels composed of AMPS monomer content of 0%, 25% and 47%, respectively, is 14–40 Pa, 10–30 Pa, and 5–18 Pa. That is to say, as the AMPS content in the polymer increases, the gel strength weakens, which is similar to the research results of Wang et al. [15].

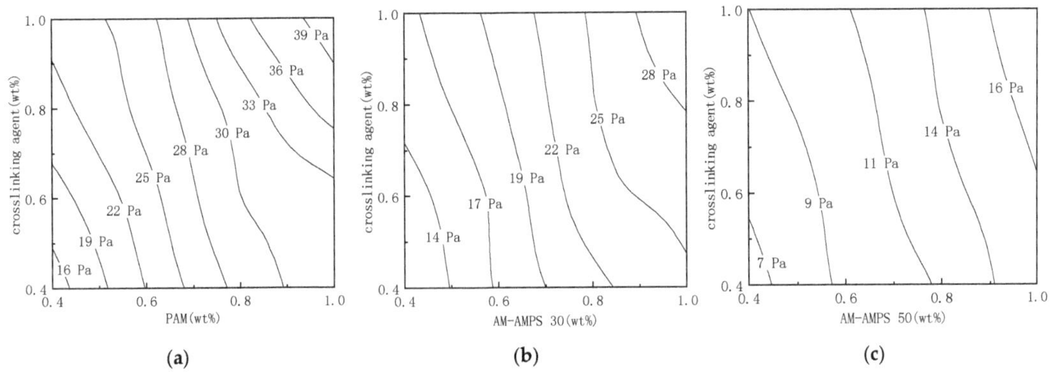

Figure 2. Storage modulus of different component gels aged at 105 °C: (**a**) Storage modulus of PAM and modified soluble phenolic resin; (**b**) Storage modulus of AM-AMPS 30 and modified soluble phenolic resin; (**c**) Storage modulus of AM-AMPS 50 and modified soluble phenolic resin.

2.1.3. Long-Term Stability of the Modified Soluble Phenolic Resin Gels

The stability of the gels was characterized by the syneresis rate at 105 °C after 90 days. The syneresis rate of the gels prepared by crosslinking copolymers of different AMPS content with the modified water-soluble phenolic resin is shown in Figure 3.

As can be observed, the stability of gels with different AMPS content is positively correlated with the mass fraction of copolymer and crosslinking agents. As the mass fraction of copolymer and crosslinking agents increases, the syneresis rate of gels decreases. When the mass fractions of the polymer and crosslinking agent are both 0.4–1.0 wt%, the syneresis rate of the gels with PAM polymer as the main agent aged at 105 °C for 90 days is between 50% and 80% (exceeding 50%). The syneresis rate of the gels mainly composed of an AM-AMPS copolymer with AMPS content of 25% varies from 10% to 70%. The gel with a high mass fraction of copolymer and crosslinking agent has good stability and a low syneresis rate. The syneresis rate of the gel prepared with 1.0 wt% copolymer and 1.0 wt% crosslinking agent is only 10%, and the syneresis rate of the gel prepared with a 0.4 wt% copolymer and a 0.4 wt% crosslinking agent is 70%. In addition, the increase in AMPS content also enhances the stability of the gel. The syneresis rate of the gels using a copolymer with AMPS content of 47% as the main agent is 0–15%, indicating a significant improvement in the stability of the gel.

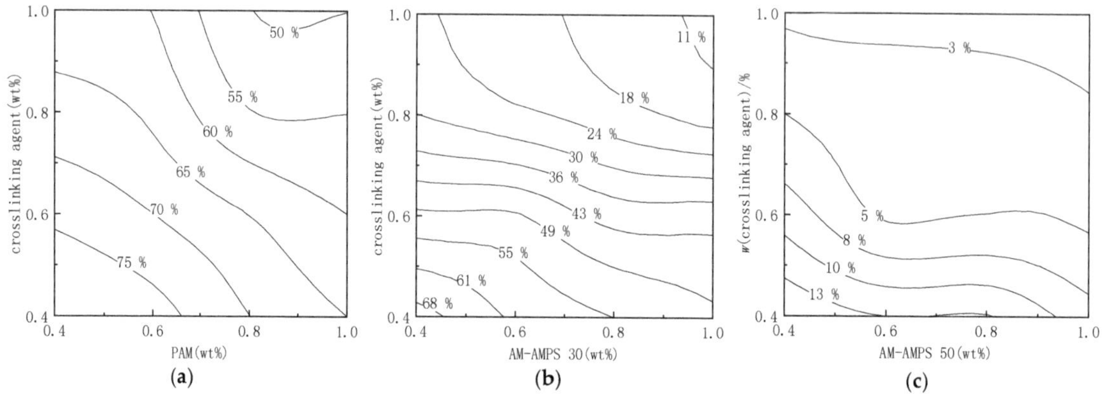

Figure 3. Syneresis rate of different component gels aged at 105 °C: (**a**) Syneresis rate of PAM and modified soluble phenolic resin; (**b**) Syneresis rate of AM-AMPS 30 and modified soluble phenolic resin; (**c**) Syneresis rate of AM-AMPS 50 and modified soluble phenolic resin.

In summary, examining the gelation performance of the gels prepared by the modified water-soluble phenolic resin and AM-AMPS copolymer, it becomes clear that with the increase in the AMPS content, the gelation time of the gels becomes longer and the long-term stability increases, but the strength of the gel decreases. This is related to the structure of the copolymer. The crosslinking mechanism of the gel is the polycondensation of phenolic resin and amide group in polymer chains [1]. As the copolymer's AMPS content rises, the amide group content correspondingly reduces, resulting in a longer crosslinking time and weaker gel strength. However, the rate of hydrolysis and degradation of the copolymer at high temperature slows down. Therefore, the long-term stability of the prepared gel is enhanced.

The performance of the prepared gel using the AM-AMPS copolymer and the phenolic aldehyde composite crosslinking agent was evaluated at 105 °C and a salinity of 22×10^4 mg/L, as shown in the following Figure 4. The copolymer with AMPS monomer ratio of 47% was selected as the gelling agent, the phenolic crosslinking agent was hydroquinone, and the aldehyde crosslinking agent was urotropine.

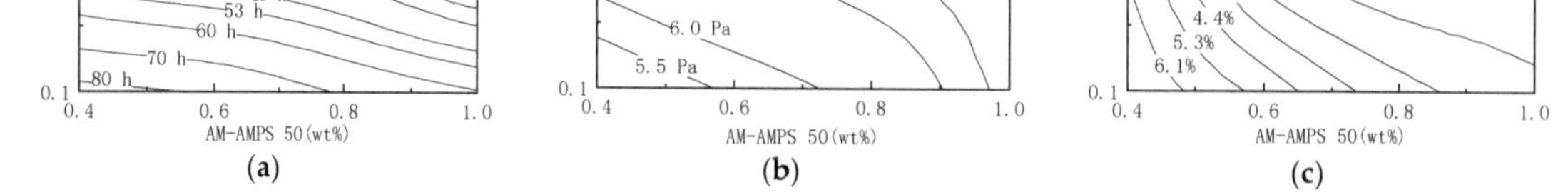

Figure 4. Performance of different component gels aged at 105 °C: (**a**) Gel formation time of AM-AMPS 50 and phenolic aldehyde composite crosslinking agent; (**b**) Storage modulus of AM-AMPS 50 and phenolic aldehyde composite crosslinking agent; (**c**) Syneresis rate of AM-AMPS 50 and phenolic aldehyde composite crosslinking agent.

It can be seen that the gelation time is 12–83 h, the strength after 5 days is 4.9–9.8 Pa, and the dehydration rate after 90 days is only 0–7%. When the mass fraction of AM-AMPS copolymer is 1.0 wt%, the hydroquinone is 0.4 wt%, and the urotropine is 0.8 wt%, the gelation time of the prepared phenolic aldehyde crosslinking agent gels can be shortened to 12 h, and the storage modulus after 5 days of aging is 9.8 Pa (Sydansk's gel strength code of "G": when the ampoule is inverted, the gel flows down to about half of the ampoule). The gel has good stability and does not dehydrate within 90 days, making it suitable for the harsh conditions of high temperature and high salinity in Tahe Oilfield.

Using a mass fraction of 1.0 wt% AM-AMPS copolymer (AMPS content is 47%), 1.0 wt% modified water-soluble phenolic resin, and 0.4 wt% thiourea as the formula, a low-cost gel with gelation time of 7.5 h, storage modulus of 18 Pa, and no syneresis at 105 °C after 90 days can be prepared. By comprehensively comparing the performance of the gel prepared by a phenolic aldehyde composite crosslinking agent and the modified water-soluble phenolic resin, it can be found that the gel constructed by the modified water-soluble phenolic resin not only reduces costs, but also has a shorter gelation time and higher gel strength, albeit with slightly poorer stability.

This may happen because the crosslinking reaction between the water-soluble phenolic resin and the polymer skips the early process of the phenol–aldehyde reaction, leading to a shortened crosslinking reaction time. In addition, the water-soluble phenolic resin has a larger conformation and a certain degree of rigidity, resulting in the formation of gels with higher strength.

2.2. Evaluation of Plugging Performance of Modified Water-Soluble Phenolic Resin Gel

2.2.1. Gel Strength after Adding Water

After the modified water-soluble phenolic resin gel is injected into the formation, it first comes into contact with a large amount of formation water. In order to investigate the effect of high-salinity Tahe water on the performance of the gel under 105 °C conditions, simulated formation water was added to the gel in a 1:1 volume ratio to observe the strength changes in the gel. The results of the experiment can be seen in Figure 5.

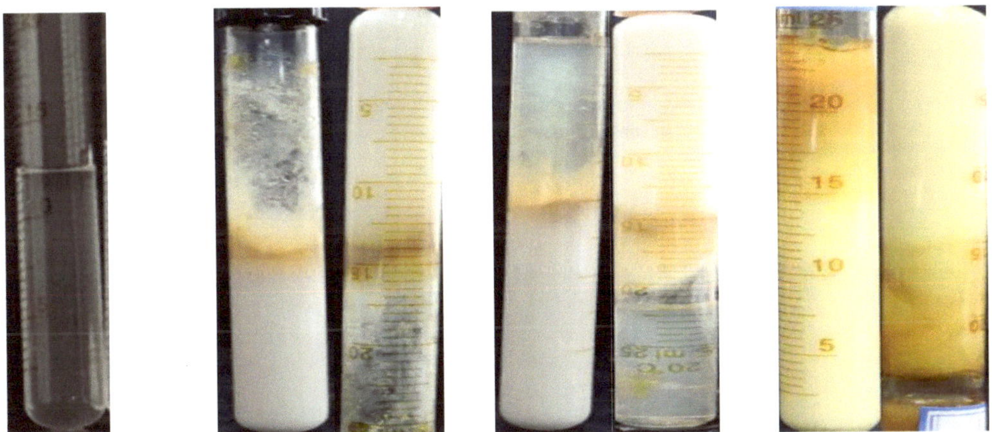

(a) (b) (c) (d)

Figure 5. State changes of the modified water-soluble phenolic resin gel after adding Tahe water: (**a**) the state of the gel for 0 h; (**b**) the state of the gel for 24 h; (**c**) the state of the gel for 7 days; (**d**) the state of the gel for 30 days.

From the results, it can be seen that for the modified water-soluble phenolic resin gel, after the Tahe simulated formation water was added, the gel strength level was H (when the ampoule is inverted, only the surface of the gel is slightly deformed) after 30 days, and it could still maintain good stability.

2.2.2. Plugging Ability of Gel in Porous Media

The performance of the most stable water-soluble phenolic resin gel (1.0 wt% AM-AMPS 50 + 1.0 wt% modified water-soluble phenolic resin + 0.4 wt% thiourea) was evaluated by conducting a displacement experiment in a self-made visual glass flat plate model filled with sand. The experimental process, which is shown in Figure 6, can be divided into three steps: first, after the model is saturated with oil, primary water flooding is carried out until the water cut reaches 94%; then, 0.3 PV of the modified water-soluble phenolic resin gelant is injected and allowed to sit for 24 h; finally, the subsequent water drive is conducted until the water cut reaches 98%.

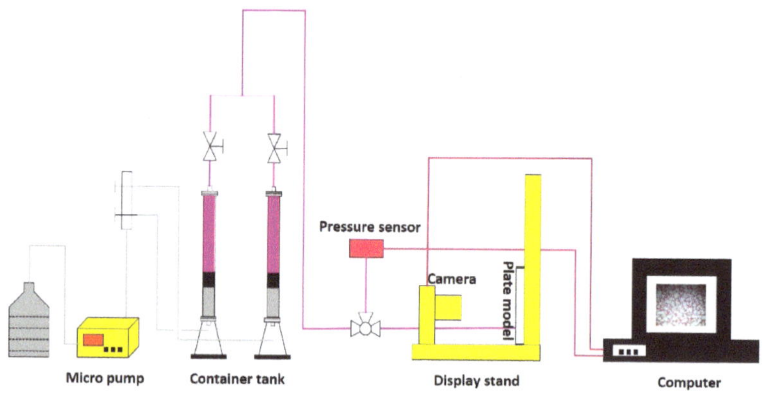

Figure 6. Visualized plate model displacement experiment.

The distributions of residual oil at different stages were observed using a heterogeneous visual sand-filling model, as shown in Figure 7. From Figure 7b, it can be seen that during water injection development under heterogeneous reservoir conditions, water gives priority to flow in the high permeability zones to form fluid channeling, and the color becomes significantly lighter because of the decrease in oil content after the primary water drive. The analysis of Figure 7d reveals that the gelling agent successfully forms a robust gel in the central region of the reservoir, thereby inducing a plugging effect. The gel formation obstructs the "dominant channel" in the high permeability zones, leading to the displacement of oil by water in the low-permeability section. Consequently, the oil sand in the low permeability zones exhibits a lighter color, indicating an enhancement in the sweep efficiency and an improvement in the recovery of crude oil due to the injection of gel.

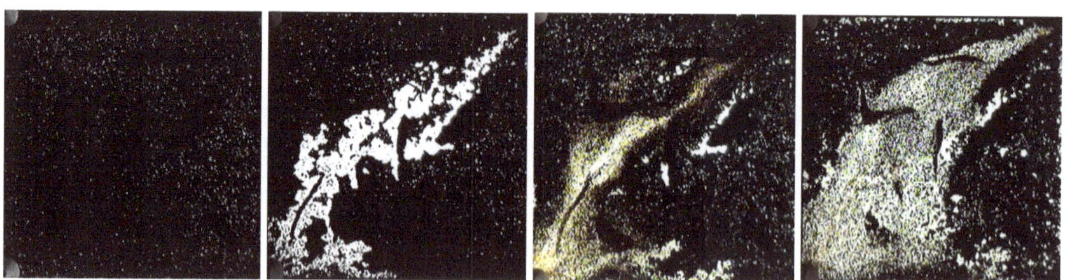

(a) (b) (c) (d)

Figure 7. Residual oil distributions at different stages of heterogeneous model: (**a**) Residual oil distribution before first water drive; (**b**) Residual oil distributions after first water drive; (**c**) Residual oil distributions after gel formation; (**d**) Residual oil distributions after subsequent water drive.

Figure 8 illustrates the correlations of oil recovery and water cut with injected volume of water in the experiment. The results indicate that an increase in water injection leads to a gradual rise in oil recovery. At water cut of 93.48%, the cumulative oil production amounted to 16.3 mL, resulting in a recovery rate of 20.45%. After the gel is injected, it occupies the pores in the highly permeable zone of the model. Subsequently, during the secondary water flooding phase, water infiltrates the low permeability layer, displacing some of the oil before creating a new water flooding path. The displacement of oil by water flooding results in a gradual improvement in recovery. After water cut of 97.84% is reached, the cumulative oil production amounts to 31.22 mL, and the recovery rate reaches 39.17%. The introduction of gel during secondary water flooding leads to an increase in oil recovery by 18.72%.

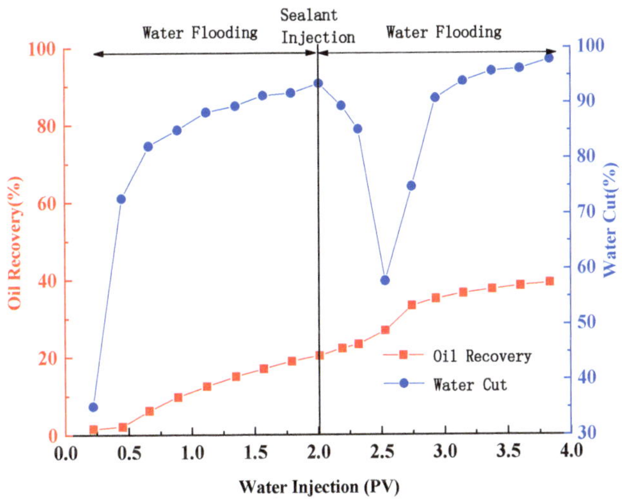

Figure 8. Production curve of heterogeneous model.

3. Conclusions

Different gels were prepared using the modified water-soluble phenolic resin and AM-AMPS copolymers with AMPS content of 0%, 25%, and 47%, respectively. The experimental results indicate that the gelation time decreases with the increase in the mass fraction of the copolymer and crosslinking agent when the content of AMPS is constant. Furthermore, with the increase in the AMPS content of the copolymer, the gelation time is significantly prolonged, the strength of the gel decreases, but Its long-term stability improves.

The optimal formulation for the gelling solution is 1.0 wt% AM-AMPS copolymer (AMPS content is 47%), 1.0 wt% modified water-soluble phenolic resin, and 0.4 wt% thiourea. The high-performance gel with a gelation time of 7.5 h, a storage modulus of 18 Pa, and no syneresis after 90 days can be prepared in the simulated Tahe water at 105 °C and a salinity of 22×10^4 mg/L. By comprehensively comparing the performance of the gel prepared by a phenolic aldehyde composite crosslinking agent and the modified water-soluble phenolic resin, it can be found that the gel constructed by the modified water-soluble phenolic resin not only reduces costs, but also has a shorter gelation time and higher gel strength, though it has slightly poorer stability.

The displacement experiment with a visual glass plate model has proved that the formed gel has good plugging ability and can improve the sweep efficiency. This research expands the application range of water-soluble phenolic resin gels, which has an important implication for profile control and water plugging in high-temperature and high-salinity reservoirs.

4. Materials and Methods

4.1. Materials

A series of analytical pure chemical reagents such as NaCl, $CaCl_2$, $MgCl_2$, $NaHCO_3$, formaldehyde, phenol, sodium hydroxide, and p-hydroxybenzoic acid were purchased from Sinopharm Chemical Reagent Co., Ltd. (Shanghai, China); hydroquinone, thiourea, and urotropine were provided by Aladdin Biochemical Technology Co., Ltd. (Shanghai, China); PAM was provided by Anhui Jucheng Fine Chemical Co., Ltd. (Huaibei, China); AM-AMPS copolymers (AMPS content is 25% and 47%, respectively) were prepared by an aqueous solution polymerization method, with the initiator being a redox system consisting of ammonium persulfate and sodium bisulfite; nitrogen was provided by Qingdao Xinkeyuan Technology Co., Ltd. (Qingdao, China)

The main instruments are shown in Table 1.

Table 1. The main instruments.

Instrument Name	Instrument Type	Manufacturer
Precision digital display mixer	JJ-1	Jintan Jincheng Guosheng Experimental Instrument Factory (Changzhou, China)
Electronic balance	ME403	Mettler Toledo International Trade Co., Ltd. (Shanghai, China)
Projector display stand	GK-8000A	Ruiying Information Technology Co., Ltd. (Guangzhou, China)
Micro injection pump	100DX	Teledyne Isco, Inc. (Lincoln, Nebraska, USA)
Alcohol blowtorch	GW-6	Subei Experimental Instrument Co., Ltd. (Taizhou, China)
Constant temperature oven	DHG-9070A	Jinghong Experimental Equipment Co., Ltd. (Shanghai, China)

4.2. Methods

4.2.1. Preparation of the AM-AMPS Copolymer Solution

The molecular weight of a water-soluble AM-AMPS copolymer is relatively large, and its dissolution in water requires a certain amount of time. The AM-AMPS copolymer with AMPS content of 47% was taken as an example, and the AM-AMPS copolymer was added to the simulated Tahe water to prepare a certain mass fraction of the copolymer solution, which was stirred slowly for 24 h, left to swell for 72 h, and then stirred for an additional 2 h.

4.2.2. Synthesis Method of the Modified Water-Soluble Phenolic Resin

At present, many scientists have synthesized soluble phenolic resins using phenol and formaldehyde in the laboratory. The two-step catalytic method is a commonly used method with similar steps [30–32]. The thermostatic water bath is controlled at 50 °C, with the molar ratio of formaldehyde to phenol set at 3:1, and 8.0% NaOH of the total mass of formaldehyde and phenol added as the catalyst. Then, the formaldehyde is slowly dropped into the chemical reactor using a separatory funnel, with the dropping time controlled at 30 min. The temperature is quickly raised to 70 °C after the synthesis of water-soluble phenolic resin, 10.0% p-hydroxybenzoic acid of the total mass of formaldehyde and phenol is added to the water-soluble phenolic resin and the mixture is stirred thoroughly for 40 min. Finally, the modified water-soluble phenolic resin with good transparency is obtained.

4.2.3. Preparation Method of Water-Soluble Phenolic Resin Gel

The gel consists of a copolymer solution, the crosslinking agent, or a certain stabilizer. Firstly, at room temperature, the prepared water-soluble phenolic resin is added to the simulated Tahe water according to the designed gel formula, and a mass fraction of 0.4 wt% thiourea is added as a stabilizer. Then, AM-AMPS copolymer is added while stirring, and a uniform gel solution is obtained after stirring. A total of 20 g of the gelling solution is weighed and injected into an ampoule, sealed with an alcohol blowtorch, and then placed in a thermostatic oven at 105 °C for heat treatment to form the gel.

4.2.4. Determination of Gelation Time and Strength of Gel

The gelation time and gel strength were qualitatively determined using the bottle test method, following Sydansk's gel code [33]. The gelation time was defined as the duration required for the gels to attain Sydansk's code of "D" (when the ampoule is inverted, only a small part of the gel is difficult to flow to the bottom). After sealing the sample, the gelation time and gel strength were observed at ambient temperature. In addition, the storage modulus of the gel was evaluated using a rheometer at a temperature of 25 °C. The setting method of the instrument can be referred to our previous article [34].

4.2.5. Method for Determining the Syneresis Rate of Gel

The evaluation of the long-term stability of a gel is mainly based on the syneresis rate of a gel as a reference indicator; when the syneresis rate of the gel is high after aging under the same temperature conditions, it can be considered that the long-term stability of the gel is poor, whereas if the syneresis rate is low or there is no syneresis phenomenon, the long-term stability of the gel is good. Thermal stability experiments were conducted at 105 °C; the gelling solution was sealed in an ampoule bottle, and after gelation, it was taken out of the oven at the set time and the ampoule bottle containing the gel was opened. The mass of gel syneresis was measured using a ME403 electronic balance; due to the initial mass of the gel forming solution being 20 g, the syneresis rate of gel was equal to the mass of the separated water weighed by the balance divided by 20 g.

Author Contributions: Conceptualization, Y.R. and G.Z.; methodology, Y.R.; formal analysis, P.J.; investigation, P.J. and H.P.; resources, G.Z.; writing—original draft preparation, Y.R.; writing—review and editing, G.Z.; supervision, H.P. All authors have read and agreed to the published version of the manuscript.

Funding: This research received no external funding.

Institutional Review Board Statement: Not applicable.

Informed Consent Statement: Not applicable.

Data Availability Statement: Data are contained within the article.

Conflicts of Interest: The authors declare no conflict of interest.

References

1. Xin, X.; Liu, Q.; Liu, S.; Yu, G.; Wan, Q. Optimization of Gel Flooding during the High Water Cut Stage in a Conglomerate Reservoir of the Xinjiang A Oilfield. *Polymers* **2023**, *15*, 1809. [CrossRef]
2. Ge, J.; Zheng, W.; Wei, K.; Deng, X.; Fang, Q. Preparation and Performance Evaluation of Phenolic Gel Using for Plugging Cracks in Honghe Reservoir. *Oilfield Chem.* **2020**, *37*, 432–437. [CrossRef]
3. Wu, Q.; Ge, J.; Zhang, G.; Guo, H.; Zhao, Q. The water shutoff simulation experiment using tube for high-strength plugging agent in fractured reservior. *Acta Pet. Sin.* **2019**, *40*, 1368–1375.
4. Zhou, R.; Zhang, D.; Wei, J. Experiment on the profile control effect of different strength gel systems in heterogeneous reservoir. *Energy Rep.* **2021**, *7*, 6023–6030. [CrossRef]
5. Ding, F.; Dai, C.; Sun, Y.; Zhao, G.; You, Q.; Liu, Y. Gelling Behavior of PAM/Phenolic Crosslinked Gel and Its Profile Control in a Low-Temperature and High-Salinity Reservoir. *Gels* **2022**, *8*, 433. [CrossRef]
6. Lv, Q.; Ge, J.; Guo, H.; Chen, J.; Fan, J.; Mao, Y. Development of Phenolic Gel for Profile Control under High Temperature and High Salt Condition. *J. Xi'an Shiyou Univ. (Nat. Sci. Ed.)* **2023**, *38*, 61–67+112.
7. Yang, Q.; Zhao, M.; Gao, M.; Song, X.; Dai, C. The experimental study of silica nanoparticles strengthened polymer gel system. *J. Dispers. Sci. Technol.* **2019**, *42*, 298–305. [CrossRef]
8. Unomah, M.; Thach, S.; Shong, R.; App, J.; Zhang, T.; Kim, D.H.; Malik, T.; Dwarakanath, V. Performance of conformance gels under harsh conditions. In Proceedings of the SPE Improved Oil Recovery Conference, Oklahoma City, OK, USA, 14–18 April 2018.
9. Juárez, J.L.; Rodriguez, M.R.; Montes, J.; Trujillo, F.D.; Monzòn, J.; Dupuis, G.; Gaillard, N. Conformance gel design for high temperature reservoirs. In Proceedings of the SPE Europec, Online, 1–3 December 2020.
10. Guo, N.; Wu, J.; Li, L.; Zhang, R.; Liao, X.; Chen, L. Study on preparation and stability mechanism of water plugging gel with high performance for Tahe Oilfield. *Appl. Chem.* **2020**, *49*, 3011–3015. [CrossRef]
11. Ge, J.; Guo, H.; Zhang, T.; Zhou, D.; Xu, Y.; Lv, Q. Development of temperature and salinity resistant phenolic gel and its performance regulation mechanism. *Acta Pet. Sin.* **2022**, *43*, 1145–1157.

12. Chen, L.; Zhang, G.; Ge, J.; Jiang, P.; Zhu, X.; Lin, Y.; Han, S. A Novel Thermal-Resistance and Salt-Tolerance Gel with Low-Concentration Crosslinkers for Water Shutoff in Tahe Oilfield. In Proceedings of the SPE Asia Pacific Unconventional Resources Conference and Exhibition, Brisbane, Australia, 9–11 November 2015.
13. Wang, J.; Luo, P. Development of the new associative polymer in profile control. *Oil Drill. Prod. Technol.* **2000**, *3*, 54–56+85. [CrossRef]
14. Liu, Y.; Dai, C.; Wang, K.; Zhao, M.; Zhao, G.; Shuai, Y.; Yan, Z.; You, Q. New insights into the hydroquinone (HQ)–hexamethylenetetramine (HMTA) gel system for water shut-off treatment in high temperature reservoirs. *J. Ind. Eng. Chem.* **2016**, *35*, 20–28. [CrossRef]
15. Wang, W.; Xu, Y.; Ge, J.; Guo, H.; Wu, Q.; Mao, Y. Phenolic resin gel suitable for medium-temperature and high-salinity reservoirs. *J. Mol. Liq.* **2022**, *364*, 119887. [CrossRef]
16. Zhang, J.; He, H.; Wang, Y.; Xu, X.; Zhu, Y.; Li, R. Gelation Performance and Microstructure Study of Chromium Gel and Phenolic Resin Gel in Bulk and Porous Media. *J. Energy Resour. Technol.-Trans. Asme.* **2014**, *136*, 042910. [CrossRef]
17. Shang, X.; Pu, C.; Wu, F.; Liu, J. Laboratory study on the high strength gel system. *Appl. Chem.* **2011**, *40*, 1911–1914. [CrossRef]
18. Cui, C.; Zhou, Z.; He, Z. Enhance oil recovery in low permeability reservoirs: Optimization and evaluation of ultra-high molecular weight HPAM/phenolic weak gel system. *J. Pet. Sci. Eng.* **2020**, *195*, 107908. [CrossRef]
19. Zhao, G.; Dai, C.; Chen, A.; Yan, Z.; Zhao, M. Experimental study and application of gels formed by nonionic polyacrylamide and phenolic resin for in-depth profile control. *J. Pet. Sci. Eng.* **2015**, *135*, 552–560. [CrossRef]
20. Gu, C.; Lv, Y.; Fan, X.; Zhao, C.; Dai, C.; Zhao, G. Study on rheology and microstructure of phenolic resin cross-linked nonionic polyacrylamide (NPAM) gel for profile control and water shutoff treatments. *J. Pet. Sci. Eng.* **2018**, *169*, 546–552. [CrossRef]
21. Yu, H.; Ma, Z.; Tang, L.; Li, Y.; Shao, X.; Tian, Y.; Qian, J.; Fu, J.; Li, D.; Wang, L.; et al. The Effect of Shear Rate on Dynamic Gelation of Phenol Formaldehyde Resin Gel in Porous Media. *Gels* **2022**, *8*, 185. [CrossRef] [PubMed]
22. Guo, H.; Ge, J.; Li, L.; Zhang, G.; Li, Z.; Wang, W.; Liu, M. New insights and experimental investigation of high-temperature gel reinforced by nano-SiO_2. *Gels* **2022**, *8*, 362. [CrossRef] [PubMed]
23. Xu, Y. Study on Phenolic Resin Gel for Plugging Fractures in Honghe Oilfield. Master's Thesis, China University of Petroleum (East China), Qingdao, China, 2022.
24. Xu, Y.; Ge, J.; Guo, H.; Wei, K. The Gelation Law of Partially Hydrolyzed Polyacrylamide-Phenol-Formaldehyde Resin Under Medium-High Temperature and Medium-High Salinity Conditions. *Acta Pet. Sin.* **2022**, *38*, 1336.
25. Qu, J.; Wang, P.; You, Q.; Zhao, G.; Sun, Y.; Liu, Y. Soft Movable Polymer Gel for Controlling Water Coning of Horizontal Well in Offshore Heavy Oil Cold Production. *Gels* **2022**, *8*, 352. [CrossRef] [PubMed]
26. Zhi, J.; Liu, Y.; Chen, J.; Bo, L.; Qu, G.; Jiang, N.; He, W. Preparation and Performance Evaluation of a Temperature and Salt Resistant Hydrophobic Associative Weak Polymer Gel System. *Molecules* **2023**, *28*, 3125. [CrossRef] [PubMed]
27. Liu, X.; Zhao, J.; Dong, R.; Wang, J.; Yang, Z. Modification of Phenolic Prepolymer Crosslinking Agent and Evaluation of Glue-forming Properties. *Shandong Chem. Ind.* **2020**, *49*, 24–26. [CrossRef]
28. Zhang, W. Study on High Temperature and High Salt Tolerance of Gel in Tahe Oilfield. Master's Thesis, China University of Petroleum (East China), Qingdao, China, 2018.
29. Guo, H.; Ge, J.; Xu, Y.; Lv, Q.; Li, Z.; Zhou, D.; Tao, Z. Preparation and Mechanism of Stability for High-Temperature and High-Salinity Gels. *SPE J.* **2022**, *27*, 3565–3578. [CrossRef]
30. Qiu, L.; Pu, W.; Zhang, J. Synthesis and Gelling Performance Evaluation of an Improved Phenolic Resin Crosslinker. *Adv. Fine Petrochem.* **2011**, *12*, 15–18.
31. Sun, L.; Li, M.; Lin, M.; Guo, J.; Peng, B.; Dong, Z. Synthesis and Hydration Properties of Water Soluble Phenol-formaldehyde Resin. *Chin. J. Appl. Chem.* **2008**, *25*, 961–965.
32. Chen, S.; Zhao, K.; Chang, D. Synthesis of phenolic resin prepolymer plugging agent with strength for oil well plugging. *China Synth. Resin Plast.* **2015**, *32*, 30–33.
33. Sydansk, R.D. A New Conformance-Improvement-Treatment Chromium (III) Gel Technology. In Proceedings of the SPE Enhanced Oil Recovery Symposium, Tulsa, OK, USA, 17–20 April 1988.
34. Ran, Y.; Zhang, G.; Jiang, P.; Pei, H. Preparation Method and Performance Evaluation of a Gel Based on AM/AMPS Copolymer. *Gels* **2022**, *8*, 802. [CrossRef] [PubMed]

gels

MDPI

Article

Synthesis and Performance Evaluation of a Novel Nanoparticle Coupling Expanded Granule Plugging Agent

Xuejiao Li [1,*], Qi Li [1], Meilong Fu [2], Li Li [1], Lingyang Su [1] and Yingyang Wang [1]

1 Hubei Key Laboratory for Processing and Application of Catalytic Materials, College of Chemistry and Chemical Engineering, Huanggang Normal University, Huanggang 438000, China; 15949506481@163.com (Q.L.); 13227313655@163.com (L.L.); m2066503389@163.com (L.S.); wyy2309704308@163.com (Y.W.)

2 College of Petroleum Engineering, Yangtze University, Wuhan 430100, China; fml202208@163.com

* Correspondence: lixuejiao202207@163.com

Abstract: This study focuses on the characteristics of fractured and vuggy high-temperature and high-salt reservoirs in the Tahe Oilfield. The Acrylamide/2-acrylamide-2-methylpropanesulfonic copolymer salt was selected as a polymer; the hydroquinone and hexamethylene tetramine was selected as the crosslinking agent with a ratio of 1:1; the nanoparticle SiO_2 was selected, and its dosage was optimized to 0.3%; Additionally, a novel nanoparticle coupling polymer gel was independently synthesized. The surface of the gel was a three-dimensional network structure, with grids arranged in pieces and interlaced with each other, and the structure was very stable. The SiO_2 nanoparticles were attached to the gel skeleton, forming effective coupling and enhancing the strength of the gel skeleton. To solve the problem of complex gel preparation and transportation, the novel gel is compressed, pelletized, and dried into expanded particles through industrial granulation, and the disadvantage of the rapid expansion of expanded particles is optimized through physical film coating treatment. Finally, a novel nanoparticle coupling expanded granule plugging agent was developed. Evaluation of the performance of the novel nanoparticle coupling expanded granule plugging agent. With an increase in temperature and mineralization, the expansion multiplier of granules decreases; aged under high-temperature and high-salt conditions for 30 days, the expansion multiplier of granules can still reach 3.5 times, the toughness index is 1.61, and the long-term stability of the granules can be good; the water plugging rate of granules is 97.84%, which is superior to other widely used particle-based plugging agents.

Keywords: temperature and salt resistant type; expanded granule; nanoparticle SiO_2; coupling; performance evaluation

Citation: Li, X.; Li, Q.; Fu, M.; Li, L.; Su, L.; Wang, Y. Synthesis and Performance Evaluation of a Novel Nanoparticle Coupling Expanded Granule Plugging Agent. *Gels* **2023**, *9*, 479. https://doi.org/10.3390/gels9060479

Academic Editor: Emiliano Carretti

Received: 24 April 2023
Revised: 5 June 2023
Accepted: 6 June 2023
Published: 12 June 2023

1. Introduction

The Tahe Oilfield is located in the Tarim Basin, a rare large ultra-deep marine seam-hole-type carbonate reservoir, with a planar well-controlled oil-bearing area of 2800 km^2. By August 2019, the Tahe Oilfield submitted a total of 13.5×10^8 t of proven petroleum geological reserves, and the cumulative oil production exceeded 1×10^8 t. The Tahe Oilfield reservoirs are divided into fracture, seam-hole, and cavern. The reservoir temperature is as high as 120–150 °C, and the total mineralization of the formation water is (200–250) g/L, which belongs to the super high temperature and high salt reservoir [1,2]. At present, the main challenges encountered in the extraction process of the Tahe Oilfield are the rapid increase in water content of oil wells, rapid production reduction of oil wells, and overall deviation of the water plugging effect; therefore, it is urgent to slow down the increasing rate of water content after water is seen in oil wells [3].

A granular plugging agent is a popular plugging agent commonly used in water plugging and dissection. It is stable, has a wide variety, has good mechanical properties, and is suitable for plugging water in oil reservoirs with large pores. Expanded granules are a type of granular plugging agent developed in the last 20 years and can be expanded when encountering water; they are cross-linked, granulated, and dried on the ground, avoiding the influence of complex conditions of the formation on the cross-linking of the system, which can better plug the reservoir with large pores [4]. Compared with cement-based inorganic granular plugging agents and curable granular plugging agents widely used in the Tahe Oilfield, the expanded granules can absorb water and swell, can plug the throat with their deformation, or penetrate the formation through deformation under certain pressure differences to seal the high-permeability layer containing seams and, large pores [5,6].

In this paper, a novel nanoparticle coupling polymer gel was independently synthesized. The novel gel is compressed, pelletized, and dried into expanded particles through industrial granulation. The disadvantage of too fast expanded particle expansion is optimized through physical film coating treatment. Finally, a novel nanoparticle coupling expanded granule-plugging agent was developed. The novel gel was chemically characterized by FT-IR analysis, SEM, and a structural diagram, and the performance of the plugging agent was evaluated from the aspects of temperature impact on expansion ratio, salinity impact on expansion ratio, and long-term thermal stability of particles. This article provides favorable technical support for applying the bulk expanded granule plugging agent in the Tahe Oilfield.

2. Results and Discussions

2.1. Synthesis of Novel Nanoparticle Coupling Polymer Gel

2.1.1. Polymerization of Polymer Gel

In this paper, Acrylamide/2-acrylamide-2-methyl propane sulfonic copolymer salt (AM/AMPS) was selected as the polymer. The AM/AMPS polymer was polymerized by acrylamide (AM) monomer and AMPS monomer; for AMPS monomer, the highly stable carbon chain was its main chain, and strong anionic hydration groups, i.e., methyl propyl sulfonate groups, were introduced into its molecule, which not only improved the hydrophilicity, but also greatly improved the salinity resistance of AMPS monomer due to the hyposensitivity of propane sulfonic acid groups to external cations; after the polymerization of acrylamide and AMPS monomer, the steric hindrance of polymer molecule significantly increased, effectively inhibiting the hydrolysis of amide groups, empowering strong stability to the polymer, and greatly improving its heat resistance and salt tolerance, which laid a solid foundation for the subsequent preparation of heat-resistant and salt-tolerant gels.

The hydroquinone and hexamethylenetetramine were selected as crosslinking agents with a ratio of 1:1. Hydroquinone and hexamethylenetetramine can form water-soluble (methyl) phenolic resin, which contains a benzene ring structure, making its temperature resistance even better; In addition, hydroquinone contains two hydroxyl groups. The ortho and para positions of the hydroxyl groups are chemical reaction active points, which is easier to form free hydroxymethyl groups so that they can be better crosslinked with AM/AMPS polymers, forming a network-like gel structure with a skeleton structure extending to four aspects.

The synthesis idea of polymer gel polymerization is as follows Figure 1:

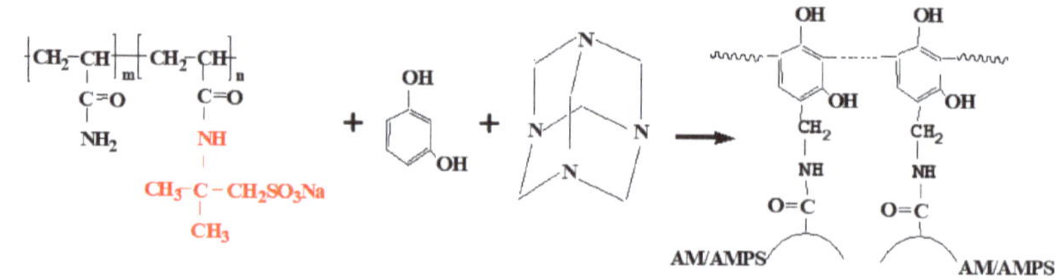

Figure 1. Schematic Diagram of Gel Structure.

2.1.2. Screening and Dosage Optimization of Nanoparticles

The size range of nanoparticles is generally 0.1–100 nm. Nanoparticles coupled with polymer gel can strengthen the viscoelasticity and rheological properties of the gel system, enhance the strength of the gel skeleton, improve the water retention capacity, and form a higher strength and more compact filling network [7].

Nine different types of nanoparticles were screened out and added to the gel system. The viscosity and dehydration rate of the gel prepared with the Tahe simulated formation water at 130 °C in 20 days were investigated. The viscosity of the gel was measured by DV2T Viscometer, and the shear rate is 5.8 rpm. The experimental results are shown in Table 1.

Table 1. Effect of different types of nanoparticles on the dehydration rate of gel.

20 Days	Nanoparticle ZrO_2	Nanoparticle SiO_2	Nanoparticle Al_2O_3	Nanoparticle Fe_2O_3	Nanoparticle ZnO	Nanoparticle MgO	Nanoparticle TiO_2	Nanoparticle $CaCO_3$	Attapulgite Soil
Gel viscosity, Pa·s	36.8	43	37	42.8	36	36.4	19.8	20.6	27.2
Dehydration rate%, %	8.4	2.8	10.2	4.3	9	12.5	36.2	30.8	24

According to Table 1, adding nanoparticles to the gel system can effectively improve the long-term stability of the gel system. The nanoparticle SiO_2 was aged for 20 days under high temperatures and high salt, the viscosity of the gel was up to 43 Pa·s, and the dehydration rate was only 2.8%. Therefore, nano SiO_2 was screened as coupling nanoparticles.

The dosage of nanoparticles was optimized by setting the dosage of nanoparticle SiO_2 as 0.05%, 0.08%, 0.1%, 0.3%, 0.5%, and 1.0%, and other conditions remain unchanged. Using the Tahe simulated formation water to prepare the gel, and place the system at 130 °C for cross-linking reaction. The investigation time is 10 days. The experimental results are shown in Figure 2.

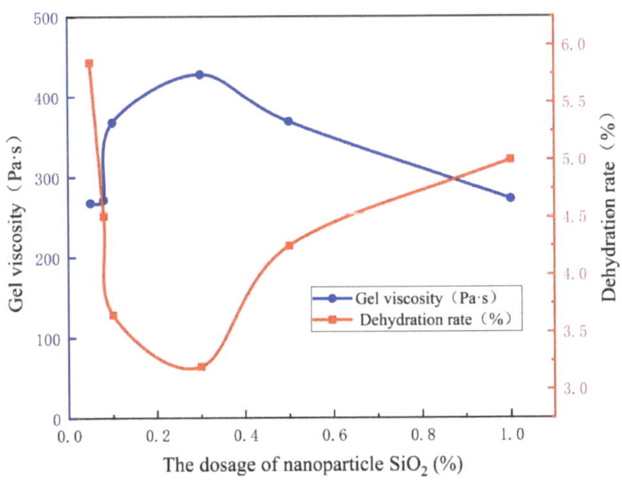

Figure 2. Optimization experiment of nanoparticle SiO_2 dosage.

As shown in Figure 2, within a certain concentration range, the gelling strength of the system gradually increases with the increase of the dosage of nanoparticle SiO_2. When the dosage of nanoparticle SiO_2 reaches 0.3%, the gel viscosity is up to 42.8 mPa·s, and the dehydration rate for 10 days is only 3.2%. If the dosage of nanoparticle SiO_2 continues to increase, the gelling strength of the system shows a downward trend, So 0.3% is the optimal dosage of nanoparticles.

Nanoparticles have a surface effect, volume effect, quantum size effect, and macro quantum tunnel effect [8] and do not participate in chemical crosslinking reactions in the whole gel system. Nanoparticles are mainly combined with gel molecular chains in the way of hydrogen bonds, forming effective coupling, increasing the bonding point of the gel system, improving the tensile and compressive strength of the system, and greatly improving the stability of the gel system structure, The three-dimensional network skeleton of the gel system is effectively strengthened. In addition, nanoparticles coupled with polymer gel can effectively change the wettability of rock surfaces, reduce the interfacial tension between oil and water, and improve oil recovery efficiency. As shown in Figure 3.

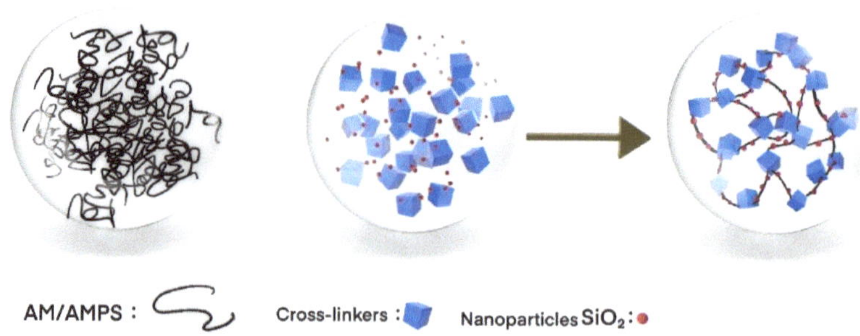

Figure 3. Structural Diagram of Nanoparticle Coupling Polymer Gel.

2.1.3. Chemical Structure Characterization of Nanoparticle Coupling Polymer Gel

The nanoparticle coupling polymer gel belongs to soft-solid-like systems with a permanent system of bonds, which has high strength. According to FT-IR analysis, as shown in Figure 4. There is an obvious vibration absorption peak at 1203 cm^{-1}, which belongs to the C-H vibration absorption peak in the benzene ring, which indicates that the aromatic ring crosslinker has been successfully introduced into the gel structure [9]. At the same time, under normal circumstances, the vibrational absorption peak of C=O in the amide group is at 1600 cm^{-1}. Still, the vibrational absorption peak of C=O in the amide group in this graph moves to 1629 cm^{-1}, with a significant shift, indicating that the amide group in AM/AMPS undergoes a crosslinking reaction with the crosslinking agent [10].

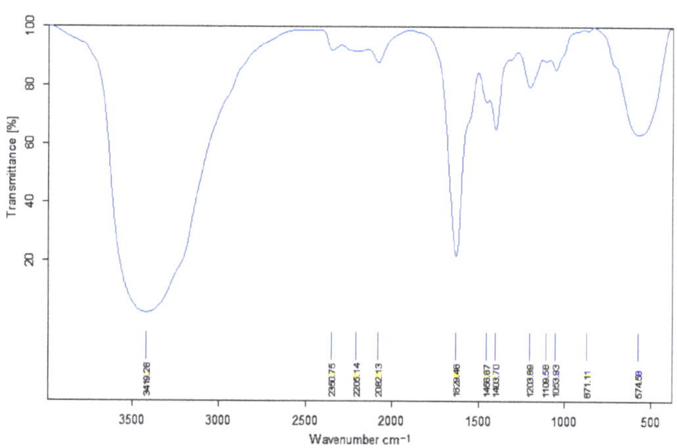

Figure 4. FT-IR analysis of nanoparticle coupled polymer gel (the concentration of SiO_2 is 0.3%).

The microstructure of the nanoparticle coupling polymer gel was analyzed by SEM, as shown in Figure 5.

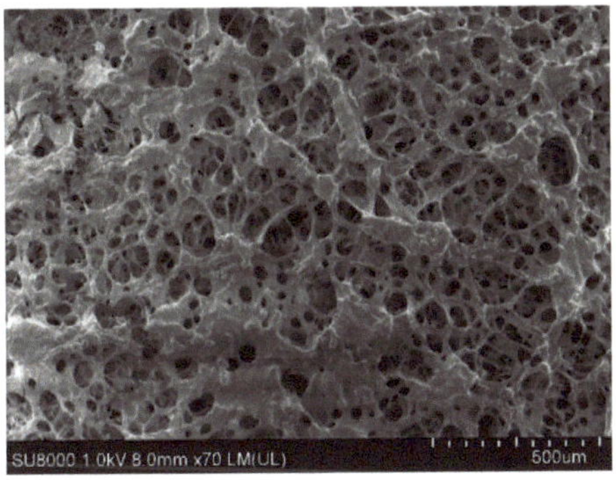

Figure 5. SEM of nanoparticle coupling polymer gel (the concentration of SiO_2 is 0.3%).

As shown in Figure 5, the surface of the novel nanoparticle coupling polymer gel is a three-dimensional network structure, with grids arranged in pieces and interlaced with each other, and the structure is very stable.

2.2. Preparation of Laminated Expanded Granules

The new nanoparticle coupled polymer gel has a stable three-dimensional network structure, good temperature, and salt resistance, and it is suitable for water plugging and profile control of fracture cavity reservoirs in high temperature and high salinity reservoirs in Tahe Oilfield. However, the preparation of the polymer gel is complex, and the transportation is tedious. To facilitate the on-site construction, the polymer gel is processed, compressed, pelletized, and dried into nanoparticle coupling expansion granules, as shown in Figure 6.

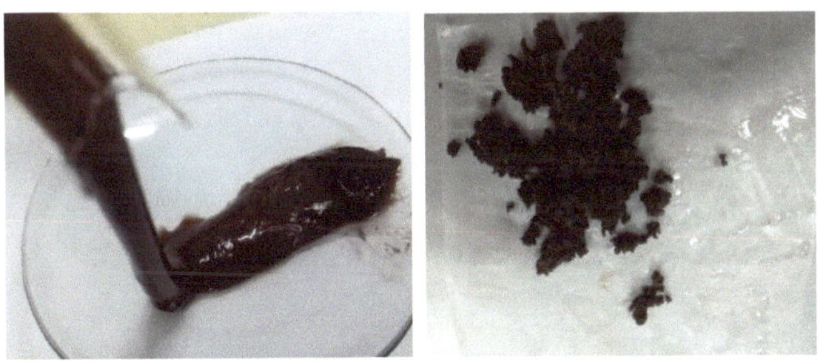

Figure 6. Morphology of nanoparticle coupling polymer gel (**left**) and nanoparticle coupling expansion granules (**right**).

The novel expansion granules have been treated professionally in industrialization, and the gel framework structure has been preserved. In a reservoir with rich water content, the granules can absorb water and expand. This is due to the main chain of the granule molecule containing more hydrophilic groups. When the granule first comes into contact with water, the hydrophilic functional group can hydrate with water and draw water molecules into the interior of the granule. The water molecules can form hydrogen bonds with the main chain of the granule molecule, making water enter the granule molecule's three-dimensional network structure smoothly. The granule network-like structure is very stable and has a good water retention capacity, which makes the osmotic pressure difference formed inside and outside the granule molecule. Under this pressure difference, the granules can be induced to absorb water continuously until the dissolution equilibrium is reached inside and outside the molecule [11]. According to the whole aspect, the granule water absorption swelling results from both physical and chemical adsorption. The absorbed water exists in the network structure mainly in two states: free water molecules and hydrated molecules formed by hydrogen bonds.

The expansion rate of the nanoparticle coupling expansion granule is too fast. The expansion multiplier can reach 4.7 times within 0.5 h in the Tahe simulated stratigraphic water. The maximum expansion multiplier is reached within 8 h, which badly affects the injection of field granules. The novel expansion granules were optimized for achieving a slow expansion effect.

This section exhibits the idea of physical coating of the granules by using reagents that do not react chemically with the granules [12] to form a coating film on its surface, thus retarding its expansion, as shown in Figure 7. Polyvinylidene chloride viscosity (average molecular weight of 3 million), unsaturated sulfide (antioxidant), and organophosphate (stabilizer) were used to formulate the laminating solution. Polyvinylidene chloride has strong intermolecular cohesiveness, high crystallinity, excellent barrier properties, and does not react chemically with the granules. It can well physically coat the granules and form a capsule-like coating on their surface, thus, the swelling gets retarding.

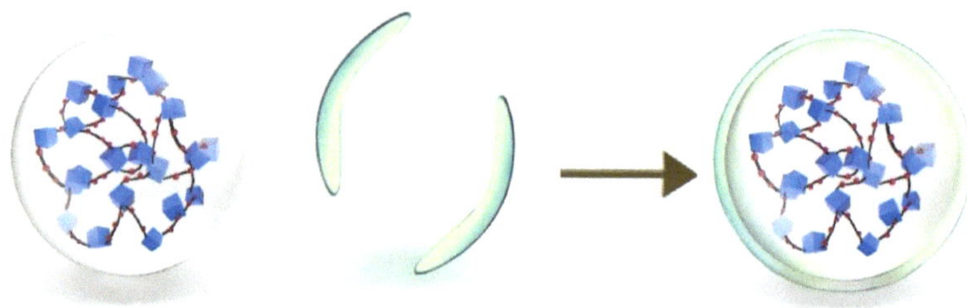

Figure 7. Structural diagram of overcoated expanded granules.

The 1% and 2% coating solutions were prepared in the laboratory, and the nanoparticle coupling expansion granules were coated with a film. Finally, two types of coating granules were prepared, which were the expanded granules with a film and the expanded granules with a thick film.

According to Figure 8, the initial expansion of the granule is rapid, reaching the maximum expansion multiplier within 8 h. In contrast, the thin and thick film of the granule begins to slowly decompose at high temperatures, and the granule slowly expands, reaching the maximum expansion multiplier of 5.31 within about 24 h, with good slow expansion performance. After the granules reached the maximum expansion multiplier, the granule began to shrink slowly and finally stabilized gradually, the expansion multiplier of 72 h stabilized at about 4 times.

Considering the effect of retarding expansion, economic costs, and other factors, using thin overcoated expanded granules is recommended.

2.3. *Evaluation of the Nanoparticle Coupling Expanded Granule*

2.3.1. Effect of Temperature on the Expansion Multiplier

There is a test to the effect of temperature on the expansion multiplier of overcoated granules, Tahe Oilfield is a high-temperature reservoir, so three higher temperature gradients, 120 °C, 130 °C, and 140 °C, are set. By comparing the expansion performance of overcoated type expansion granules in 1–10 days, the experimental results are shown in Figure 9.

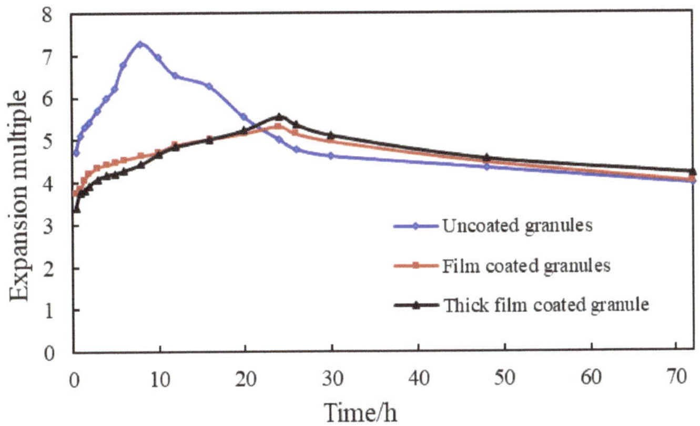

Figure 8. Evaluation Experiment of Expansion Performance of Overcoated Expanded Granules.

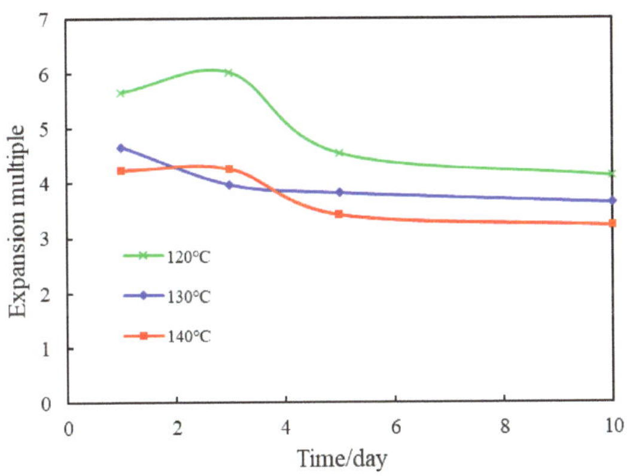

Figure 9. Effect of Temperature on the Expansion Multiplier of the Laminated Expanded Granules.

According to Figure 10: the granule swelling multiplier of 1 day is 5.66 under 120 °C condition, and the swelling multiplier of 10 days can reach 4.13. Under the condition of 140 °C, the granule swelling multiplier declines to 4.23 after 1 day, and the swelling multiplier after 10 days is 3.21. The granule swelling multiplier is between the above temperatures under 130 °C. It shows that the temperature has a great influence on the granule swelling multiplier, and the granule swelling multiplier tends to decrease as the temperature increases. The results indicate that the high temperature reduces the performance of water-absorbing functional groups on the surface of the granules, the water molecules entering the molecular network structure of the granules are reduced, which has a certain inhibitory effect on the water absorption capacity of the granules [13,14].

2.3.2. Effect of Mineralization Degree on Expansion Multiplier

The mineralization of the simulated formation water in Tahe River was 220 g/L. The simulated formation water was diluted with distilled water by 20% and 60%, and the water samples with mineralization of 44 g/L and 132 g/L were prepared, respectively. 3% NaCl was added to the simulated formation water and water samples with mineralization of 250 g/L were prepared to investigate the changes in granule swelling times under different mineralization. The evaluation time was 10 days and the experimental results are shown in Figure 10.

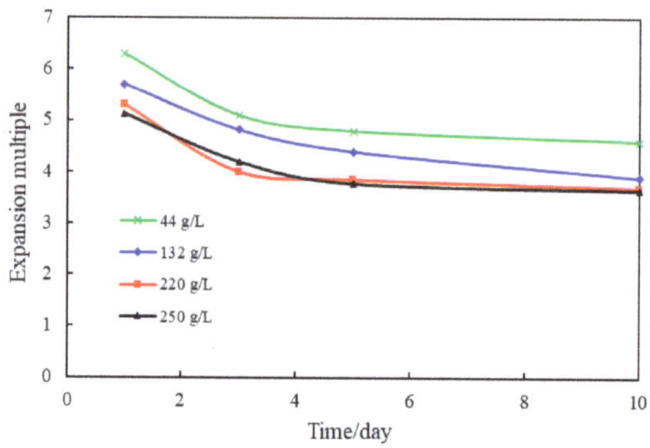

Figure 10. Effect of mineralization on the expansion multiplier of the laminated expanded granules.

From the figure, we could know that when the mineralization of the water sample rises from 44 g/L to 220 g/L, the swelling multiplier of the granules shows an obvious decreasing trend. There are several main reasons for this. First, according to the theory of absorption thermo dynamics, the size of the ionic strength of the external solution can affect the absorption capacity of the body's expansion particles. With the increase of mineralization, the ionic strength in the solution increases. The osmotic pressure inside and outside the granules' molecules also decreases, inhibiting the expanded granules' water absorption capacity, and the swelling multiplier decreases accordingly. In addition, the Ca^{2+} ion content in the Tahe model formation water is high, and Ca^{2+} ions affect the functional groups' activity on the granule macromolecules' surface, making the swelling multiplier decrease. When the mineralization of the water sample increased from 220 g/L to 250 g/L, the change in the swelling multiplier was very small, indicating that the change in the granule swelling multiplier decreased when the mineralization exceeded a certain range [15].

2.3.3. Long-Term Thermal Stability of the Granules

The 7 groups of granule suspensions of the same granule size and 10% mass concentration were prepared with simulated formation water (mineralization of 220 g/L) from the Tahe River and placed at 130 °C for 30 days. They were removed at 2 h, 1 day, 2 days, 5 days, 10 days, 20 days, and 30 days. The granule swelling multiples and the granule strength were measured to evaluate the long-term stability performance of the granules under high temperatures and high salt.

Here, J. E. Smith's toughness index method was used to measure the strength of the granules, quantifying the tensile and compressive strength of the granules. The toughness index can be defined as follows: selecting a sieve with a suitable mesh (the mesh should be smaller than the granule size after expansion), and measuring the pressure difference between the granules passing through the sieve twice (P1, P2). The multiplier of the two pressure differences is the toughness index, that is toughness index = P1/P2, the closer the toughness index is to 1.0, the better the strength of the granules. The pressure difference between the granules passing through the screen twice before and after different aging times was recorded and the toughness index value was calculated, as shown in Table 2.

To summarize and compare the experimental results of the expansion multiplier and toughness index measured by aging the granules under high temperature and high salt for 2 h–30 days, as shown in Figure 11.

According to Figure 11, under the conditions of high temperature and high salt, the granules swell slowly from 0 to 1 day, and the swelling multiplier increases linearly, while the toughness index decreases, reaching the maximum swelling multiplier of 5.31 and the minimum toughness index 1.14 within 1 day. This indicates that the granule strength increases with the increase of the swelling multiplier during the water absorption stage, the granule swelling multiplier and strength both reach the maximum in 1 day. This is mainly due to the high temperature and high salt, which have a certain inhibitory effect on the granules' water absorption capacity and surface functional group activity [16]. After 5 days, the curves of the expansion multiplier and toughness index all tend to be stable, and after 30 days of aging, the expansion multiplier of the granules can reach 3.5 times. The toughness index goes to 1.61, which can meet the water plugging requirements of the Tahe Oilfield, indicating that the granules have good long-term stability under high temperature and high salt conditions.

Table 2. Toughness Index of Granules at Different Aging Times.

Times	2 h	1 Day	2 Days	5 Days	10 Days	20 Days	30 Days
P_1/MPa	0.06	0.058	0.069	0.072	0.07	0.074	0.074
P_2/MPa	0.045	0.051	0.056	0.051	0.046	0.047	0.045
Toughness index	1.33	1.14	1.23	1.41	1.52	1.57	1.61

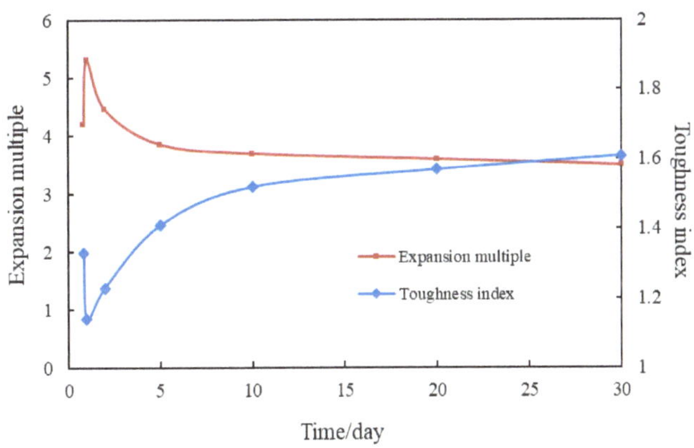

Figure 11. Experiment to evaluate the long-term stability performance of the granules.

2.3.4. Granule Sealing Reservoir Performance

The core flow experiments were conducted to evaluate the water-blocking ability of the granules in the reservoir. In the experiment, the saturated oil was dehydrated, and degassed crude oil from Tahe Oilfield (with a viscosity of 55.1 mPa·s), and the injected water was simulated with Tahe formation water at an injection rate of 0.5 mL/min, as shown in Figure 12.

The particles slowly absorb water and expand in the cracks and pores, and the expanded particles are prone to accumulate and bridge at the junction of the cracks and pores, forming a sealing effect. Based on the pressure difference data in Figure 13, the water phase permeability K_w before particle injection and the water phase permeability Kw after injection of the sealing agent is calculated; the particle water blocking rate is $\Phi_W = (K_W - K_{W'})/K_W \times 100\% = 97.84\%$. It can be seen that particles can effectively seal cracks and holes, with a sealing rate superior to other widely used particle-based plugging agents.

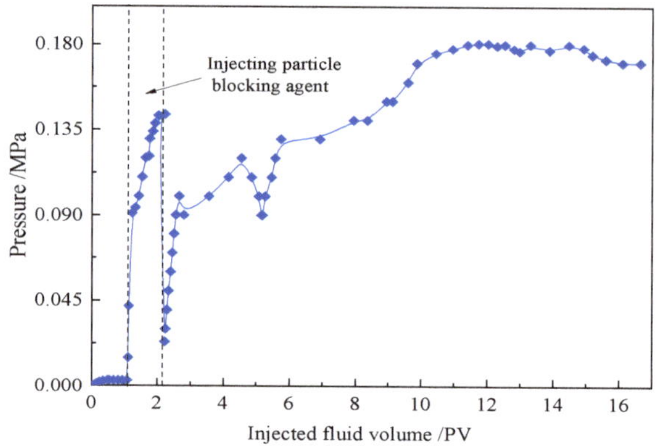

Figure 12. Pressure variation curve of core flow experiment.

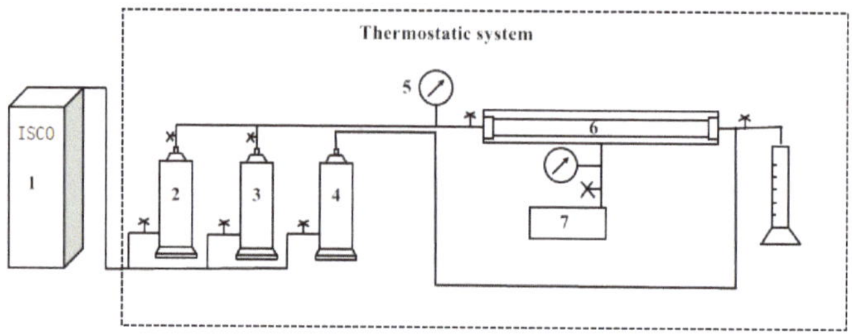

Figure 13. Core flow experimental device. 1- High pressure plunger pump; 2- Simulated oil; 3- Formation water; 4- Plugging agent solution; 5- Pres sure gauge; 6- Core gripper; 7- Hand pump.

3. Conclusions

1. This article focuses on high-temperature and high-salt reservoirs in the Tahe Oilfield. Selecting Acrylamide/2-acrylamido-2-methylpropane sulfonic acid copolymer as a polymer; selecting hydroquinone and hexamethylene tetramine as the crosslinking agent with a ratio of 1:1. To enhance the stability of the system, nanoparticles are added to the system. The nanoparticle SiO_2 was selected from nine different types of nanoparticles, and its dosage was optimized to 0.3%; A novel nanoparticle coupling polymer gel was synthesized.
2. To solve the problem of complex gel preparation and complicated transportation, the novel gel is compressed, pelletized, and dried into expanded particles through industrial granulation, and the disadvantage of too fast expansion of expanded particles is optimized through physical film coating treatment. Finally, a novel nanoparticle coupling expanded granule plugging agent is developed. The granules have the characteristics of slow expansion.
3. Evaluation of the performance of the novel nanoparticle coupling expanded granule plugging agent. After 30 days of high temperature and high salt aging conditions, the granule expansion performance is good, and the granules have high strength and good long-term stability performance. With the increase of temperature or salinity, the expansion ratio of particles shows a decreasing trend, indicating that excessive temperature and salinity inhibit the water absorption ability of particles. The water plugging rate of particle plugging agents is as high as 97.84%, which is superior to other widely used particle-based plugging agents.

4. Materials and Methods

4.1. Experimental Materials

Experimental water: Simulated formation water prepared according to the ion composition of the produced water from the Tahe Oilfield, as shown in Table 3.

Table 3. Simulated formation water ion composition.

Ion Content/(mg·L^{-1})					Total Salinity /(mg·L^{-1})	pH Value
Cl^-	HCO_3^-	Ca^{2+}	Mg^{2+}	$Na^+ + K^+$		
138,000	200	11,000	1500	73,000	223,700	6.8

Experimental chemicals: hydroquinone, thiourea, hexamethylene tetramine, nano zirconium dioxide, nano silicon dioxide, nano aluminum oxide, nano iron oxide, nano zinc oxide, nano magnesium oxide, nano titanium dioxide, nano calcium carbonate, attapulgite soils, shown in Table 4. All drugs were analytically pure and provided by Shanghai Macklin Biochemical Technology Co., Ltd., Shanghai, China. The polymers used in this paper is Acrylamide/2-acrylamido-2-methylpropanesulfonic copolymer salt, belongs to block copolymers, with a solid content of >88%, AMPS content of 58%, and viscosity average molecular weight of 5 million; polymers were purchased from Shandong Baomo Biochemical Co., Ltd. (Dongying, China).

Table 4. Nanoparticle properties.

	Nanoparticle ZrO_2	Nanoparticle SiO_2	Nanoparticle Al_2O_3	Nanoparticle Fe_2O_3	Nanoparticle ZnO	Nanoparticle MgO	Nanoparticle TiO_2	Nanoparticle $CaCO_3$	Attapulgite Soil
Particle size, nm	25 ± 5	20 ± 5	30 ± 5	30	30 ± 5	30	25	30	30 ± 5
Surface Area (m^2/g Gsurface)	150–500	45–150	500–700	400	500–700	450	150	450	500–800

4.2. Experimental Apparatus

FT-IR Hoffen-10, Shanghai ZHUJIN Analytical Instrument Co., Ltd., Shanghai, China; EM-30AX scanning electron microscope (SEM), COXEM, Korea; DV2T Viscometer, Brookfield, WI, USA (belongs to the plate/cone type); precision aging tank (6-bore screws, 500 mL volume), Jiangsu Lianyou Scientific Research Instrument Co., Ltd., Nantong, China; 85–2 thermostatic magnetic stirrer, Changzhou Longhe Instrument Manufacturing Co., Ltd., Changzhou, China; Core Flow System Device, Jiangsu Lianyou Scientific Research Instrument Co., Ltd., Nantong, China.

4.3. Experimental Methods

4.3.1. Determination of Gel Dehydration Rate

Put the prepared gelling solution into an ampoule bottle, which is recorded as m_1; After gelling, take it out of the incubator, open the ampoule, and weigh the mass of gel dehydrated water, which is recorded as m_1. The ratio of this mass to the mass of the initial gelling solution is the dehydration rate. Dehydration rate = $(m_1/m) \times 100\%$.

4.3.2. Determination of Expansion Multiplier

Characterize the water absorption and swelling performance of the granules by the mass multiplier of water absorption of the granules: weigh the mass of the dried granules m_0, add them into the simulated stratigraphic water of Tahe, stir them into a dispersion system, then filter the granule solution through a filter, absorb the free water on the surface of the granules with filter paper, weigh the mass of the granules m_1 after water absorption and swelling, and the swelling multiplier is $S_w = m_1/m_0$.

4.3.3. Determination of Granule Strength

The *method* of foreign scholar J. E. Smith [17] was used to determine the granule strength. The experimental steps are as follows: (1) Screen out the granules with a granule size of 0.4–0.6 mm, and age the granules under the set high temperature and high salt conditions; (2) Prepare the aged granules into a granule suspension with a mass concentration of 10%; (3) Install a 10-mesh screen at the outlet of the sand-filling tube and drive the granule suspension at a high flow rate of 25 mL/min until it completely flows out of the sand-filling tube, and record the first differential pressure P_1; (4) Collect the outgoing suspension and flow through the screen again under the same conditions, and record the second differential pressure P_2. Toughness Index = P_1/P_2.

4.3.4. Measurement of Particle Blocking Rate in Core Flow Experiment

(1) Saturate the core with formation water, raise the temperature to 130 °C, establish bound water saturation through oil flooding, and let it stand for 24 h; (2) Simulate the displacement of formation water to 98% water content in the core, record the pressure after liquid production, and calculate the K_W before plugging agent injection; (3) Reverse injection of a 10% particle solution 1 PV and aging at 130 °C for a while; (4) Forward water drive again, record the breakthrough pressure and calculate the K_w' after core plugging after the pressure stabilizes.

Author Contributions: Writing—Review and Editing and Methodology, X.L.; Validation, L.L.; Formal Analysis, Q.L.; Resources, M.F. and L.S.; Data Curation, Y.W. All authors have read and agreed to the published version of the manuscript.

Funding: Funding organization: the 2022 Scientific Research Program Funded by the Hubei Provincial Education Department; Funding project number: Q20222902.

Institutional Review Board Statement: Not applicable.

Informed Consent Statement: Not applicable.

Data Availability Statement: No new data were created.

Acknowledgments: Financial support by the 2022 Scientific Research Program Funded by the Hubei Provincial Education Department (Project No Q20222902), is gratefully acknowledged.

Conflicts of Interest: The authors declare no conflict of interest.

References

1. Yu, B.; Ruan, Z.; Zhang, C.; Pan, Y.; Lin, C.; Wang, L. Tectonic evolution of Tarim basin in Cambrian–Ordovician and its implication for reservoir development, NW China. *J. Earth Syst. Sci.* **2016**, *125*, 285–300.
2. Li, Y. *Quantitative Characterization of Internal Heterogeneity of Carbonate Fracture Cave Unit in Tahe Oilfield*; China University of Petroleum (Beijing): Beijing, China, 2017; pp. 8–10.
3. Davletbaev, A.; Baikov, V.; Doe, T.; Emchenko, O.; Zainulin, A.; Igoshin, A.; Fedorov, A. Fracture-Based Strategies for Carbonate Reservoir Development. In Proceedings of the SPE Russian Oil and Gas Conference and Exhibition. Society of Petroleum Engineers, Moscow, Russia, 26–28 October 2010.
4. Alkhorshid, N.R.; Araujo, G.L.; Palmeira, E.M. Consolidation of soft clay foundation improved by geosynthetic-reinforced granular columns: Numerical evaluation. *J. Rock Mech. Geotech. Eng.* **2021**, *13*, 1173–1181. [CrossRef]
5. Kumar, A.; Mahto, V.; Sharma, V.P. Development of fly ash reinforced nanocomposite preformed particle gel for the control of excessive water production in the mature oil fields. *Oil Gas Sci. Technol.-Rev. D Ifp Energ. Nouv.* **2019**, *74*, 1–10. [CrossRef]
6. Ahmadi, M.A.; Ahmad, Z.; Phung, L.T.K. Experimental investigation the effect of nanoparticles on micellization behavior of a surfactant: Application to EOR. *Pet. Sci. Technol.* **2016**, *34*, 1055–1061. [CrossRef]
7. Gorbacheva, S.N.; Yadykova, A.Y.; Ilyin, S.O. Rheological and tribological properties of low-temperature greases based on cellulose acetate butyrate gel. *Carbohydr. Polym.* **2021**, *272*, 118509. [CrossRef] [PubMed]
8. Albanese, A.; Tang, P.; Chan, W. The effect of nanoparticle size, shape, and surface chemistry on biological systems. *Annu. Rev. Biomed. Eng.* **2012**, *14*, 1–16. [CrossRef] [PubMed]
9. Chen, L.; Wang, J.; Yu, L.; Zhang, Q.; Fu, M.; Zhao, Z.; Zuo, J. Experimental Investigation on the Nanosilica-Reinforcing Polyacrylamide/Polyethylenimine Hydrogel for Water Shutoff Treatment. *Energy Fuels* **2018**, *32*, 6650–6656. [CrossRef]

10. Nazari Moghaddam, R.; Bahramian, A.; Fakhroueian, Z.; Karimi, A.; Arya, S. Comparative study of using nanoparticles for enhanced oil recovery: Wettability alteration of carbonate rocks. *Energy Fuels* **2015**, *29*, 2111–2119. [CrossRef]
11. Lan, W.U. Absorbing Mechanism and Preparation Methods of Superabsorbent Polymers. *Chem. Adhes.* **2006**, *3*, 11–28.
12. Wei, F. Study on microencapsulation expansion retarding method of super absorbent resin particle plugging agent. *Oil Gas Geol. Recovery Effic.* **2005**, *12*, 74–80.
13. Goudarzi, A.; Zhang, H.; Varavei, A.; Taksaudom, P.; Hu, Y.; Delshad, M.; Bai, B.; Sepehrnoori, K. A laboratory and simulation study of preformed particle gels for water conformance control. *Fuel* **2015**, *140*, 502–513. [CrossRef]
14. Jayakumar, S.; Lane, R.H. Delayed Crosslink Polymer Gel System for Water Shutoff in Conventional and Unconventional Oil and Gas Reservoirs. In Proceedings of the SPE International Symposium on Oilfield Chemistry, The Woodlands, TX, USA, 8–10 April 2013.
15. Koohi, A.D.; Seftie, M.V.; Ghalam, A. Rheological characteristics of sulphonated polyacrylamide/chromium triacetate hydrogels designed for water shut-off. *Iran. Polym. J.* **2010**, *19*, 757–770.
16. Almohsin, A.; Bai, B.; Imqam, A.; Wei, M.; Kang, W.; Delshad, M.; Sepehrnoori, K. Transport of nanogel through porous media and its resistance to water flow. In Proceedings of the SPE Improved Oil Recovery Symposium, Tulsa, OK, USA, 12–16 April 2014.
17. Smith, J.E. The Transition Pressure: A Quick Method for Quantifying Poly-acrylamide Gel Strength. In Proceedings of the SPE International Symposium on Oilfield Chemistry, Houston, TX, USA, 8–10 February 1989.

gels

MDPI

Article

Injection of Gelling Systems to a Layered Reservoir for Conformance Improvement

Konstantin Fedorov [1], Alexander Shevelev [1,*], Alexander Gilmanov [1], Andrey Arzhylovskiy [2], Denis Anuriev [2], Ivan Vydysh [2] and Nikita Morozovskiy [3]

1 Department of Modeling of Physical Processes and Systems, University of Tyumen, 6 Volodarskogo Str., 625003 Tyumen, Russia
2 Tyumen Petroleum Research Center, 79/1 Osipenko Str., 625003 Tyumen, Russia
3 Rosneft Oil Company, 26/1 Sofiyskaya Naberezhnaya Str., 625000 Moscow, Russia
* Correspondence: alexandershevelev@mail.ru; Tel.: +7-912-991-9014

Abstract: The paper describes the introduction and estimation of performance criteria for the gelling agent injection technology based on a general approach to modeling physical and chemical enhanced oil recovery (EOR) methods. The current mathematical models do not include performance criteria for the process of gelling agent injection and do not allow for assessing the level of success of a treatment job in production wells. The paper introduces such criteria for the first time. To simulate the effect on injection wells, the mass conservation laws and the generalized flow law are used, and closing relations for the gelling rate are taken into account. A conformance control coefficient is introduced which characterizes the positive effect of well treatments and injectivity drop which characterizes the negative effect. The performance criteria allow for identifying the wells where the treatment jobs were the most successful. The model verification, based on the comparison of post-treatment injectivity estimated in the developed model, with Rosneft's field data showed a satisfactory match. The developed correlations can be used as the basis for a surrogate model that allows for avoiding building sector geological and flow simulation models of the treated zone.

Keywords: gelling agents; incremental oil production; mass conservation law; flow characteristics; conformance control coefficient; injectivity drop

Citation: Fedorov, K.; Shevelev, A.; Gilmanov, A.; Arzhylovskiy, A.; Anuriev, D.; Vydysh, I.; Morozovskiy, N. Injection of Gelling Systems to a Layered Reservoir for Conformance Improvement. *Gels* **2022**, *8*, 621. https://doi.org/10.3390/gels8100621

Academic Editors: Qing You, Guang Zhao and Xindi Sun

Received: 10 September 2022
Accepted: 27 September 2022
Published: 29 September 2022

1. Introduction

Flooding of oil reservoirs is a conventional technology of secondary oil recovery. It addresses two main issues: maintenance of reservoir pressure (and, therefore, the flow rates of production wells), and the displacement of oil from the rock pores [1]. However, due to vertical and lateral heterogeneities of reservoirs, the behavior of the water–oil displacement front is rather complex. Such complex behavior leads to an early breakthrough of the injected water into production wells. Oil production with high water content shows low performance and incomplete sweep efficiency.

The term "water control" includes defining the field zones with minimal fluid cross-flows, measures to control water injection into various injection wells, and controlling the voidage replacement by injecting water in certain zones [2]. Recently, a new term has been widely used, i.e., "flooding control" or "conformance control". This term implies comprehensive measures to redistribute flows in a reservoir vertically and laterally and reduce the water cut of recovered fluid resulting from such well interventions [3].

Water-flood management covers two focus areas. The first is stimulating the bottom-hole zone of injection wells and ensuring conformance control between layers with various permeabilities [4]. The response of the nearest producers is determined by a change in lateral piezo-conductivity when forming low-permeable barriers in high-permeable interlayers.

However, the presence of vertical cross-flows between interlayers leads to flow forming around these barriers and the return of the flows to their original state. The difference in the rate of vertical conformance and lateral transmission of pressure disturbances leads to a short-term (several months) positive effect on producers (reduction of water-cut, increase in oil rate) [5]. Examples of conformance control technologies or stimulating the zone closest to wells are the injection of cross-linked polymer agents [6–8] and sodium-silicate-based gelling agents [9–13], sediment-forming technologies [14], the use of thermo-gels [15,16] and injection of suspensions [17], etc. The examples are shown in Figure 1 below.

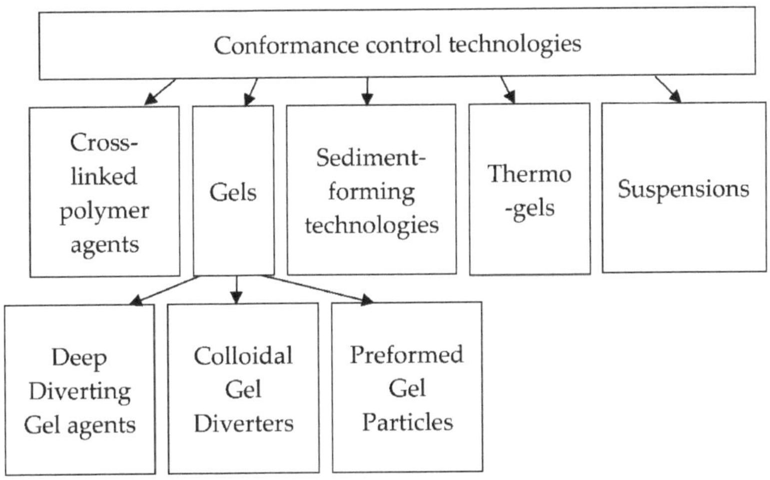

Figure 1. Various conformance control technologies.

The second focus area is related to reservoirs with local highly permeable channels between injection and production wells. Examples of such irregularities are "super-reservoirs" with permeabilities that differ from the average field values by tens or hundreds of times, regional or induced fracturing of formations. Examples include the local occurrence of a super-reservoir at the Talinskoye Field [18] or the development of induced or so-called auto-fractures at the Priobskoye Field, where the water breaks through into fractured producers due to the convergence of two types of fractures [19]. Manipulating these causes of reduced reservoir sweep is called flow-diverting technologies, although the agent bases of conformance control and injected water diverting technologies are very close, as a rule, only the content and injection volumes differ.

The flow-diverting technologies allow the injected agents to be pushed deep enough into the formation or effectively block the fractures without a noticeable impact on the formation matrix. For this purpose, the same suspension agents are used, Deep Diverting Gel agents [6], such as Colloidal Gel Diverters [20], and Preformed Gel Particles [21,22].

The considered approach includes a model of "deep-bed suspension migration" [23] and a general compositional model with chemical reactions [24]. In the first model, the "active" component is suspended particles; in the compositional simulator, the principal components are dissolved substances reacting in porous media. Several analytical solutions were obtained for 1D problems in the framework of deep-bed suspension migration [25].

The 1D objective of oil displacement from a homogeneous reservoir by polymer solution is a "classical" problem [26] and has an analytical solution, which was used to solve an inverse problem for adsorption parameters determination in this paper. For relatively low agent injection volumes in a layered reservoir, vertical fluid cross-flows are negligible in comparison with lateral flows. Therefore, the 1D approximation approach can be applied. This approach allows for considering the conformance improvement problem as a set of 1D objectives in each interlayer without cross-flows [27].

With relatively high treatment performance (more than 1500 m^3 of incremental oil per well treatment), the success rate for the analyzed sample does not exceed 70%. The reason for this is likely due to the insufficient attention to the geological and physical features of the wells selected for treatment and the non-optimal process parameters of treatment jobs. The solution to this problem is the preliminary designing of treatment jobs based on mathematical models of the process.

Two reasons hinder the use of special reservoir models for treatment design purposes. The first is the high time and financial costs, the second is the fact that the objectives of agent injection to form low-permeable barriers and the response of production wells are of different scales. To solve this problem, the authors suggest applying a combined approach or a "surrogate" model.

This paper discusses only the conformance control technologies. Such an approach allows us to estimate and optimize the technologies to maximize the flow redistribution effect. Another application of such models includes evaluating and comparing the treatment performance under certain geological and physical conditions.

2. Results and Discussion

2.1. The Influence of Slug Volume

The influence of slug volume on the introduced coefficients was calculated for a particular well described in Section 4.1 and the treatment parameters, which are mentioned in the Materials and Methods section. The results are presented in Figure 2. For comparison, the same calculations were made for a vertical well without a fracture and are given in Figure 2. For the generality of the results, a dimensionless volume of the gel was introduced: $M = \beta\, V/V_r$, where β is the rock damage coefficient, V is the volume of the injected slug and V_r is the volume of the treated reservoir zone.

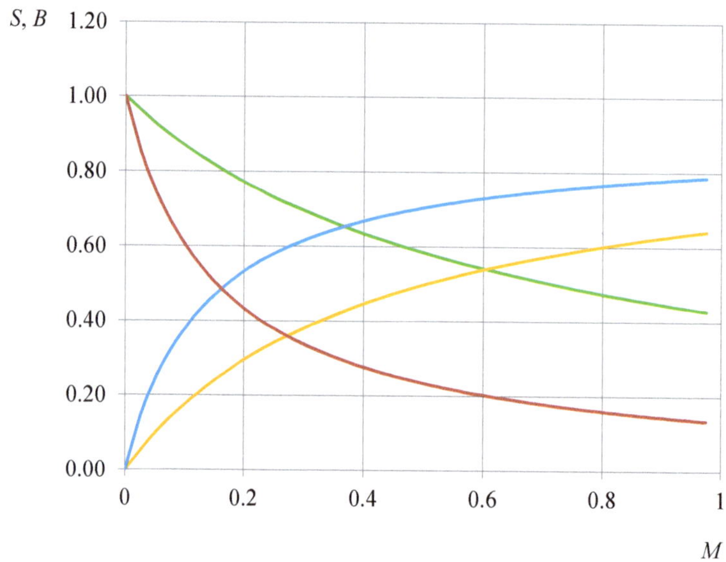

Figure 2. The correlation of CC performance criteria and the dimensionless mass of the injected gelling agent (green line—injectivity drop S for linear inflow, red line—injectivity drop S for radial inflow, yellow line—CC coefficient B for linear inflow, blue line—CC coefficient B for radial inflow).

The analysis of the resulting correlations shows that with the growth of the injected agent mass, the conformance control (CC) coefficient increases, the flows are redistributed more efficiently, and the injectivity profile is improving; however, this leads to a decrease in the overall well injectivity.

2.2. Analysis of Field Experience in Gelling Agents Injection

We analyzed the available data on the use of gelling agents based on HPAA and sodium silicate in recent years at Rosneft's license blocks in West Siberia. The field data analysis allowed us to identify a few features of applying gelling agents. First of all, we concluded that basically such treatments are used in vertical and directional wells both with and without hydraulic fracturing, as well as in horizontal wells without hydraulic fracturing. Next, we will talk only about vertical and directional wells with and without hydraulic fracturing.

As mentioned in Materials and Methods section, the CC technology performance for injection wells is characterized by two parameters: the CC coefficient, which is an "internal" parameter that is not determined in the field, and the injectivity drop coefficient, which is measured after the treatment. The internal parameter characterizes the processes inside the formation, which can not be accurately measured by production logging tools describing the flow rates through perforation intervals, which poorly correlate with flows in layers with various permeabilities. The injectivity drop coefficient is a well-measured parameter (the rate related to drawdown).

Figure 3 shows a cross-plot of the estimates, modeled using the described model, and actual injectivities after the treatment for all well completion types. The cross-plot is described with sufficient accuracy (characterized by the correlation coefficient *D*) by a straight line leaving zero and having an inclination of about 45°, which validates the assumptions of the developed approach. The linear trend is chosen because the field and the calculated data should match. This corresponds to a straight line with an angle of inclination of 45°. The low accuracy of the approximation is associated with the low precision of the field data.

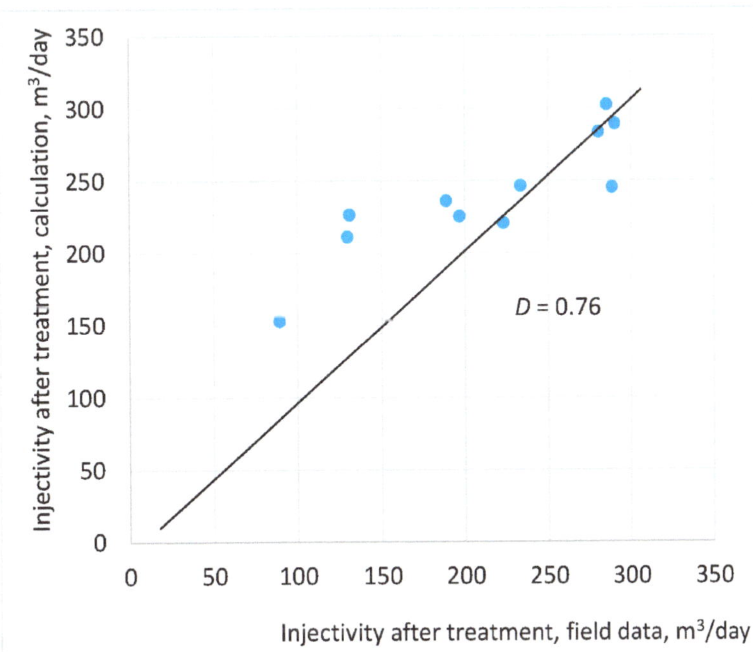

Figure 3. Comparison of post-treatment injectivity, calculated in the developed mathematical model, with its actual value.

The main point of analyzing the reliability of the ideas embedded in the developed line of gelling mathematical models is to confirm the influence of the formulated criteria on incremental oil production as a result of applying these technologies. For the generality of the results, the processing of dimensionless parameters should be used to characterize incremental oil production from the previously introduced dimensionless criteria for the process's performance. As such a parameter, it is advisable to use a dimensionless ratio of the incremental volume to the injected volume. We can denote it as G.

The developed approach is based on the description of the efficiency of flow redistribution in the interlayers through the CC coefficient. This criterion should have a positive effect on incremental oil production, respectively. To confirm this idea, the incremental production data on vertical and directional injection wells were compared with the estimated CC coefficient (B). These calculations are based on well-logging interpretations, as well as the treatment parameters (volumes, compositions, and injection rates). The results of data statistical processing are shown in Figure 4. The resulting function demonstrates a satisfactory correlation coefficient $D = 0.7$.

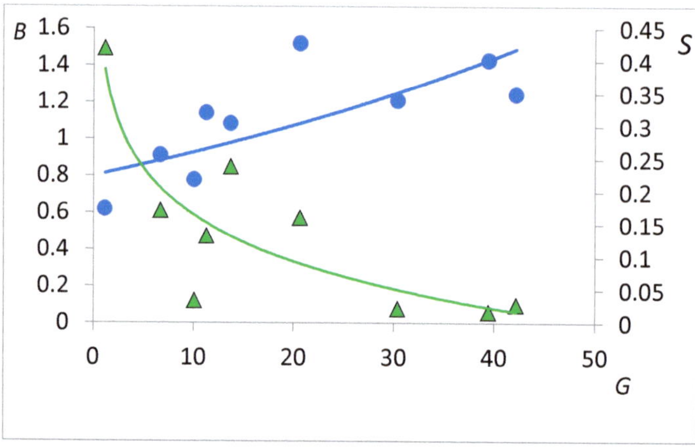

Figure 4. The results of statistical processing of the correlation between the specific incremental production in producers responding to gelling agents injection and the CC coefficient (blue dots and trend) and injectivity drop (green dots and trend).

A negative consequence of the use of gelling agents is a decrease in the injectivity of injection wells, which forces an increase in the BHP to maintain the level of water injection into the reservoir. Statistical processing of data on the effect of the introduced injectivity drop coefficient S on dimensionless incremental oil production from applying the gelling technology is also shown in Figure 4. The correlation coefficient is $D = 0.68$. Figure 4 analysis shows that for the considered process, we can set and solve an optimization problem of maximizing incremental oil production with a minimum reduction in well injectivity. Today, the developed approach for the gel and suspension treatment designs is considered in the Rosneft Company. Executed designs are recommended but the field confirmation has not yet been obtained.

3. Conclusions

The paper presents a new approach to modeling the conformance improvement processes in the application of gelling technologies in a porous medium with the formation of gel barriers leading to the redistribution of flows in an inhomogeneous reservoir. The approach identifies two related tasks: predicting the processes of forming gel barriers in the bottom-hole zone of a heterogeneous reservoir and the response of production wells to treatment jobs. Analysis of the characteristic times of various effects shows that the first problem can be considered in the framework of a one-dimensional flow in non-connected interlayers. The second task can be considered on the basis of a statistical analysis of the field experience in applying gelling technologies.

In the first problem, the efficiency of flow redistribution in the bottom-hole zone can be characterized by the difference in the root-mean-square deviation of the interlayers' permeability and the average value before and after the treatment or by the CC coefficient, which was introduced for the first time. It is established that the CC coefficient increases with the increase in the slug volume. The negative effect of the well treatment is manifested through a decrease in the reservoir injectivity. These coefficients describe the integral effect of near-wellbore flow distribution.

The experimental procedures for the model's parameter determination are considered. The solution of inverse problems was used to find the common time of the gelling reaction, polymer adsorption parameters, and flow characteristics. Examples of such procedures are presented. Unlike the existing approaches, the developed experimental method does not require destructing the core sample to determine the adsorption constants.

The specific effect in the nearest producers from an injector can be characterized by the specific value or the ratio of incremental oil production to the volume of the injected gelling agent. Existing approaches do not contain such characteristics. An analysis of the field experience in applying HPAA/sodium-silicate-based gelling agents showed that the specific incremental oil production will also be determined by the CC coefficient and a decrease in the injectivity coefficient. The developed correlations can be used as the basis for a surrogate model that does not require building geological sector models and flow simulation sector models of the treated zone.

4. Materials and Methods

4.1. General Approach to Modeling

With all the variety of agents recommended for conformance control (CC) applications, three main groups can be distinguished by the sedimentation/gelling mechanism. The most widely used group includes injection of polymers and cross-linkers. These are partially hydrolyzed polyacrylamides (HPAA) [28,29] cross-linked by polyvalent anions that change valence under reservoir conditions, or organic cross-linkers having a low kinetic constant [24]. For some cases, biopolymers [30] or other types of polymers are recommended.

The second group includes gels formed at high reservoir temperatures [14,15] or preformed gels that swell at elevated temperatures [6,22,31,32]. Sometimes the temperature triggers a primary reaction, the products of which create conditions for gelling [15].

The third group includes agents that form an insoluble precipitate during reaction under reservoir conditions. Most of those agents contain sodium silicate as the main agent reacting with calcium chloride [4]. Somewhat apart in this group is the use of suspended particle systems, the main difference of which is that the particles are captured in a porous medium much faster than chemical actions occur and the particles can break away into the flow due to a change in the liquid flow rate [23].

Thus, the main feature of these processes is the presence of two principal components dissolved in water. For suspensions, the principal component in the mixture is particles that are trapped in a porous medium to form a precipitate. Polymers or gels in suspensions have an auxiliary function of stabilizing the injected system and preventing its gravitational segregation.

The conformance control technology consists of injecting into an injection well a slug of agents that form a gel or sediment in a porous medium and form a low-permeable barrier. The slugs used for such purposes have small volumes of up to 2000 m^3. Such volumes are several orders of magnitude smaller than the volumes of injection sites and surrounding production wells. Therefore, from the point of view of mathematics, the problem of a production wells response is a small-parameter problem.

It is impossible to find a solution to such a problem in the general formulation, even with a sector flow simulation model. In such cases, the general task is divided into two components. The first is to analyze the distribution of gel or sediment in the bottom-hole zone of injection wells and determine the quantitative characteristics of the redistribution of flows in the bottom-hole zone. The second is the response of production wells taking into account these characteristics.

Before solving these problems, it is advisable to estimate the time of lateral transmission of the pressure disturbance within the reservoir system and the time of gravitational vertical flows. The lateral transmission of pressure disturbances is described by the piezo-conductivity equation, where we can assume that the piezo-conductivity coefficient is determined by the average lateral reservoir permeability k_x. The viscosity of a reservoir fluid mixture is denoted as μ_{ef}, and the compressibility of a saturated porous medium is β_l. Then, the common time scale of pressure transfer or flow redistribution is estimated as

$$\tau_p \sim \frac{L^2 \beta_l \mu_{ef}}{k_x}, \tag{1}$$

where L is the typical size of a problem.

The time of the gravitational flow of fluid between the layers is determined by the Buckley–Leverett equation where the pressure drop is due to the difference in the fluid densities $\Delta\rho g$, where $\Delta\rho$ is the difference in the densities of water and oil, and g is the acceleration of gravity. The flow time can be estimated as

$$\tau_S \sim H\varnothing \frac{\mu_{ef}}{k_y \Delta\rho g}, \tag{2}$$

Here, $\varnothing$ is the average reservoir porosity, k_y is vertical reservoir permeability, and H is reservoir thickness; the effective viscosity is determined by the formula:

$$1/\mu_{ef} = (f_w/\mu_w + f_o/\mu_o)\partial F/\partial S, \tag{3}$$

where f_w, f_o, μ_w, μ_o are water and oil relative permeability and water and oil viscosity and F is the Buckley–Leverett function.

Let us consider the tasks set for the reservoir with the following parameters: L = 1000 m, H = 50 m, $\varnothing$ = 0.2, anisotropy $a = \sqrt{k_y/k_x} = 0.1$, $\beta_l = 10^{-7}$ Pa^{-1}, μ_{ef} = 1 mPa · s, $g\Delta\rho$ = 1000 Pa/m. The time of gravitational cross-flows for such a reservoir will be 1–2 years and the time of the pressure disturbance propagation is 10 times less. Thus, in the first task of injecting agents into the reservoir, it is possible to neglect the cross-flows between the layers, i.e., consider them isolated, and consider the flow radial near vertical or directional wells and linear near hydraulic fractures. In the second task, it is possible to estimate the time during which the effect of the redistribution of flows will be noticeable, which is about a year.

As a result, we can consider the first task in the simplified formulation of single-phase water flow at residual oil saturation as a one-dimensional statement (radial or linear, depending on the well completion type) for all layers of the reservoir separated from well-logging interpretations. Let us first consider this problem for a single interlayer.

As noted above, for the principal gelling components, the continuity equations that determine the kinetics of reactions may be written, as well as the continuity equation for immobile gel or sediment. The generalized Darcy's law is used as the momentum conservation equation for slow flows. The closing conditions are the laws of the chemical reaction kinetics. Let us consider these equations using an example of two components flowing to form a gel. The mass conservation equations for a multicomponent mixture are written for the mass concentrations of the components in question:

Compositional or multiphase: multicomponent flow models are a common tool for the simulation of fluid flow in reservoirs. These models use the standard specification of specific phase density in terms of saturations and component mass concentrations. Heterogeneous mechanics introduce the concepts of volumetric content of phases and mass concentration of components dissolved in a phase, as well as a pseudo homogeneous mixture in which the velocities of the carrier phase and suspended particles coincide. The latter is often called advective flow. As a result, under the condition of the incompressibility of fluid and rock matrix, the laws of mass conservation are reduced to equations for mass concentration of the components in question:

$$\frac{\partial(\varnothing-\sigma)(1-S_{or})c_i\rho_w+a_i(1-\varnothing)\rho_r}{\partial t}+\operatorname{div}\left(\vec{U}c_i\rho_w\right)=-K_i\rho_w j, \tag{4}$$

where c_i is the mass concentration of the i-th component in the carrier phase (© = 1, 2); a_i is the concentration of the adsorbed i-th component of the polymer; σ is the volume fraction of gel in a single volume of a saturated porous medium; S_{or} is residual oil saturation of the porous medium; ρ_w, ρ_r are the density of the water phase and porous medium matrix; U is the flow rate; K_i is the mass fraction of the i-th component necessary for the formation of 1 kg of sediment or gel; j is the rate of precipitate formation/gelling; and t is time.

The profile of the change in the gel mass per unit volume of the entire saturated porous medium is written as:

$$\frac{\partial(\sigma\rho_g)}{\partial t}=\rho_g j, \tag{5}$$

where ρ_g is gel density.

The generalized flow law takes into account the change in flow resistance due to the formed gel:

$$\vec{U}=-\frac{k\,k_r(S_{or})}{\mu R_f(\sigma,a_i\,)}\operatorname{grad}P, \tag{6}$$

where k is the absolute permeability; $k_r(S_{or})$ is the phase permeability of water at residual oil saturation; μ is the viscosity of the injected solution; $R_f(\sigma,a_i\,)$ is the resistance factor in the zone of formation of a low–permeable barrier; and P is the pore pressure.

The closing relations for the gelling rate in the approximation of an elementary homogeneous one-sided chemical reaction in a closed system proceeding at constant volume and temperature will be written in the form of Guldberg and Waage's Law of Mass Action and the reaction constant through the Arrhenius law [33]:

$$j=Z\prod\nolimits_i c_i^{n_i},\quad Z=Z_0\exp\left(-\frac{E}{RT_0}\right), \tag{7}$$

where the $c_i^{n_i}$ function determines the probability of finding the number of molecules of the i-th component necessary for the reaction at a given point in space; n_i is the stoichiometric reaction coefficient (the number of molecules involved in the reaction) for the i-th component; Z is the reaction constant; Z_0 is the kinetic coefficient; E is the reaction activation energy; RT_0 is the energy of molecules thermal motion; R is the universal gas constant; and T_0 is the temperature. Note that the complex nature of the reaction can manifest itself in fractional values of n_i and the need to determine the empirical value of the reaction constant. A similar approach to describe the polycondensation reaction is presented, for example, in [21].

The functional relationship of the resistance factor on the sediment or gel in a porous medium is described by different functions. This is due to the fact that there is extremely little special research in this area. Here are just a few examples of the resistance factor as a function of the volume content of precipitate in the rock [5]:

$$R_f=(1+\beta\sigma),\qquad R_f=(1+\beta\sigma)^{\gamma}, \tag{8}$$

where γ is the exponent and β is the rock damage coefficient.

The description of suspension flow also fits into the framework of the proposed approach. Suspension consists of micron particles of clay, chalk or wood flour in water, stabilized from gravitational segregation by additives of polymer or weak gel. The migration of particles in a porous medium is described in the framework of a homogeneous multiphase medium in which the velocities of the carrier and dispersed phases coincide. The fraction of particles in the flow is recorded through the mass concentration of the suspension and the particles trapped in the porous medium through the volume fraction in a single volume of the entire porous medium. Such a model is called deep bed suspension migration. The main conditions for its application are a small particle size (smaller than the pore size), and small suspension concentrations (significantly less than the porosity value).

The mechanism of particle capture may vary, but its nature is not chemical, so the trapping kinetics is usually considered proportional to the particle flow modulus:

$$j = \lambda U c, \tag{9}$$

where λ is the flow coefficient and c is the concentration of suspension particles.

This model of suspension flow is widely used to describe the processes of propagation of small particles in a porous medium [23].

When gelling agents are injected into a layered reservoir, they are unevenly distributed over the productive interval: a larger volume of agents enters the highly permeable layers, and a smaller volume, respectively, enters the less permeable ones. Therefore, larger low-permeable barriers will be formed in high-permeable interlayers.

As an example, let us consider the process of HPAA and chrome acetate injection into a vertical well with a fracture with l length in layered strata with reservoir net pay H_{ef}. The reservoir consists of N layers with corresponding thicknesses h_n, porosities $\varnothing_n$ and permeabilities k_n, where n identifies the layer's number.

The governing equations in each layer are as follows:

$$\varnothing_n \frac{\partial c_{1n}}{\partial t} + \operatorname{div}(U_n c_{1n}) = -c_{10} c_{20} Z c_{1n} \tag{10}$$

$$\frac{\partial \sigma_n}{\partial t} = c_{20} z c_{1n} \tag{11}$$

$$U_n = -\frac{k_n\, k_r(S_{or})}{\mu(1 + \beta\sigma_n)} \operatorname{grad} P \tag{12}$$

Here, c_{10} and c_{20} are the initial polymer and chrome acetate concentrations.

The initial conditions for the equation set are the absence of components in a reservoir, the boundary conditions are constant injection rate Q_0 and agents concentrations c_{10} and c_{20} in the well. The analytical solution for the initial and boundary conditions of the model (10)–(12) is as follows:

$$\sigma_n = \frac{Z c_{10} c_{20} V_{inj}}{Q_0} \exp\left(\frac{2HlZ c_{10} c_{20} \sum_m k_m h_m}{Q_0 k_n h_n} x\right), \qquad x > 0 \tag{13}$$

where V_{inj} is the agent slug volume; Q_0 is the injection rate; x is the linear distance from the fracture; and n and m are the layer's indicators.

The method and procedure of the governing equations' solution are presented in the authors' previous paper [27].

The values of layer permeability after treatment k_n^* are calculated through the provided analytical solution [27] of gel distribution $\sigma_{©}$). For the linear flow near the fracture, for example, this formula has the view:

$$k_n^* = \frac{k_n r_c}{\int_0^L (1 + \beta\sigma_n) dx} \tag{14}$$

where r_c is the mid-length between the injector and the nearest producer.

The methodology for assessing the redistribution of the injectivity profile is given, for example, in [27]. Analytical solutions on the distribution of gel or precipitations in each layer of the reservoir system allow for estimating additional hydraulic resistance. Based on the additional resistance, it is possible to estimate the mathematical expectation of the redistribution of flows near an injection well.

The authors recommend characterizing the reservoir heterogeneity through the mean-square deviation of the interlayers permeability:

$$B_0 = \sqrt{\frac{1}{N}\sum_n \left(\frac{k_n H_{ef}}{\sum_m k_m h_m} - \frac{k_a H_{ef}}{\sum_m k_m h_m}\right)^2};\ B_1 = \sqrt{\frac{1}{N}\sum_n \left(\frac{k_n^* H_{ef}}{\sum_m k_m^* h_m} - \frac{k_a^* H_{ef}}{\sum_m k_m^* h_m}\right)^2}, B = B_0 - B_1, \quad (15)$$

where k_a, k_a^* is the arithmetic mean of permeability before and after the treatment job, B_0 and B_1 are the difference in the mean-square deviations of the interlayers permeabilities before and after the treatment job k_n and k_n^* are permeabilities before and after the treatment job; h_m is the thickness of the m-th interlayer; and N is the number of interlayers. For brevity, the difference between these indicators B is called the CC (conformance control) coefficient. The growth of parameters characterizing the reservoir heterogeneity shows an increase in the CC performance.

The negative effect of treatments is associated with a drop in injectivity of injection wells due to deteriorating reservoir properties after the treatment due to partial blocking of the porous medium channels by gel. Therefore, the second important parameter characterizing the negative effect of treatments should be an injectivity drop:

$$S = \frac{PI_1}{PI_0},\ PI_1 = \frac{Q_1}{\Delta P_1}, PI_o = \frac{Q_1}{\Delta P_o}, \quad (16)$$

where PI_1 and PI_0 are injectivities after and before the treatment; Q_1 and Q_0 are injection rates before and after the treatment; and ΔP_1 ΔP_0 are pressure differences between the bottom-hole and the reservoir (at the distance r_c).

As an example of the algorithm for calculating the parameters of flow redistribution in the bottom-hole zone of an injection well, let us consider a treatment job in a vertical well with a fracture of length l = 20 m intersecting a reservoir with net pay H = 4.7 m, containing seven interlayers with the following parameters: k_n = 5 mD, 100 mD, 15 mD, 50 mD, h_n = 1 m, 0.8 m, 0.7 m, 2.2 m, $k(S_{or})$ = 0.2. HPAA and chromium acetate solution with initial concentrations $c_{10} = 0.1$, $c_{20} = 0.091$ were injected as slug V_o = 600 m^3 into the well, the initial viscosity of the solution was close to the water viscosity of μ = 1 cP. The following values were accepted for the HPAA adsorption and reaction constants: a_{imax} = 10^{-7}, Z = 0.002 s^{-1}, β = 21.

Application of the described procedure resulted in a CC coefficient of 0.34 and an injectivity drop factor of 0.74. The redistribution of layers' permeabilities is shown in Figure 5.

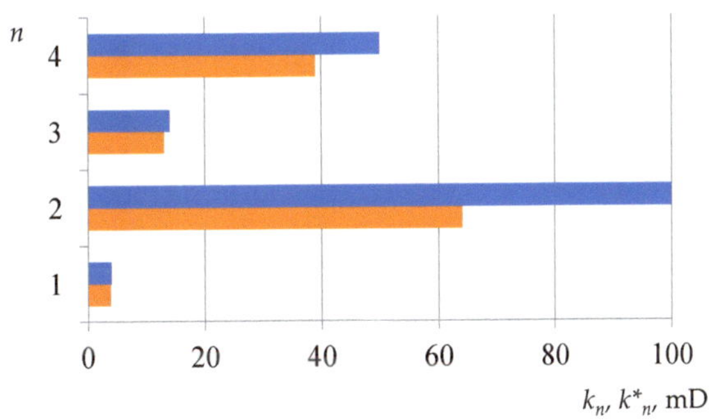

Figure 5. Calculated data on permeability k_n, k_n^* change due to gel treatment for concrete well (blue line stands for k_n, orange line stands for k_n^*).

The schematic view of the key objectives and the whole solution process are demonstrated in Figure 6.

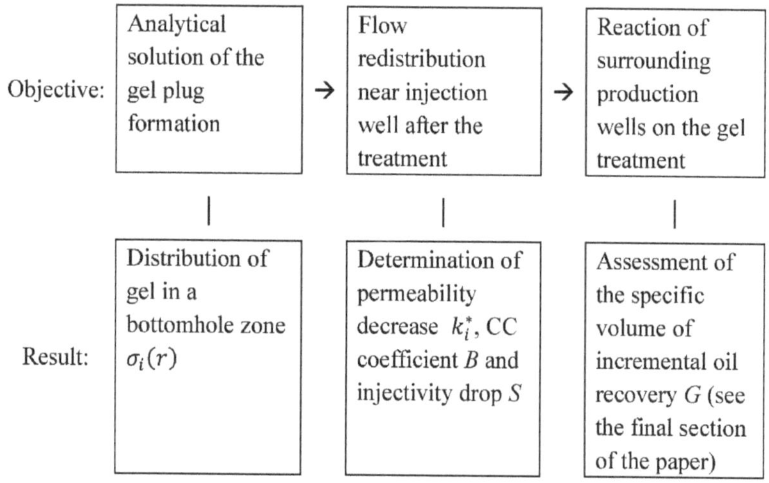

Figure 6. The general workflow to estimate the treatment process parameters.

4.2. Experimental Determination of Constants Closing the Described Models

Let us consider schematically the gelling reaction of partially hydrolyzed polyacrylamide and, for example, a chromium-acetate-based cross-linker:

$$n_1 \cdot M_p + n_2 \cdot M_{Cr} + n_3 \cdot M_w \to n_4 \cdot M_g, \tag{17}$$

where M_p, M_{Cr}, M_w and M_g are the averaged analogs of the masses of polymer, chromium acetate, water, and gel molecules. It is assumed here that the polymerization reaction or the formation of the gel mesh structure occurs when interacting with polyvalent metal anions with water absorption into the mesh structure. The mass ratio of the reacting components cannot be calculated or determined from this equation. However, a classical theory of homogeneous reactions allows us to model the process under certain assumptions that simplify further reasoning.

In physical chemistry, the gel point is defined as the time of the beginning of rheological changes in the system. This time is associated with the transition from the formation of linear or branched oligomers or sol to polycondensation of oligomers into two or three-dimensional meshes [34].

For the analysis of the gelling process kinetics, a more important time is the characteristic gelling time, which determines the completeness of the gelling reaction [35]. The characteristic gelling time is determined under laboratory conditions and its value is usually estimated by the following: (a) by the ability to hold a stick in an upright position in a test tube with the initial agents, (b) by forming a 15% immobile gel layer on the walls of the glass when pouring the mixture out of it [24], (c) by forming a 5 cm gel column when dipping and lifting the chemical spatula at the mixture level [36]. All methods allow only for estimating the characteristic time. We will conditionally call these methods "bottle tests".

We can consider the continuity Equations (11) and (12) in the conditions of chemical experiments (in a beaker, test tube, etc.). There is no movement of components in the vessel and the porosity is equal to unity. Then, the defining system of equations using the gelling reaction as an example will take the form:

$$\frac{dc_i}{dt} = -K_i j, \quad \frac{d\sigma}{dt} = j, \quad j = Zc_1^{n_1} c_2^{n_2} c_3^{n_3}, \tag{18}$$

where it is assumed that the water phase density is close to the gel density $\rho_g \approx \rho_w$.

An example of a partially hydrolyzed polyacrylamide (PAA) cross-linker, as mentioned earlier, is chromium acetate, and a change in its valence under reservoir conditions is the triggering mechanism of the reaction. Sometimes organic cross-linkers are used to form gels that are more resistant to reservoir conditions [24].

Let us estimate the mass fractions of the components in the Equation (13) reaction. The average molecular weight of polymer molecules is thousands of grams of molecules and the proportion of water in gelling process is high (more than 90% of the gel consists of water), therefore, the mass fractions of these agents in the reaction significantly exceed the proportion of the cross-linker.

Thus, we can assume that the reaction proceeds with an excess of water ($c_3 \approx 1$, $c_1 \approx 0.001$, and $c_2 \approx 0.001$), and the consumption of the cross-linker is insignificant ($K_2 \to 0$, $c_2 \approx c_{20}$). We will also consider the reaction to be of the first order for the polymer. Thus, the rate of the chemical reaction will be approximated by the expression: $j = Zc_1c_{20}$. Analysis of the mass balance of the chemical reaction shows that this approach can also be applied to sodium-silicate-based gelling systems [37]. These approximations make it possible to obtain analytical solutions for estimating the kinetic constant.

Let us consider the bottle test results for the studied system, described in [24]. The initial agents were as follows: partially hydrolyzed PAA (25% degree of hydrolysis) with a molecular weight of 20 molecular-mass distribution (MMD), with a mass concentration of 0.003, polyethyleneimine was used as an organic cross-linker that allows for increasing the temperature of gel destruction, with a concentration of 0.0005 in water. In the above study, the results obtained by various methods were compared and the most accurate method was identified with the outflow of reaction products from a chemical beaker and forming of 15% gel on the beaker walls.

The solution of the system of Equation (4) with a simplified setting of the chemical reaction rate is transformed to:

$$c_1 = c_{10}\ \exp(-(K_1 Z c_{20})t), \qquad \sigma = \frac{c_{10}}{K_1}(1 - \exp(-(K_1 Z c_{20})t)). \tag{19}$$

Here, the initial conditions are used when the polymer concentration is equal to $c_1 = c_{10}$ and the gel concentration is $\sigma = 0$. According to the solution, for a sufficiently long time, the volume fraction of the gel is equal to unity, provided the reaction is complete. In this case, the condition for the reaction completeness is equality: $K_1 = c_{10}$. Usually, the initial composition of the solution is selected from the condition of reaction completeness, so further we will assume that the polymer mass fraction in the reaction is determined by its initial concentration.

The solutions of Equation (15) are called kinetic curves. With a volume fraction of the gel equal to 0.15, the characteristic reaction time τ or the reaction constant $Z = (c_{10}\ c_{20}\ \tau)^{-1}$ is determined. Using the data from [24], the following values were obtained: $\tau = 52$ h and $Z = 0.579\ s^{-1}$. In the absence of kinetic experimental data, it is possible to obtain the considered constants from the history matching data [38].

The flow characteristics of the gelling agent are also considered by the example of determining the flow and adsorption characteristics of a polymer–hydrolyzed polyacrylamide (HPAA) [39]. These characteristics are determined from core flow experiments with polymer solution with measuring differential pressure drop and, if possible, measuring the polymer concentration in the outgoing flow.

After pumping a sufficient volume of polymer solution (pressure stabilization), the resistance factor $R(a)$ is determined, where a is the concentration of the adsorbed polymer. The transient flow regime is compared with the classical solutions for water displacement by the polymer solution, which allows for determining the adsorption characteristics of a porous medium. Analysis of the flow of the tail and leading edges of the polymer slug also allows us to conclude that the pore volume is inaccessible to the polymer [36].

An example of solving these inverse problems is schematically shown in Figure 7, where ΔP is the pressure drop, T is the number of pumped pore volumes, T_b is the break-point, and α is the inclination angle. The formulation of inverse problems and their solution are presented in [26,39].

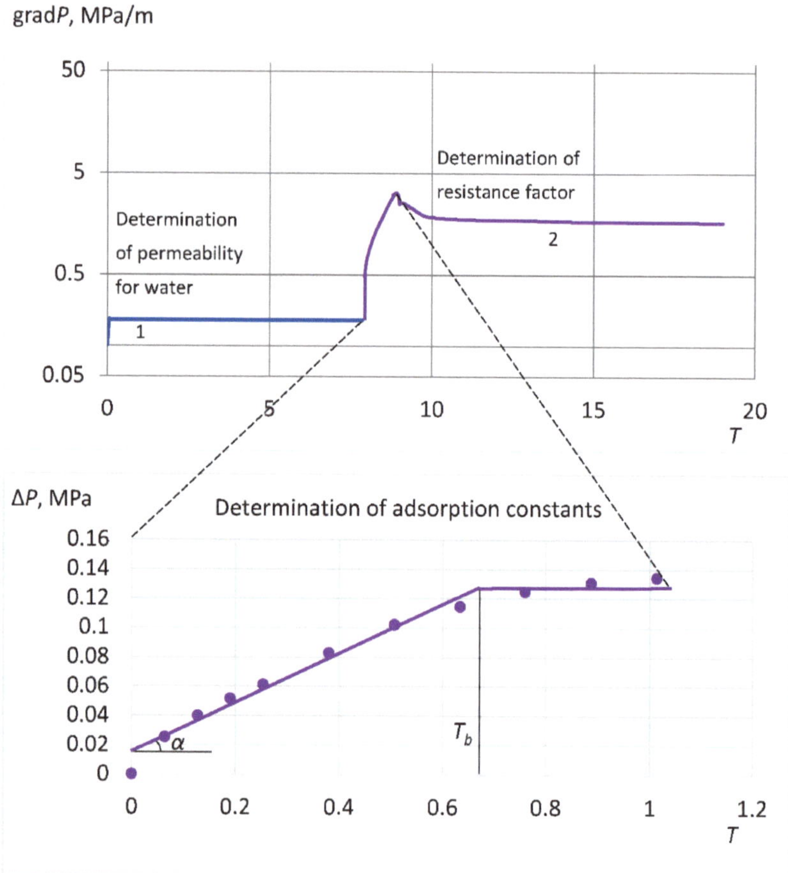

Figure 7. Graphical solution of inverse problems to determine the polymer adsorption parameters (stage 1 and blue line—determination of water permeability, stage 2 and purple line—determination of the resistance factor).

The paper [38] describes a similar approach to determine the flow parameters of the suspension λ and β. In the absence of experimental data, characteristics can be obtained from solving a more complex inverse problem of treating the bottom-hole zone of an injection well for a reservoir consisting of non-connected interlayers with various flow parameters [40].

Author Contributions: Conceptualization, K.F.; methodology, A.S.; software, I.V.; validation, A.G. and N.M.; formal analysis, K.F.; investigation, A.S. and D.A.; resources, N.M.; data curation, D.A. and N.M.; writing—original draft preparation, K.F.; writing—review and editing, A.S.; visualization, N.M.; supervision, A.A.; project administration, D.A. All authors have read and agreed to the published version of the manuscript.

Funding: This research received no external funding.

Institutional Review Board Statement: Not applicable.

Informed Consent Statement: Not applicable.

Data Availability Statement: Not applicable.

Conflicts of Interest: The authors declare no conflict of interest.

References

1. Willhite, G.P. *Waterflooding*; Society of Petroleum Engineers: Richardson, TX, USA, 1986; pp. 1–326.
2. Wolcott, D. *Applied Waterflood Field Development*; Energy Tribune Publishing: Houston, TX, USA, 2009; pp. 1–417.
3. Sydansk, R.D.; Laura, R.-Z. *Reservoir Conformance Improvement*; Society of Petroleum Engineers: Richardson, TX, USA, 2011; pp. 1–138.
4. Zemtsov, Y.V.; Mazaev, V.V. *The Current State of Physical and Chemical Enhanced Oil Recovery Methods (Literature and Patent Review)*; LLC "Izdatelskiye resheniya": Yekaterinburg, Russia, 2021; pp. 1–240. (In Russian)
5. Ruchkin, A.A.; Yagafarov, A.K. *Optimization of Conformance Improvement Technologies Application for Samotlor Oil Field*; Vector Buk: Tyumen, Russia, 2005; pp. 1–165. (In Russian)
6. Okeke, T.; Robert, L. Simulation and Economic Screening of Improved Oil Recovery Methods with Emphasis on Injection Profile Control Including Waterflooding, Polymer Flooding and a Thermally Activated Deep Diverting Gel. In Proceedings of the Society of Petroleum Engineers (SPE) Western Regional Meeting, Bakersfield, CA, USA, 19–23 March 2012. [CrossRef]
7. Kang, W.; Kang, X.; Lashari, Z.A.; Li, Z.; Zhou, B.; Yang, H.; Sarsenbekuly, B.; Aidarova, S. Progress of polymer gels for conformance control in oilfield. *Adv. Colloid Interface Sci.* **2021**, *289*, 102363. [CrossRef] [PubMed]
8. Ding, F.; Dai, C.; Sun, Y.; Zhao, G.; You, Q.; Liu, Y. Gelling behavior of PAM/Phenolic crosslinked gel and its profile control in a low-temperature and high-salinity reservoir. *Gels* **2022**, *8*, 433. [CrossRef] [PubMed]
9. Kan, V.A.; Poddubniy, Y.A.; Sidorova, I.A. Hydrogels from sodium silicate solutions. *Oil Ind.* **1984**, *10*, 44–46. (In Russian)
10. Hatzignatiou, D.G.; Askarinezhad, R.; Giske, N.H.; Stavland, A. Laboratory testing of environmentally friendly sodium silicate systems for water management through conformance control. *SPE Prod. Oper.* **2016**, *31*, 337–350. [CrossRef]
11. Krumrine, P.H.; Boyce, S.D. Profile Modification and Water Control with Silica Gel-Based Systems. In Proceedings of the Society of Petroleum Engineers (SPE) Oilfield and Geothermal Chemistry Symposium, Phoenix, AZ, USA, 9–11 April 1985. [CrossRef]
12. Tang, X.; Kang, W.; Zhou, B.; Gao, Y.; Cao, C.; Guo, S.; Iqbal, M.W.; Yang, H. Characteristics of composite microspheres for in-depth profile control in oilfields and the effects of polymerizable silica nanoparticles. *Powder Technol.* **2020**, *359*, 205–215. [CrossRef]
13. Tang, X.; Zhou, B.; Chen, C.; Sarsenbekuly, B.; Yang, H.; Kang, W. Regulation of polymerizable modification degree of nano-SiO2 and the effects on performance of composite microsphere for conformance control. *Colloids Surf. A Physicochem. Eng. Asp.* **2020**, *585*, 124100. [CrossRef]
14. Brilliant, L.S.; Antipov, V.S.; Starkova, N.R.; Gordeev, A.O.; Chernavskikh, S.F.; Chernyshov, A.V.; Negomedzyanov, V.R. *Technologist Handbook*; JSC "SIBINKOR": Tyumen, Russia, 1998; pp. 1–91. (In Russian)
15. Altunina, L.K.; Kuvshinov, V.A. Physical and chemical enhanced oil recovery methods. *St.-Petersburg Univ. Her.* **2013**, *4*, 46–76. (In Russian)
16. Ghaddab, F.; Kaddour, K.; Tesconi, M.; Brancolini, A.; Carniani, C.; Galli, G. A Tertiary Method for Enhanced Oil Recovery for a Mature Field. In Proceedings of the Society of Petroleum Engineers (SPE) Production and Operations Conference and Exhibition, Tunis, Tunisia, 8–10 June 2010. [CrossRef]
17. Gazizov, A.S.; Nizamov, R.K. Evaluation of the effectiveness of the polymer-dispersed system application technology based on the results of field studies. *Oil Ind.* **1990**, *7*, 49–52. (In Russian)
18. Volkov, V.P.; Brilliant, L.S. Geological aspects of the reservoir of Shercalinskaya suite of Talinskaya area. *Oil Ind.* **2013**, *1*, 18–22.
19. Shel, E.V.; Kabanova, P.K.; Tkachenko, D.R.; Bazyrov, I.S.; Logvinyuk, A.V. Modeling of initiation and propagation of hydraulic fractures at an injection well for non-fractured terrigenous rocks on the example of the Priobskoye field. *Prof. Oil* **2020**, *2*, 36–42. (In Russian)
20. Manrique, E.; Reyes, S.; Romero, J.; Aye, N.; Kiani, M.; North, W.; Thomas, C.; Kazempour, M.; Izadi, M.; Roostapour, A.; et al. Colloidal Dispersion Gels (CDG): Field Projects Review. In Proceedings of the Society of Petroleum Engineers (SPE) EOR Conference at Oil and Gas West Asia, Muscat, Oman, 31 March–2 April 2014. [CrossRef]
21. Bai, B.; Liu, Y.; Coste, J.-P.; Li, L. Preformed particle gel for conformance control: Transport mechanism through porous media. *SPE Reserv. Eval. Eng.* **2007**, *10*, 176–184. [CrossRef]
22. Caili, D.; Qing, Y.; Fulin, Z. In-depth profile control technologies in China—A review of the state of the art. *Pet. Sci. Technol.* **2010**, *28*, 1307–1315. [CrossRef]
23. Bedrikovetsky, P.; Siqueira, F.D.; Furtado, C.A. Modified particle detachment for colloidal transport in porous media. *Transp. Porous Media* **2011**, *86*, 353–383. [CrossRef]
24. Al-Anazi, A.; Al-Kaidar, Z.; Wang, J. Modeling Gelation Time of Organically Crosslinked Polyacrylamide Gel System for Conformance Control Applications. In Proceedings of the Society of Petroleum Engineers (SPE) Russian Petroleum Technology Conference, Moscow, Russia, 22–24 October 2019. [CrossRef]

25. Vaz, A.; Bedrikovetsky, P.; Fernandes, P.; Badalyan, A.; Carageorgos, T. Determining model parameters for non-linear deep-bed filtration using laboratory pressure measurements. *J. Pet. Sci. Eng.* **2017**, *151*, 421–433. [CrossRef]
26. Barenblatt, G.I.; Entov, V.M.; Ryzhik, V.M. *Theory of Fluid Flows through Natural Rocks*; Springer Science and Business Media LLC: Berlin/Heidelberg, Germany, 1990; pp. 1–396.
27. Fedorov, K.M.; Gilmanov, A.Y.; Shevelev, A.P.; Kobyashev, A.V.; Anuriev, D.A. A theoretical analysis of profile conformance improvement due to suspension injection. *Mathematics* **2021**, *9*, 1727. [CrossRef]
28. Al Brahim, A.; Bai, B.; Schuman, T. Comprehensive review of polymer and polymer gel treatments for natural gas-related conformance control. *Gels* **2022**, *8*, 353. [CrossRef]
29. Rozhkova, Y.A.; Burin, D.A.; Galkin, S.V.; Yang, H. Review of microgels for enhanced oil recovery: Properties and cases of application. *Gels* **2022**, *8*, 112. [CrossRef]
30. Agzamov, F.A.; Morozov, D.V. Application of biopolymers for reservoir waterproofing. *Neftegazov. Delo* **2002**, *1*, 1–8. (In Russian)
31. Aqcheli, F.; Salehi, M.B.; Taghikhani, V.; Pahlevani, H. Synthesis of a custom-made suspension of preformed particle gel with improved strength properties and its application in the enhancement of oil recovery in a micromodel scale. *J. Pet. Sci. Eng.* **2021**, *207*, 109108. [CrossRef]
32. Farasat, A.; Younesian-Farid, H.; Sadeghnejad, S. Conformance control study of preformed particle gels (PPGs) in mature waterflooded reservoirs: Numerical and experimental investigations. *J. Pet. Sci. Eng.* **2021**, *203*, 108575. [CrossRef]
33. Stromberg, A.G. *Physical Chemistry*; Vysshaya Shkola: Moscow, Russia, 1999; pp. 1–527. (In Russian)
34. Tcharkhtchi, A.; Nony, F.; Khelladi, S.; Fitoussi, J.; Farzaneh, S. Epoxy/amine reactive systems for composites materials and their thermomechanical properties. In *Advances in Composites Manufacturing and Process Design*, 1st ed.; Boisse, P., Ed.; Woodhead Publishing: Sawstone, UK, 2015; Volume 1, pp. 269–296. [CrossRef]
35. Frolov, Y.G.; Shabanova, N.A.; Savochkina, T.V. Kinetics of gelation and spontaneous dispersion of silicic acid gel. *Colloid. J.* **1980**, *42*, 1015–1018. (In Russian)
36. Safarov, F.E.; Gusarova, E.I.; Karazeev, D.V.; Arslanov, I.R.; Telin, A.G.; Dokichev, V.A. Synthesis of polyacrylamide gels for restricting the water inflow in development of oil fields. *Russ. J. Appl. Chem.* **2018**, *91*, 872–876. [CrossRef]
37. Khamees, T.K.; Flori, R.E.; Fakher, S.M. Numerical Modeling of Water-Soluble Sodium Silicate Gel System for Fluid Diversion and Flow-Zone Isolation in Highly Heterogeneous Reservoirs. In Proceedings of the Society of Petroleum Engineers (SPE) Trinidad and Tobago Section Energy Resources Conference, Port of Spain, Trinidad and Tobago, 25–26 June 2018. [CrossRef]
38. Fedorov, K.M.; Shevelev, A.P.; Kobyashev, A.V.; Zakharenko, V.A.; Kochetov, A.V.; Neklesa, R.S.; Usoltsev, A.V. Determination of Suspension Filtration Parameters from Experimental Data. In Proceedings of the Society of Petroleum Engineers (SPE) Russian Petroleum Technology Conference, Moscow, Russia, 26–29 October 2020. [CrossRef]
39. Fedorov, K.M.; Pospelova, T.A.; Kobyashev, A.V.; Gilmanov, A.Y.; Kovalchuk, T.N.; Shevelev, A.P. Determination of Adsorption-Retention Constants and Inaccessible Pore Volume for High-Molecular Polymers. In Proceedings of the Society of Petroleum Engineers (SPE) Russian Petroleum Technology Conference, Moscow, Russia, 12–15 October 2021. [CrossRef]
40. Fedorov, K.M.; Zubkov, P.T. Placement of gels in stratified reservoirs using a sequential injection technique. *J. Pet. Sci. Eng.* **1996**, *15*, 69–80. [CrossRef]

gels

Article

Experimental Evaluation of the Rheological Properties and Influencing Factors of Gel Fracturing Fluid Mixed with CO_2 for Shale Gas Reservoir Stimulation

Mingwei Wang [1], Wen Wu [2], Shuyang Chen [3], Song Li [4,*], Tao Li [5], Gensheng Ni [2], Yu Fu [1] and Wen Zhou [6]

1 School of Oil & Natural Gas Engineering, Southwest Petroleum University, Chengdu 610500, China
2 Development Division, PetroChina Southwest Oil and Gasfield Company, Chengdu 610041, China
3 Sinopec Northwest Oilfield Company, Urumqi 830000, China
4 Engineering Research Institute, PetroChina Southwest Oil and Gasfield Company, Chengdu 610017, China
5 Shunan Gas Mine, PetroChina Southwest Oil and Gasfield Company, Luzhou 646000, China
6 State Key Laboratory of Oil and Gas Reservoir Geology and Exploitation, Chengdu University of Technology, Chengdu 610059, China
* Correspondence: lisong03011640@163.com

Abstract: Foam gel fracturing fluid has the characteristics of low formation damage, strong flowback ability, low fluid loss, high fluid efficiency, proper viscosity, and strong sand-carrying capacity, and it occupies a very important position in fracturing fluid systems. The rheological properties of gel fracturing fluid with different foam qualities of CO_2, under different experimental temperatures and pressures, have not been thoroughly investigated, and their influence on it was studied. To simulate the performance of CO_2 foam gel fracturing fluid under field operation conditions, the formula of the gel fracturing fluid was obtained through experimental optimization in this paper, and the experimental results show that the viscosity of gel fracturing fluid is 2.5 mPa·s (after gel breaking at a shear rate of 500 s^{-1}), the residue content is 1.3 mg/L, the surface tension is 25.1 mN/m, and the interfacial tension is 1.6 mN/m. The sand-carrying fluid has no settlement in 3 h with a 40% sand ratio of 40–70-mesh quartz sand. The core damage rate of foam gel fracturing fluid is less than 19%, the shear time is 90 min at 170 s^{-1} and 90 °C, the viscosity of fracturing fluid is >50 mPa·s, and the temperature resistance and shear resistance are excellent. The gel fracturing fluid that was optimized was selected as the base fluid, which was mixed with liquid CO_2 to form the CO_2 foam fracturing fluid. This paper studied the rheological properties of CO_2 foam gel fracturing fluid with different CO_2 foam qualities under high temperature (65 °C) and high pressure (30 MPa) and two states of supercooled liquid (unfoamed) and supercritical state (foamed) through indoor pipe flow experiments. The effects of temperature, pressure, shear rate, foam quality, and other factors on the rheological properties of CO_2 foam gel fracturing fluid were considered, and it was confirmed that among all the factors, foam quality and temperature are the main influencing factors, which is of great significance for us to better understand and evaluate the flow characteristics of CO_2 foam gel fracturing fluid and the design of shale gas reservoir fracturing operations.

Keywords: shale gas reservoir; CO_2 foam fracturing; gel fracturing fluid; foam quality; rheological property

Citation: Wang, M.; Wu, W.; Chen, S.; Li, S.; Li, T.; Ni, G.; Fu, Y.; Zhou, W. Experimental Evaluation of the Rheological Properties and Influencing Factors of Gel Fracturing Fluid Mixed with CO_2 for Shale Gas Reservoir Stimulation. *Gels* **2022**, *8*, 527. https://doi.org/10.3390/gels8090527

Academic Editors: Mario Grassi, Qing You, Guang Zhao and Xindi Sun

Received: 25 July 2022
Accepted: 17 August 2022
Published: 23 August 2022

1. Introduction

The use of foam gel fracturing fluid is a great achievement of liquid technology. Foamed fracturing fluid is formed by dispersing N_2 or CO_2 in water, acid, methanol/water mixture, or hydrocarbon liquid as bubbles, and is usually a two-phase mixture of 70%–80% dryness gas (N_2 or CO_2) and fracturing of fluid (water-based polymer solution). Foam gel fracturing fluid is essentially a kind of gas-in-liquid emulsion, and bubbles provide high viscosity and excellent proppant-carrying capacity. Because it has the characteristics of low reservoir damage, strong flowback ability, low fluid loss, high fluid efficiency, proper viscosity, and strong sand-carrying capacity, it occupies a very important position in fracturing fluid systems.

To solve the shortcomings and defects of conventional fracturing technology, researchers began to study foam fracturing technology in the 1970s [1]. Since foam fracturing was first completed in Lincoln County, West Virginia, USA, foam fracturing technology has developed from the initial N_2 foam fracturing to the present CO_2 foam fracturing. In 1986, in the Federal Republic of Germany, 60% CO_2 foam gel fracturing fluid was used in the carboniferous gas reservoir in Fez Dolf, which was buried 3400−3650 m underground. The fracturing was successful, and the natural gas production increased by nearly 12 times after fracturing [2]. By the 1990s, about 90% of gas wells and 30% of oil wells in the United States and Canada had adopted CO_2 foam fracturing technology [3]. Nowadays, it is very common to use foam fracturing technology for fracturing worldwide. In the United States, about 3600 foam fracturing operations are carried out every year, which not only has a high success rate but also has an obvious effect on increasing production. Foams have been considered the most attractive and preferred fluid for fracturing unconventional reservoirs due to their ability to reduce formation damage and improve the recovery of injected fluid [4,5]. Gel and foam systems, as the two most widely used plugging agents for lost circulation control, have achieved positive progress in both laboratory experiments and field applications in recent decades [6,7]. Wang et al. (2022) developed a composite gel foam plugging system, which is used to plug and control flooding for heterogeneous reservoirs, and it showed better plugging and recovering performance for field applications [8].

The characteristics of CO_2 foam fracturing fluid can also influence the propagation of hydraulic fractures. Several authors reported that the high performance of CO_2 foam is attributed to its unique and favorable rheological characteristics [9–13]. However, due to the complex nature of foam, it is difficult to understand and model its flow behavior, especially under operating conditions. The versatility and uniqueness of foam are attributed to its enormously high viscosity profile compared to its base fluids and the efficiency of foam fracturing is dictated by the complex non-Newtonian behavior of foam [14–18]. Numerous authors agree that the design of fracturing treatments highly depends on foam rheology and it governs the overall process performance [19–26]. Foam rheology also determines the properties of the fracture network that may help in obtaining the required fracture geometry. At present, due to the limitation of equipment conditions and research methods, the research on the rheological properties of fracturing fluid under simulated field construction conditions has not been reported. The prediction of foam rheological behavior is a complex task and the direct determination of foam rheology under operating conditions is still considered a challenge [27]. Fu et al. (2021) investigated the rheology and stability of nanoparticle-stabilized CO_2 foams under reservoir conditions (high temperature and high pressure) for fracturing applications [28]. Li et al. (2022) investigated the rheology properties of thickened liquid CO_2 by measuring the viscosity of thickened liquid CO_2 in different physical parameters of this prepared thickener and explained the causes of the rheological changes [29]. Kadafur et al. (2022) investigated the rheology of a CO_2 foamed chelating agent, L-glutamic acid-N, N-diacetic acid (GLDA), which was conducted at 100 C, 1000 psi, 3.5 pH level, and various water salinities, resembling harsh reservoir conditions [30]. Tariq et al. (2022) established a data pool, which was analyzed using four machine learning techniques: Artificial Neural Network (ANN), Decision Trees (DT), Random Forest Regressor (RFR), and K-Nearest Neighbor (KNN), and it provides a simplified ANN-based model which can be used on the fly to predict the effective bulk foam viscosity in both laboratory and field conditions [31].

Presently, the limit pressure of the experimental system for studying the rheological properties of CO_2 foam gel fracturing fluid is only 2000 psi (13.8 MPa). Under the condition of simulating tubing or formation temperature (30−50 °C), the gas phase of the foam gel fracturing fluid is in a gas state, so the foam gel fracturing fluid in the experiment is in a gas-liquid two-phase flow, while for the actual fracturing technology, the pumping pressure is extremely high, reaching tens of MPa. At the same time, the fluid temperature in the wellbore or formation fracture is also high. For CO_2 foam gel fracturing fluid, CO_2 is in a supercritical fluid state. The physical property of the supercritical fluid is closer to that of liquid, and the rheological property at this time is closer to that of a liquid–liquid emulsion. Most of the literature has investigated the rheological performance of the CO_2 foam gel fracturing fluid unfoamed and ignored the different foam qualities' effects in the foaming process, and the experimental temperature and pressure are so low that they are unable to simulate actual field fracturing conditions. Therefore, it is very important to study the rheological properties of foam gel fracturing fluid in the two states of supercooled liquid (unfoamed) and supercritical state (foamed) under simulated actual construction conditions—high pressure (tens of MPa) and high shear rate—for the effective implementation of fracturing technology, the selection of reasonable fracturing parameters, more accurate fracturing prediction, and the evaluation of fracturing effects.

To simulate the rheological performance of CO_2 foam fracturing fluid in the two states of foamed and unfoamed under field operation conditions, the formula of the gel fracturing fluid is obtained through experimental optimization firstly, and the viscosity, static sand setting performance, and rheological performance of the foam gel fracturing fluid are experimentally evaluated. The goal is to obtain a foaming gel fracturing fluid with good performance parameters and which is mixed with liquid CO_2. This paper selects the foam fracturing fluid formed by CO_2 and gel fracturing fluid and studies the rheological properties of CO_2 foam fracturing fluid under high temperature (65 °C) and high pressure (30 MPa), considering the two states of supercooled liquid and supercritical through indoor pipe flow experiments. The effects of temperature (15–90 °C), pressure (10, 20, 30 MPa), shear rate (100–3000 s^{-1}), foam quality (0, 45, 55, 65, 75%), and other factors on the rheological properties of fracturing fluid are investigated, which is of great significance for better understanding and evaluating the flow characteristics of CO_2 foam gel fracturing fluid and on-site fracturing construction design.

2. Results and Discussion

2.1. Experimental Study on the Rheological Characteristics of CO_2 Foam Gel Fracturing Fluid

The effects of temperature, pressure, shear rate, and foam quality on the rheological properties of fracturing fluid are considered. In this experiment, the inner diameter of the test instrument pipeline was 12 mm. The effective viscosity of the CO_2 foam gel fracturing fluid changed with the shear rate at 20 MPa, 30 MPa, and 40 MPa, and the temperature changed from 0 to 80 °C.

2.1.1. Effect of Shear Rate on the Effective Viscosity of CO_2 Foam Gel Fracturing Fluid

In actual fracturing construction, high-pressure supercooled liquid CO_2 is often mixed with guanidine gum and then injected into the formation by tubing for fracturing. As the fracturing fluid enters the formation, the temperature gradually rises, and the high-pressure CO_2 completely changes from supercooled liquid to a supercritical state, and the effects of CO_2 in the two states on the effective viscosity of CO_2 foam gel fracturing fluid are completely different. Therefore, the research on the influence of various factors on the effective viscosity of CO_2 foam gel fracturing fluid is divided into two processes for analysis. The CO_2 foam fracturing fluid with CO_2 in the liquid state is defined as the fracturing fluid system under unfoamed conditions, and the CO_2 foam fracturing fluid with CO_2 in the supercritical state is defined as the fracturing fluid system under foamed conditions.

Figure 1 is the curve of the effective viscosity of the fracturing fluid under the unfoamed condition, with a pressure of 10 MPa and a temperature of 20 °C changing with the shear rate. It can be seen from the figure that the effective viscosity of the fluid decreases exponentially with the increase in the shear rate at the same temperature, which fully shows that the unfoamed fracturing fluid is a typical shear-thinning non-Newtonian fluid, and the changing trend when the shear rate is lower than 500 s^{-1} is other than that when the shear rate is higher than 500 s^{-1}. The shear-thinning characteristics of the foam system in the unfoamed state are mainly due to the influence of shear on the base liquid of the gel fracturing fluid. Linear guanidine gum is a long-chain polymer without a cross-linking structure. Increasing the shear rate will reduce the intermolecular interaction force caused by polymer molecular entanglement and hydrogen bonding, which will lead to a decrease in effective viscosity.

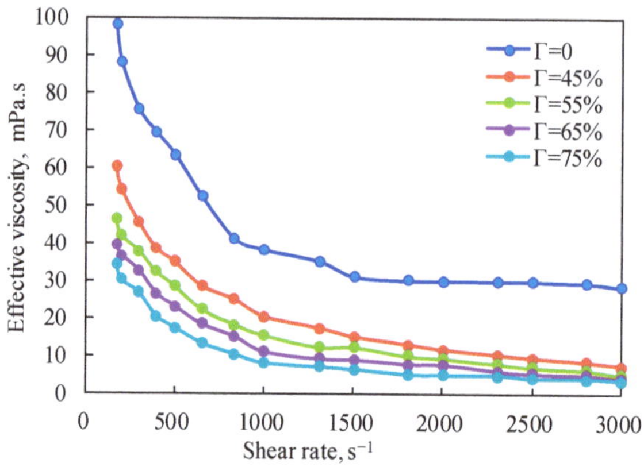

Figure 1. Variation curve of effective viscosity with shear rate (unfoamed, T = 20 °C).

Figure 2 is the curve of the effective viscosity of CO_2 foam gel fracturing fluid with a shear rate at a pressure of 10 MPa and a temperature of 65 °C. It can be seen from the figure that the effective viscosity of CO_2 foam gel fracturing fluid decreases exponentially with the increase in the shear rate at the same pressure, indicating that the CO_2 foam gel fracturing fluid system is a typical shear-thinning non-Newtonian fluid, with a changing trend when the shear rate is lower than 1000 s^{-1}. It can be seen from the figure that for CO_2 foam gel fracturing fluid during foaming, CO_2 is in a supercritical state, and its physical properties are increasingly close to those of gas. At this time, the emulsion formed by guanidine gum solution and supercritical CO_2, which are two limited miscible fluids, is closer to the traditional foam system. The weakening effect of shearing on the viscosity of CO_2 foam fluid is mainly reflected in two aspects: on the one hand, the shearing mentioned above will reduce the intermolecular interaction force caused by polymer molecular entanglement and hydrogen bonding; on the other hand, it is due to the destruction of the internal-phase CO_2 foam structure by shearing.

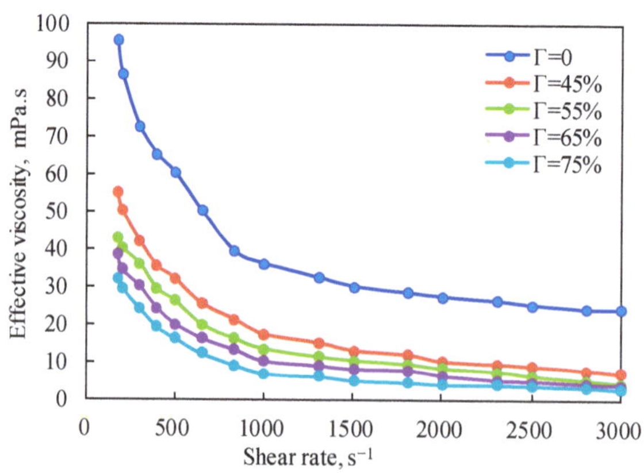

Figure 2. Variation curve of effective viscosity with shear rate (foamed, T = 65 °C).

2.1.2. Effect of Foam Quality on the Effective Viscosity of CO_2 Foam Gel Fracturing Fluid

Figure 3 is the curve of the variation of effective viscosity of fracturing fluid with a CO_2 volume fraction when the pressure is 10 MPa and the temperature is 20 °C. It can be seen from the figure that the effective viscosity of unfoamed fracturing fluid decreases with the increase in the CO_2 volume fraction, and the change range is large. The main reason is that the unfoamed CO_2 is in the supercooled liquid form, similar to Newtonian fluid in this mixed system. At this time, CO_2 has little significance for the viscosity increase in the whole system. On the contrary, the increase in the CO_2 volume fraction will dilute the guanidine gum base liquid. When the volume shares of CO_2 increase to a certain extent, the fluid-structure will suddenly change, from the previous guanidine gum base liquid as the continuous phase to the liquid one. When the guanidine gum base liquid changes from the external phase to the internal phase, the continuous phase of the fluid becomes the liquid CO_2, which greatly reduces the viscosity of the whole system.

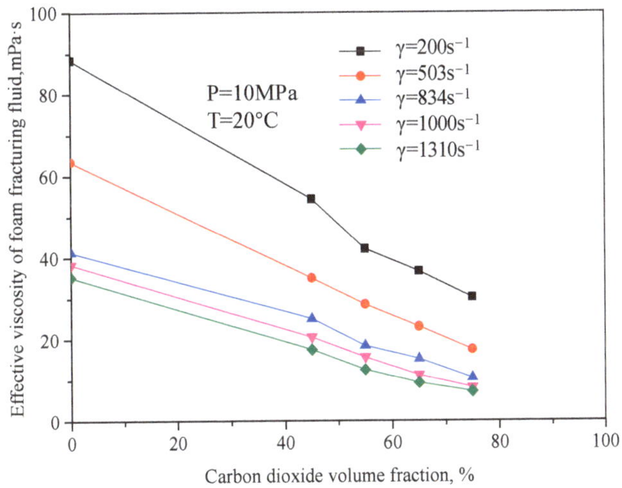

Figure 3. The curve of effective viscosity changes with CO_2 volume fraction (Unfoamed, T = 20 °C).

Figure 4 is the variation law curve of the effective viscosity of CO_2 foam gel fracturing fluid with foam quality under foaming conditions, with a pressure of 10 MPa and temperature of 65 °C. It can be seen from the figure that the change rule of effective viscosity with foam quality is opposite to that without foam. The increase in foam quality makes the effective viscosity of the whole system increase, and the increased range is large. For example, when the shear rate is 834 s^{-1}, the viscosity of foam gel fracturing fluid increases from 26.45 mPa·s to 56.32 mPa·s, with an increasing range of 113%. When the foam mass is 75%, the effective viscosity reaches the maximum value, and when the foam mass is more than 75%, the viscosity of foam gel fracturing fluid decreases obviously. In the research of this system, it can be seen that when the foam mass is more than 55%, with the increase in foam mass, the number of bubbles in the foam system increases, the mutual interference, and deformation among bubbles increase, the bubble structure becomes denser, and the viscosity of the foam system continues to increase.

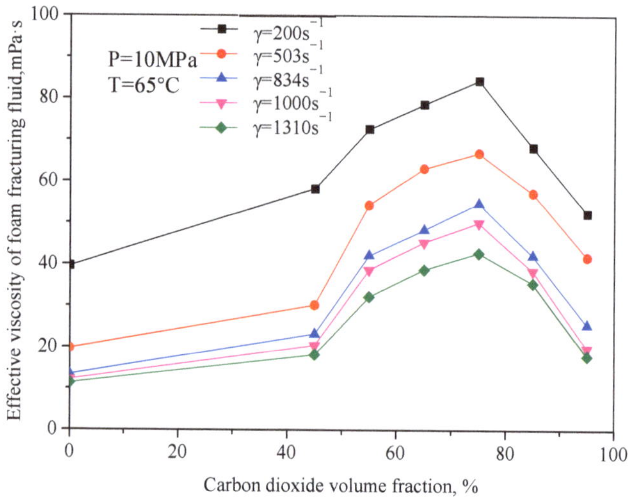

Figure 4. Variation curve of effective viscosity with foam quality (Foamed, T = 65 °C).

2.1.3. Effect of Temperature on Effective Viscosity of CO_2 Foam Gel Fracturing Fluid

The effective viscosity of CO_2 foam gel fracturing fluid with a foam mass of 45–75% was tested at elevated temperatures, and the influence of temperature on the effective viscosity of CO_2 foam gel fracturing fluid was analyzed. In the experiment, a pipe diameter of 12 mm, a shear rate of 170 s^{-1}, a heating rate of 1 °C/min, and a temperature of 80 °C were selected, and the viscosity–temperature characteristics of CO_2 foam gel fracturing fluid under conditions of 10–40 MPa were tested.

Figures 5 and 6 are the curves of the effective viscosity of the fracturing fluid with temperature in an unfoamed state, pressure 20 MPa, and different shear rates and CO_2 volume fractions. The temperature ranges from 5 to 25 °C, the shear rate ranges from 503 s^{-1} to 1500 s^{-1}, and the volume fraction of CO_2 from 45 to 75%. It can be seen from the figure that the effective viscosity of unfoamed fluid decreases with the increase in temperature, showing an exponentially decreasing trend. The main reason is the influence of temperature on the rheological properties of the guanidine gum base liquid: with the increase in temperature, the movement activity of guanidine gum molecules increases, and the thermal fracture of the hydrogen bonds of linear guanidine gum in the n liquid system is accelerated, so that the activation energy of guanidine gum base liquid decreases, which comprehensively shows that the effective viscosity of the solution decreases.

Figures 7 and 8 are the curves of the effective viscosity of CO_2 foam gel fracturing fluid with temperature under the foaming condition, with a pressure of 20 MPa, different shear rates, and foam quality. The temperature ranges from 35 to 75 °C, the shear rate ranges from 503 s^{-1} to 1500 s^{-1}, and the foam mass ranges from 45 to 75%. It can be seen from the figure that the effective viscosity–temperature characteristics of CO_2 foam gel fracturing fluid during foaming are consistent with those of unfoamed fracturing fluid, showing an exponentially decreasing trend. At the same time, when the temperature is greater than 55 °C, the variation ranges of the effective viscosity of foam gel fracturing fluid with temperature becomes smaller.

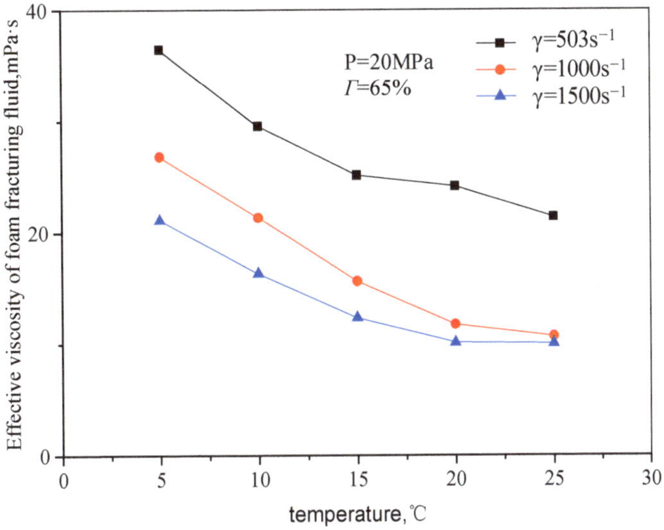

Figure 5. Variation curve of effective viscosity with temperature (unfoamed, 65% by volume).

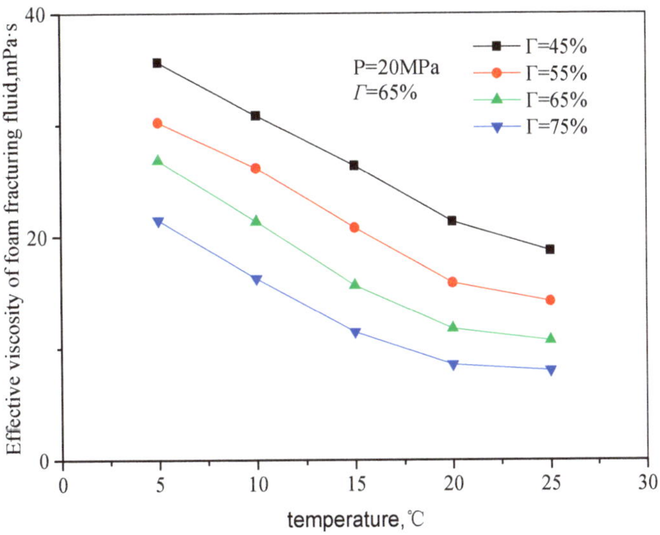

Figure 6. Variation curve of effective viscosity with temperature (unfoamed, shear rate 1000 s^{-1}).

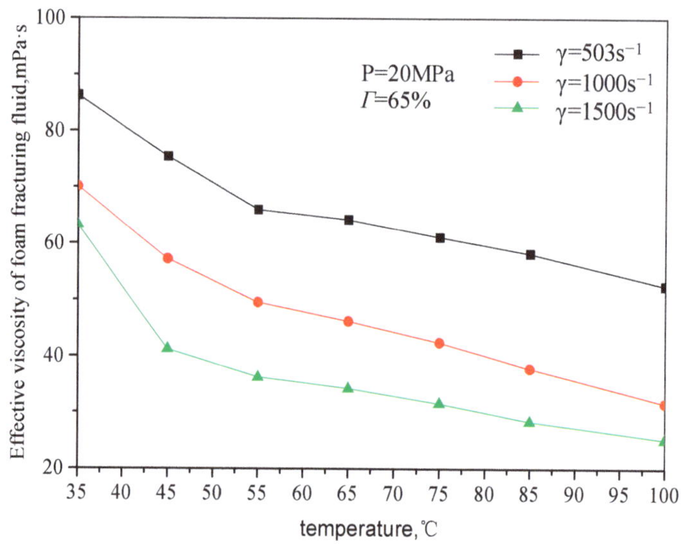

Figure 7. Variation curve of effective viscosity with temperature (foamed, foam mass is 65%).

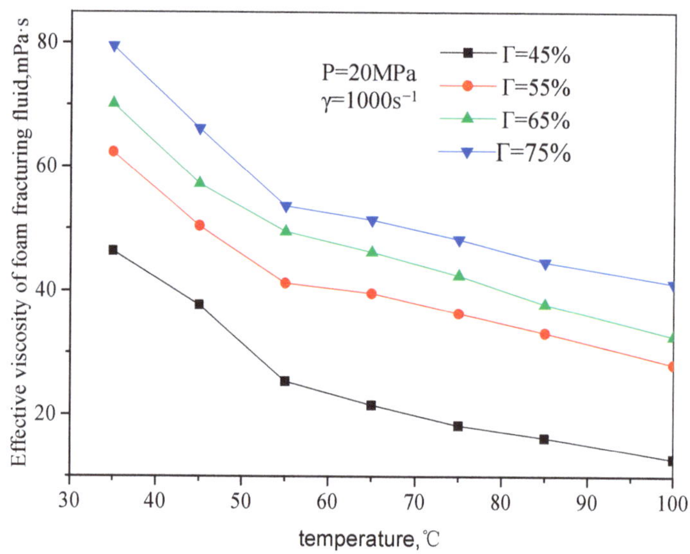

Figure 8. Variation curve of effective viscosity with temperature (foamed, the shear rate is 1000 s^{-1}).

2.1.4. Effect of Pressure on the Effective Viscosity of CO_2 Foam Gel Fracturing Fluid (Shear Rate)

Figures 9 and 10 are the curves of the effective viscosity of CO_2 foam gel fracturing fluid with a shear rate under different pressure conditions and under unfoamed and foamed conditions at 25 °C and 65 °C, respectively. It can be seen from Figure 9 that the effective viscosity of the unfoamed fracturing fluid system is affected very little by pressure, and the effective viscosity increases slightly with the increase in experimental pressure. The influence of pressure on the effective viscosity of fracturing fluid without foaming is mainly that pressure can effectively change the interaction between guar gum molecules, making the linear structure change and improving the stability of the linear structure, which shows that the increase in pressure strengthens the long-chain linear structure of guar gum, slows down the damage of shear to the linear structure, and makes the viscosity of the solution increase to a certain extent. Figure 10 shows the variation law of the effective viscosity of CO_2 foam gel fracturing fluid with pressure under the conditions of foaming at 65 °C, 55% foam mass, and 10 MPa, 20 MPa, and 30 MPa respectively. Similar to the unfoamed condition, the effective viscosity is less affected by pressure due to the pressure's influence on bubble size and distribution in the foam gel fracturing fluid. The results show that the diameter of bubbles in the foam system gradually decreases with the increase in pressure, and the higher the pressure, the more uniform the bubble size distribution. On the one hand, the stability of bubbles is increased, while on the other hand, the nonlinear interaction between bubbles is enhanced, and the viscosity of the foam gel fracturing fluid is improved as a whole.

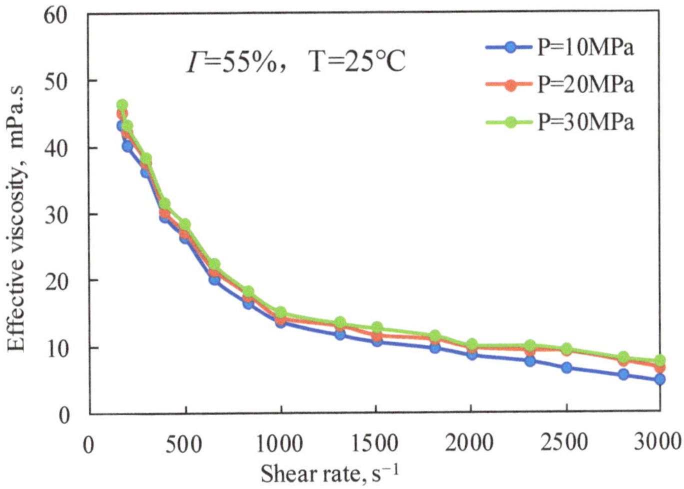

Figure 9. Variation of effective viscosity with shear rate under different pressures (unfoamed).

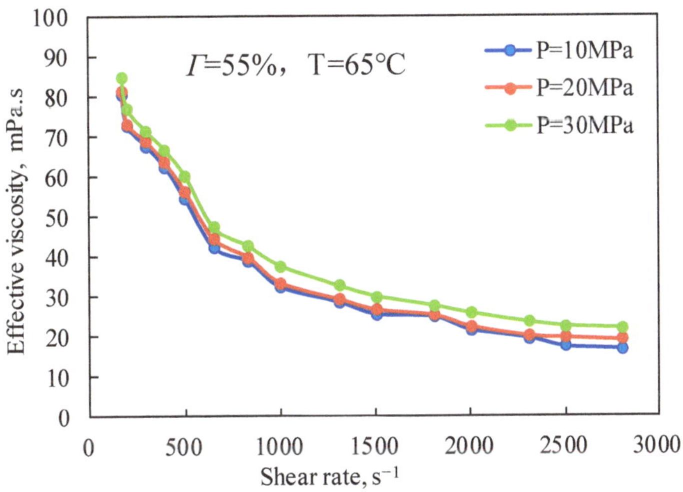

Figure 10. Variation of effective viscosity with shear rate under different pressures (foamed).

2.2. Changes in the Rheological Parameters of CO_2 Foam Gel Fracturing Fluid with Various Factors

Figures 11 and 12 are curves of the flow index and consistency index of fracturing fluid with CO_2 volume fraction when the temperature is 25 °C and the pressure is 10 MPa, 20 MPa, and 30 MPa, respectively. It can be seen from the figure that, when unfoamed, the flow index gradually increases with the increase in CO_2 volume fraction, while the consistency index decreases. Under experimental conditions, the change range of the flow index is 0.37−0.72; the variation range of the consistency index is 0.23−0.39.

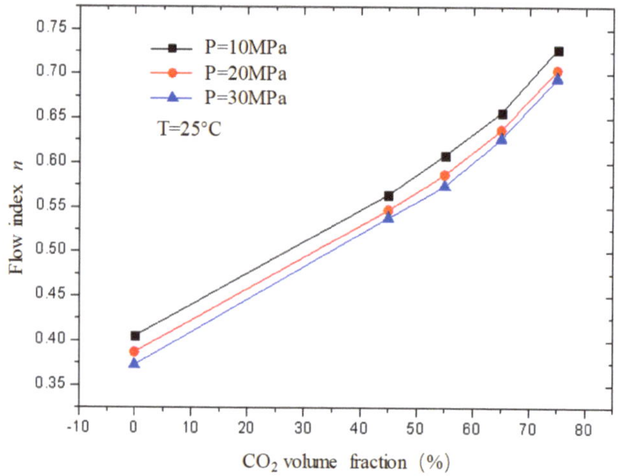

Figure 11. Variation curve of flow index n with CO_2 volume fraction under different pressures (unfoamed).

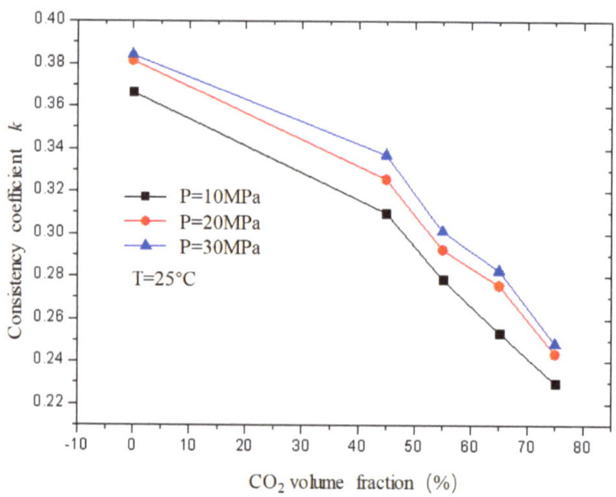

Figure 12. Variation curve of consistency index with CO_2 volume fraction under different pressures (unfoamed).

The flow index is a parameter used to describe the non-Newtonian property of the fluid. It can be seen from the figure that the flow index n of the unfoamed fracturing fluid is less than 1, indicating that the fluid is a shear-thinning non-Newtonian fluid, and the value of n is closer to 1 with the increase in CO_2 volume fraction, indicating that the fluid property of the fracturing fluid gradually changes to a Newtonian fluid with the increase in the CO_2 volume fraction, which is precise because the unfoamed CO_2 exists as a supercooled liquid similar to a Newtonian fluid, and its non-Newtonian property weakens with the increase in the CO_2 volume fraction.

2.2.1. Effect of Pressure on the Effective Viscosity of CO_2 Foam Gel Fracturing Fluid (Foam Quality)

Figures 13 and 14 are the curves of the flow index and consistency index of CO_2 foam gel fracturing fluid with foam quality when the temperature is 65 °C and the pressure is 10 MPa, 20 MPa, and 30 MPa, respectively. As can be seen from the figure, with the increase in foam quality, the flow index gradually decreases, and the consistency index increases, and from the perspective of the change range, it starts to increase substantially when the foam quality is 55%. This shows that the CO_2 foam gel fracturing fluid in the foaming stage still belongs to the shear-thinning non-Newtonian fluid, and with the increase in foam quality, the non-Newtonian fluid properties of the foam gel fracturing fluid gradually increase.

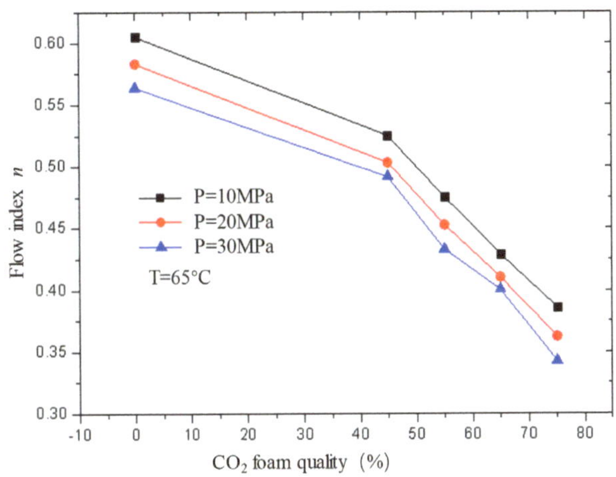

Figure 13. Variation curve of flow index n with foam quality under different pressures (foamed).

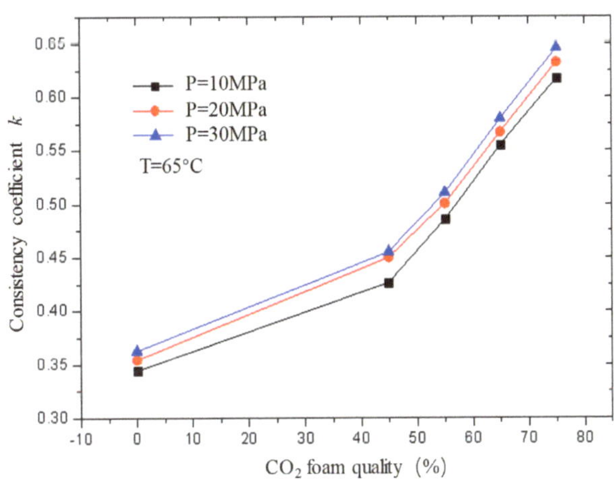

Figure 14. Variation curve of consistency index with foam quality under different pressures (foamed).

2.2.2. Effect of Temperature on the Rheological Parameters of CO_2 Foam Gel Fracturing Fluid

Figure 15 is the curves of the flow index and consistency index of fracturing fluid with temperature when the pressure is 20 MPa and the volume fraction of CO_2 is 65% without foaming. It can be seen from the figure that with the increase in temperature, the flow index of the fluid increases, and the consistency index decreases.

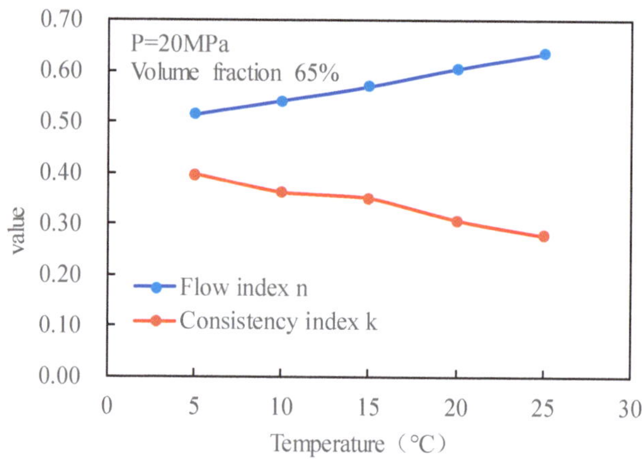

Figure 15. Variation curves of flow index *n* and consistency index k with temperature (unfoamed).

Figure 16 is the curves of the flow index and consistency index of the CO_2 foam gel fracturing fluid with temperature when the pressure is 20 MPa and the foam mass is 65%. It can be seen that the change rule of rheological parameters of foamed fracturing fluid with temperature is consistent with that of unfoamed fracturing fluid, which shows that with the increase in temperature, the flow index of fluid increases and the consistency index decreases. From the action mechanism, it is the same as that of the previous temperature on the viscosity characteristics of CO_2 foamed fracturing fluid.

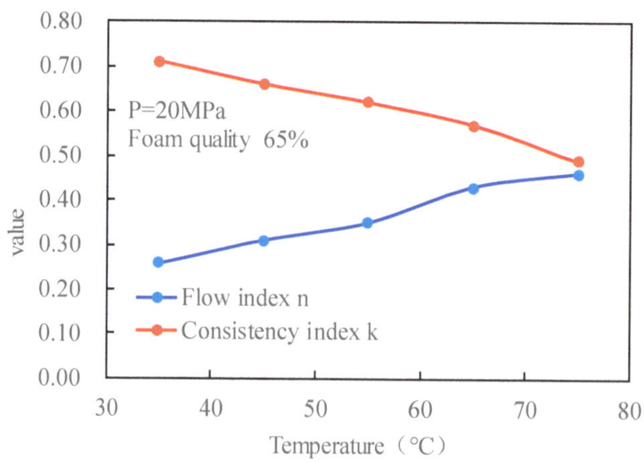

Figure 16. Variation curves of flow index *n* and consistency index *k* with temperature (foamed).

2.2.3. Effect of Pressure on the Rheological Parameters of CO_2 Foam Gel Fracturing Fluid

Figures 17 and 18 are curves of the rheological parameters of CO_2 foam gel fracturing fluid changing with pressure when the temperature is 65 °C and the foam quality is 45%,

55%, 65%, and 75% respectively. It can be seen from the figures that with the increase in pressure, the change of fluid flow index and consistency index is very small, which shows that the influence of pressure on the rheological properties of the foam gel fracturing fluid is almost negligible compared with temperature and foam quality. Meanwhile, the influence of pressure on the rheological properties of the foam gel fracturing fluid is negligible.

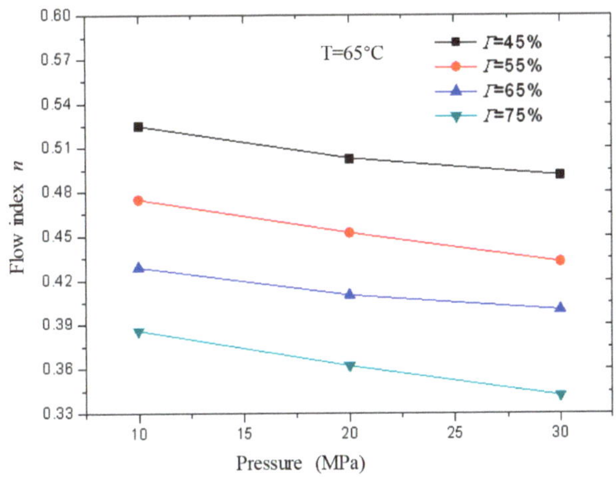

Figure 17. Variation curve of flow index n of foam gel fracturing fluid with pressure under foaming.

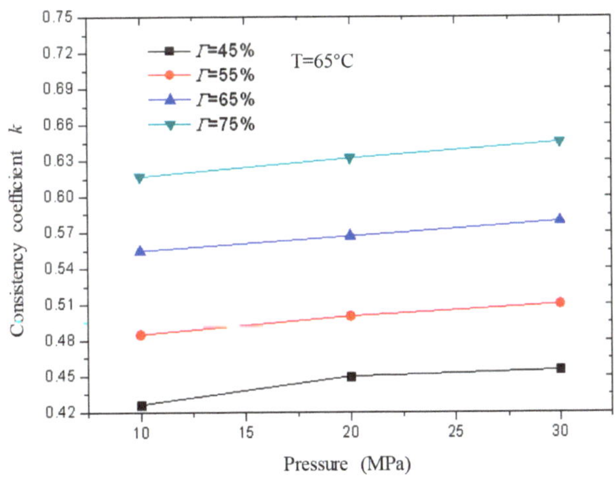

Figure 18. Variation curve of consistency index k of foam gel fracturing fluid with pressure under foaming.

2.3. Sensitivity Analysis

Sensitivity analysis is a method to analyze system stability in system analysis. There is a system, and its system characteristics, P, are mainly determined by n factors $a = \{a_1, a_2, \ldots, a_n\}$ and $P = f(a_1, a_2, \ldots, a_n)$. In a certain reference state, $a^* = \{a_1^*, a_2^*, \ldots, a_n^*\}$, the system characteristic is P^*. Let each factor change within its possible range, and analyze the trend and degree of the deviation of the system characteristic P from the benchmark state P^* due to these factors. This analysis method is called sensitivity analysis.

We define the dimensionless sensitivity function and sensitivity factor. That is, the ratio of the relative error of the system characteristic P to the relative error of the parameter a_k is defined as the sensitivity function $S_k(a_k)$ of the parameter a_k.

$$\frac{S_k(a_k)\left(\frac{|\Delta P|}{P}\right)}{\left(\frac{|\Delta a_k|}{a_k}\right)} = \left|\frac{\Delta P}{\Delta a_k}\right|\frac{a_k}{a_k}\ k = 1, 2, \ldots\ldots, n \quad (1)$$

When $|\Delta a_k|/a_k$ is small, $S_k(a_k)$ can be approximately expressed as:

$$S_k(a_k) = \left|\frac{d\ \varphi\ _k a_k}{da_k}\right|\frac{a_k}{P}\ k = 1, 2, \ldots\ldots, n \quad (2)$$

$S_k{}^*$, $k = 1, 2, \ldots, n$, is a set of dimensionless non-negative real numbers. The larger the $S_k{}^*$ value, the more sensitive P is to a_k in the reference state. Through the comparison of $S_k{}^*$, the sensitivity of system characteristics to various factors can be compared and evaluated.

To understand the sensitivity of the rheological properties of foam gel fracturing fluid to various factors, the above sensitivity analysis method was used to analyze the main factors affecting the rheological properties of foam gel fracturing fluid, and the sensitivity of each factor was compared. The characteristics of the system, that is, the rheological characteristics of foam gel fracturing fluid, are characterized by the effective viscosity of the system. The parameters for sensitivity analysis are shear rate, pressure, foam quality, and temperature. Observe the shape of the rheological curve and establish the functional relationship between effective viscosity and shear rate, temperature, and other parameters to obtain the sensitivity function, and then calculate the sensitivity factor. After analysis, the sensitivity values of each parameter are obtained (Table 1), which are ranked by size, followed by foam quality, temperature, shear rate, and pressure. It can be seen that among all the factors, foam quality and temperature are the main influencing factors. Therefore, the performance of CO_2 foam gel fracturing can be mainly regulated by the two parameters, which is helpful for application in shale gas reservoir fracturing.

Table 1. Sensitivity factors of influencing factors.

Influencing Factors	Shear Rate	Pressure	Foam Quality	Temperature
Sensitivity factor	0.685	0.092	0.973	0.735
Sensitivity grade	III	IV	I	II

3. Conclusions

The formula of the gel fracturing fluid was obtained through experimental optimization, which evaluated experimentally the viscosity, residue, surface tension, sand-carrying capacity, and interfacial tension of the gel fracturing fluid. The core damage rate of the gel fracturing fluid is less than 19%, the shear time is 90 min at 170 s^{-1} and 90 °C, and the viscosity of the fracturing fluid is >50 mPa·s.

The rheological properties of CO_2 foam gel fracturing fluid and its influencing factors were studied experimentally. For the foam gel fracturing fluid with CO_2 in supercooled liquid and at a supercritical state, the effects of temperature, pressure, shear rate, and foam quality on the rheological properties of the gel fracturing fluid were considered. The main conclusions are as follows:

In the unfoamed state, the effective viscosity of foam gel fluid decreases exponentially with the increase in shear rate, gradually decreases with the increase in CO_2 volume fraction, and the effective viscosity of fluid decreases with the increase in temperature. The effective viscosity is little affected by pressure. In the foaming state, the change rule of effective viscosity with shear rate is the same as that without foaming, the change rule with foam quality is opposite to that without foaming, and the effective viscosity–temperature characteristics of fluid are the same as that without foaming.

Without foaming, the foam quality increases, the flow index gradually increases, and the consistency index decreases. With the increase in temperature, the flow index of fluid increases, and the consistency index decreases. When foaming, the foam quality increases, the flow index gradually decreases, and the consistency index increases. When the foam quality is about 50%, a sudden change begins. With the increase in temperature, the flow index of fluid increases, and the consistency index decreases. With the increase in pressure, the flow index and consistency index of fluid change very little.

Based on the sensitivity analysis method, the influencing factors of the rheological behavior of CO_2 foam gel fracturing fluid are foam quality, temperature, shear rate, and pressure, in turn, which provides a theoretical basis for CO_2 foam fracturing technology.

4. Experiments and Methods

4.1. Experimental Optimization of Gel Fracturing Fluid

4.1.1. Formula Design

In recent years, the CO_2 foam gel fracturing fluid system, which has been applied successfully to reservoir stimulations, is a cross-linked fracturing fluid system with guanidine gum as the thickener.

The commonly used foam sealing and channeling systems are mainly composed of a foaming agent, foam stabilizer, and gas phase [32,33]. Therefore, guanidine gum fluid is used as the basic liquid phase in this study, the thickener and foaming agent are optimized by experiments, and the gel fracturing fluid that is needed to mix with liquid CO_2 in this research is formed. According to the characteristics of foam gel fracturing fluid, the high-speed mixing method (Waring Blender method) was selected to evaluate the foaming ability and foam stability of the foaming agent.

Single Agent Optimization

(1) Thickener optimization

To select a liquid that can form foam with good stability at a low dosage, the frequently used thickeners were evaluated and optimized by experiments, including CT5-7, CT5-7WI, hydroxypropyl guar gum (CMHPG), instant carboxymethyl cellulose (CMC), and polyanionic cellulose (PAC), and the results of the thickeners' foam stability are shown in Table 2. The experimental results showed that the foam stabilizing performance of CT5-7 is much better than the others. Therefore, the CT5-7 thickener is selected as the thickener of the gelled fracturing fluid formula.

Table 2. Experimental results of thickening agent foam stability.

Thickening Agent	Modulators	Foaming Agent	V_0/mL	Foam Quality/%	The Half-Life of Foam/h
0.3% CT5-7	0.3% CT5-7U	0.3% CT5-7B	280	70.4	90
0.5% CT5-7 WI			280	68.3	75
0.5% CMHPG			280	61.4	65
0.5% instant CMC			280	60.6	18
0.5% CMC			280	60.2	28
0.5% PAC			280	60.3	16

The CT5-7 is a thickener, which is prepared from acrylamide, vinylpyrrolidone, maleic acid, anionic functional monomer, initiator, inorganic salt, etc. It includes acrylamide 25~30%, vinylpyrrolidone 3~5%, maleic acid 4~6%, anionic functional monomer 5~7%, initiator 0.08~0.15%, potassium hydroxide 1~8%, inorganic salt 5~10%, and the rest is water.

(2) Foaming agent optimization

The foaming ability and foam half-life of 0.3% CT5-7B, CT5-7S, sodium dodecylbenzene sulfonate (ABS), alkyl betaine (DSB), CT5-7C and other foaming agents in CT5-7 thickener solution were experimentally evaluated. These are the foaming agent, and their main component is alcohols; CT5-7B is fusel, CT5-7S is methyl isobutyl methanol, and CT5-7C is triethoxybutane. The results are shown in Table 3. The calculation formula of foam quality Γ is:

$$\Gamma = (V_0 - 100)/V_0 \quad (3)$$

where V_0 is the volume of gel fracturing fluid, mL.

Table 3. Evaluation of foaming and foam stabilizing properties of the foaming agent.

Thickener	Foaming Agent	V_0/mL	Foam Quality/%	The Half-Life of the Foam/h
0.5% CT5-7	0.3% CT5-7B	270	60.6	110
	0.3% CT5-7S	270	64.9	120
	0.3% DSB	270	64.3	95
	0.3% CT5-7C	270	64.2	100
	0.3% ABS	270	63.7	96

The foaming power and foam stability data of five foaming agents were analyzed, and CT5-7S with good foaming power and foam stability was selected as the foaming agent of the foam gel fracturing fluid formula.

Optimization of a Single Dosage

(1) Thickener dosage

To optimize the amount of thickener in gel fracturing fluid, an experimental evaluation was carried out on the sand suspension (settling velocity of 40–70 mesh quartz sand) of fracturing fluid with 0.2, 0.3, 0.4, and 0.5% of thickener at 30~90 °C, and the experimental results are shown in Table 4.

Table 4. Experimental results of gel fracturing fluid-suspended sand at different temperatures (mm/s).

CT5-7 Dosage/%	15 °C (Room Temperature)	30 °C	50 °C	70 °C	90 °C
0.2	12.5	25.0	33.3	50.0	100.0
0.3	0.70	1.29	2.32	5.56	9.52
0.4	0.019	0.057	0.11	0.70	0.98
0.5	Almost motionless	Almost motionless	0.046	0.16	0.46

The experimental results show that the sedimentation rate of quartz sand in the gel fracturing fluid increases with the increase in temperature. When the amount of CT5-7 is 0.2% and 90 °C, the suspension capacity of quartz sand is poor. When the amount is 0.3–0.4%, the sedimentation rate of quartz sand is 1–10 mm/s at 90 °C, but when the amount is increased to 0.5%, the quartz sand hardly sinks.

The analysis shows that after the gel fracturing fluid forms foam, its apparent viscosity increases, and the ability to suspend and carry solid particles increases significantly.

(2) Stability of foam with different thickener dosage

The stability (half-life) of gel fracturing fluid with thickener dosages of 0.2, 0.3, 0.4%, and 0.5% was experimentally evaluated at room temperature and up to 90 °C The dosage of 0.3% CT5-7S foaming agent was 0.3%, as shown in Table 5. It can be seen from Table 5 that with the increase in temperature, the half-life of foam decreases. At a dosage of 0.2%, the half-life is 0.2 h at 90 °C; at a dosage of 0.3~0.4%, the half-life of foam is 1.4~2.5 h at 90 °C; and at a dosage of 0.5%, the half-life of foam is 4.2 h at 90 °C.

Table 5. The half-life of foam in gel fracturing fluid at different temperatures.

CT5-7/%	15 °C (Room Temperature)		30 °C		50 °C		70 °C		90 °C	
	Foam Quality/%	Half-Life/h	Foam Quality/%	Half-Life/h	Foam Quality/%	Half-Life/h	Foam Quality/%	Half-Life/h	Foam Quality/%	Half-Life/h
0.2	75.21	89	79.13	75	79.62	40	82.32	21	93.13	0.2
0.3	71.60	92	76.27	76	76.76	60	80.53	37	92.25	1.4
0.4	69.57	109	71.43	105	74.21	70	79.16	56	91.01	2.5
0.5	66.04	133	67.03	119	71.17	90	77.45	68	89.32	4.2

Based on the experimental data on viscosity, sand suspension, and the half-life of foam in gel fracturing fluid, the thickener dosage of gel fracturing fluid with a foam formula is determined to be 0.5%.

(3) Determination of Foaming agent dosage

To determine the dosage of the foaming agent, the half-life of foam when the dosage of CT5-7S foaming agent is 0.1, 0.3, and 0.5%, the experimental temperature is 30 °C, and the experimental data are shown in Table 6. It can be seen from Table 6 that as the dosage of CT5-7S foaming agent increases from 0.1% to 0.5%, for CT5-7 thickener, the half-life of foam shows an increasing trend. Considering the performance and cost factors, the dosage of the CT5-7S foaming agent is determined to be 0.3%.

Table 6. The half-life of foam in gel fracturing fluid with different dosages of CT5-7S.

Foaming Agent	Thickening Agent	V_0/mL	Foam Quality/%	The Half-Life of Foam/h
0.1% CT5-7S		288	63.5	90
0.3% CT5-7S	0.5% CT5-7	292	65.7	110
0.5% CT5-7S		292	65.8	113

The formula of Gel Fracturing Fluid with Foam

Through the experimental evaluation of the single agent of gel fracturing fluid with foam such as foaming agent and thickener, the experimental evaluation data analysis of foam half-life, foam quality, and the viscosity of gel fracturing fluid is carried out, and the formula of gel fracturing fluid is determined: 0.5% CT5-7 thickener + 0.3% CT5-7S foaming agent + 0.3% CT5-7D high-temperature stabilizer + 0.3% CT5-7U regulator.

4.1.2. Performance Testing

Gel Breaking Performance

The gel fracturing fluid volume of 1000 mL was prepared in the mixer, and then the 600 ppm gel breaker (ammonium persulfate) was added, which may break the gel fracturing fluid at 90 °C. The gel fracturing fluid after the gel was broken was centrifuged and dried with the separated residue at 105 °C ± 1 °C. Then, the residue content in the gel fracturing fluid was determined. The results are shown in Table 7.

Table 7. Gel breaking performance of gel fracturing fluid.

Breaker Viscosity /mPa·s	Residue Content/mg·L^{-1}	Surface Tension/mN/m	Surface Tension/mN/m
2.5	1.3	1.6	25.1

The residue content in the gel breaker is calculated according to the following formula:

$$\eta = \frac{m}{V} \times 1000 \tag{4}$$

where η is the residue content in the gel breaker, mg/L, m is the residue mass in mg, and V is the volume of gel fracturing fluid in mL.

Sand Setting Performance at the Static Station

It is necessary to test the sand carrying performance of gel fracturing fluid by a static suspended sand experiment. The gel fracturing fluid with foam quality (Q = 65%) was prepared according to the liquid formula and 200 mL of fluid was poured into the beaker and then placed in a 90 °C water bath at a constant temperature for 20 min. Then, the liquid was poured into the mixer, and 40–70 mesh quartz sand was added according to the 40% sand ratio and stirred evenly. Then, the mixed liquid with sand was poured into a 250 mL measuring cylinder and put into an oven at 90 °C (formation temperature is usually 85~100 °C). The volume of clear liquid separated from the upper layer was recorded at regular intervals. The static suspended sand test results of the gel fracturing fluid are shown in Figure 19. The experiment shows that after the gel fracturing fluid forms a stable foam, the proppant is evenly dispersed in the foam fluid. Due to the interface effect between foam, the proppant has a wrapping and supporting effect. From room temperature up to 90 °C, the sand mixing fluid remains uniform for 3 h, there is no obvious stratification phenomenon, and the sand carrying performance is good.

Figure 19. The static suspended sand test results of gel fracturing fluid.

Formation Damage Evaluation

The core of the shale reservoir is used to evaluate the damage performance of gel fracturing fluid. The test results are shown in Table 8. It can be seen that the damage rate of foam gel fracturing fluid to shale is less than 19%.

Table 8. Core damage performance of gel fracturing fluid.

Core No.	Gas phase Permeability before Fracturing Fluid Injection/10^{-3} μm^2	Gas Phase Permeability after Fracturing Fluid Injection/10^{-3} μm^2	Damage Rate/%	Average Value/%
1#	0.12863	0.10821	15.87	15.05
2#	0.09756	0.08602	11.83	
3#	0.03584	0.03023	15.65	
4#	0.02424	0.02016	16.83	

Rheological Properties

Evaluation method: take 70 mL of gel fracturing fluid, add regulator according to the proportion of 0.3% (*v/v*), adjust the pH value of the liquid to 5.8, then add foaming agent according to the design proportion, mix it evenly, transfer it to the closed system of the RS6000 high-temperature rheometer, connect the CO_2 gas source to pressurize 10 bar and ensure that the gel fracturing fluid is in the CO_2-saturated state, and test the temperature resistance and shear resistance of the fracturing fluid. The experimental results show that the fracturing fluid can still maintain a high apparent viscosity after a long time of shearing at 90 °C at a shear rate of 170 s^{-1} and the apparent viscosity is greater than 50 mPa·s, indicating that it has excellent temperature and shear resistance.

The formula of the gel fracturing fluid is obtained through experimental optimization, and the viscosity, static sand setting performance, and rheological performance of the foam gel fracturing fluid are experimentally evaluated. The experimental results show that the viscosity of the fracturing fluid is 2.5 mPa·s (after gel breaking and at a shear rate of 500 s^{-1}), the residue content is 1.3 mg/L, the surface tension is 25.1 mN/m, and the interfacial tension is 1.6 mN/m. The 40–70 mesh quartz sand commonly used in shale gas fracturing is used for static sand setting experiments and the sand ratio is 40%. The static sand carrying experiment was carried out. The sand-carrying fluid had good flow performance, and the proppant showed no settlement in 3 h. The core of the shale reservoir was selected for the damage evaluation experiment of the foam gel fracturing fluid. The test results are shown in Figure 20. It was observed that the core damage rate of foam gel fracturing fluid is less than 19%, the shear time is 90 min at 170 s^{-1} and 90 °C, and the viscosity of the fracturing fluid is >50 mPa·s, and the temperature resistance and shear resistance are excellent.

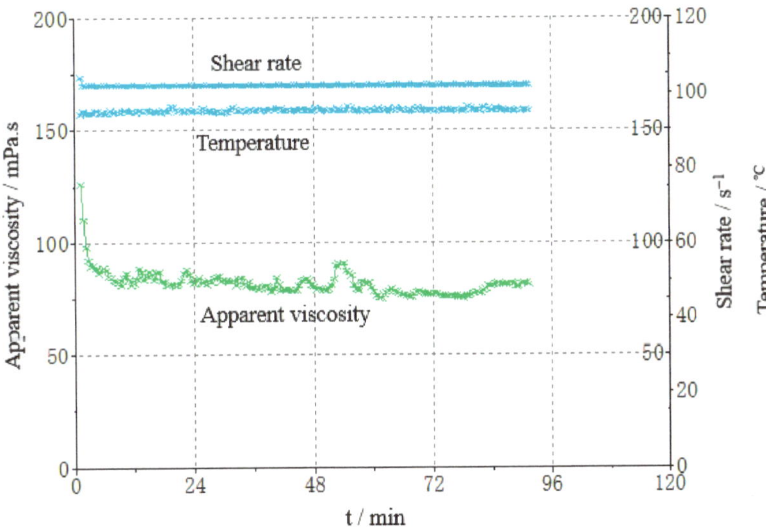

Figure 20. Evaluation of rheological properties of gel fracturing fluid with foam.

4.2. Preparation Methods of Gel Fracturing Fluid with Different CO_2 Foam Mass

According to the formula of the gel fracturing fluid optimized in Section 2.1, the gel fracturing fluid was obtained in the laboratory and stirred well to obtain the liquid-phase CO_2 foam gel fracturing fluid (Figure 21).

The proportion of gas phase in foam gel fracturing fluid is usually described by foam quality or foam dryness, which indicates the phase's percentage in the total volume of the foam gel fracturing fluid; that is, the gas volume contained in a unit volume of foam gel fracturing fluid.

Usually expressed by Γ:

$$\Gamma = \frac{V_G}{V_G + V_L} \times 100\% = \frac{V_G}{V_F} \times 100\% \tag{5}$$

where: Γ is the foam quality; V_G is the gas volume, m^3; V_L is the liquid volume, m^3; and V_F is the total volume of the foam, m^3.

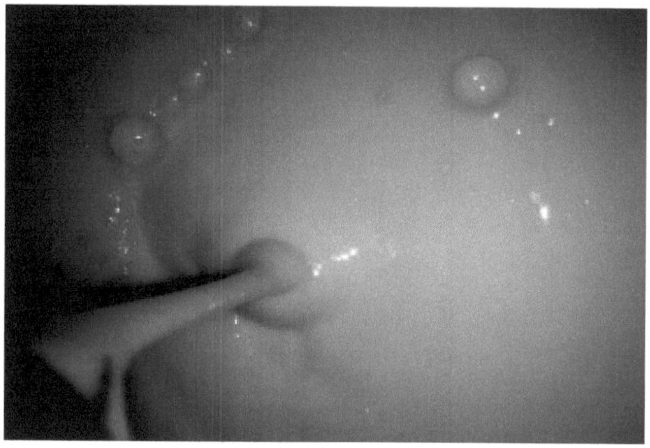

Figure 21. CO_2 foam gel fracturing fluid.

Because CO_2 is in a supercooled liquid state and supercritical state in this experiment, the foam quality is defined as the percentage of CO_2 volume in the whole foam volume under certain temperature and pressure conditions.

4.3. Experimental Principle of the Rheological Properties of CO_2 Foam Gel Fracturing Fluid

For CO_2 foam gel fracturing fluids, within the range of practical shear rate, they are close to power-law fluids, and a set of n and k values can be used to characterize their rheological characteristics.

A thin-tube rheometer calculates the relationship between shear stress and shear rate through the measured pressure drop and flow rate of fluid in the tube, to determine the rheological characteristics of the fluid. The flow of fluid in a narrow tube better meets the following conditions: ① viscous laminar flow; ② constant flow; ③ uniform flow; ④ no-slip pipe wall.

The basic formula for the laminar flow of viscous fluid in a circular tube is:

$$\overline{U} = \frac{D}{2\tau_w^3} \int_0^{\tau_w} f(\tau)\tau^2 d\tau \tag{6}$$

where $\overline{U}$ is the average velocity of the fluid in the pipe, D is the diameter of the pipe, and τ_w is the wall shear stress.

Transforming Formula (6) into Formula (7):

$$\frac{1}{4}\frac{8\overline{U}}{D}\tau_w^3 = \int_0^{\tau_w} f(\tau)\tau^2 d\tau \tag{7}$$

Formula (7) is the basic formula for the rheometer.

For power-law fluid, the constitutive formula is:

$$f(\tau) = \left(\frac{\tau}{k}\right)^{\frac{1}{n}} \tag{8}$$

Substituting Formula (8) into Formula (7) of tube rheometer to obtain Formula (9):

$$\frac{1}{4}\frac{8\overline{U}}{D}\tau_w^3 = \int_0^{\tau_w}\left(\frac{\tau}{k}\right)^{\frac{1}{n}}\tau^2 d\tau \tag{9}$$

Integrate the above formula to obtain Formula (10):

$$\tau_w = k\left(\frac{8\overline{U}}{D}\right)^n\left(\frac{3n+1}{4n}\right)^n \tag{10}$$

Take the logarithm on both sides of the formula and obtain Formula (11):

$$\lg\tau_w = \lg k\left(\frac{3n+1}{4n}\right)^n + n\lg\left(\frac{8\overline{U}}{D}\right) \tag{11}$$

Type, the wall shear stress τ_w is:

$$\tau_w = \frac{\Delta pD}{4L} \tag{12}$$

After the pressure drop and flow, Q is measured, and the relationship curve $\lg(\Delta pD/4L) \sim \lg(8\overline{U}/D)$ is compiled, as shown in Figure 22. From the slope $tg\theta$ and intercept B of the straight line, the rheological characteristic parameters n and k of the power-law fluid can be determined.

$$n = tg\theta \tag{13}$$

$$k = B/\left(\frac{3n+1}{4n}\right)^n \tag{14}$$

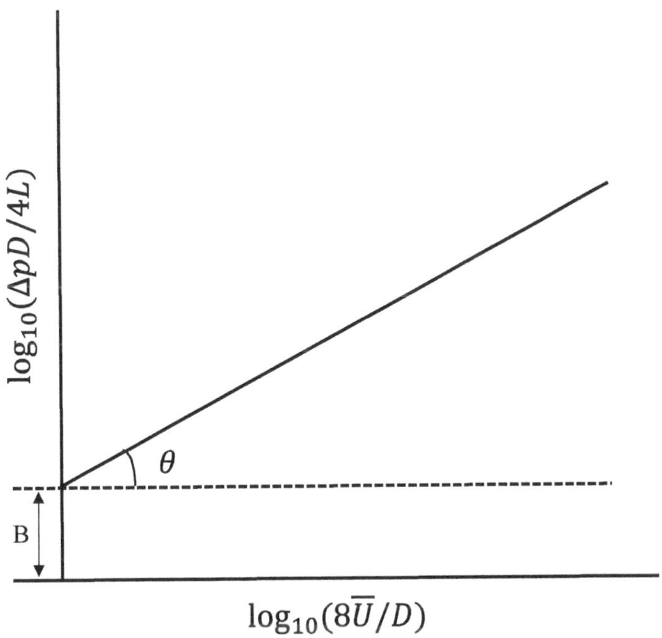

Figure 22. Power-law fluid flow curve (logarithmic coordinates). (Reprinted/adapted with permission from Ref. [9], 2014, copyright X. Sun et al.).

Author Contributions: M.W.: Conceptualization, funding acquisition, project administration, resources, and software. W.W. and S.L.: Data curation, formal analysis, methodology, writing—original draft, and writing—review and editing. S.C.: Project administration, resources. T.L.: Investigation, methodology, software, and visualization. G.N. and Y.F.: Conceptualization, funding acquisition, methodology. W.Z.: Investigation, methodology, project administration. All authors have read and agreed to the published version of the manuscript.

Funding: This work was supported by the National Natural Science Foundation of China: Study on the dynamic characteristics of methane/carbon dioxide in shale heterogeneous reservoir under multi-field coupling (Program No. 41772150).

Institutional Review Board Statement: Not applicable.

Informed Consent Statement: Not applicable.

Data Availability Statement: Not applicable.

Acknowledgments: The authors are grateful for the support of the National Natural Science Foundation of China. Thanks to Xiangrong Luo of Xi'an University of Petroleum for his guidance and suggestions, as well as reviewers and editors for their careful review of this manuscript.

Conflicts of Interest: The authors declare no conflict of interest.

References

1. Chen, Y.; Pope, T.L. Novel CO_2-emulsified viscoelastic surfactant fracturing fluid system. In Proceedings of the SPE European Formation Damage Conference, Scheveningen, The Netherlands, 25–27 May 2005. SPE Paper 94603.
2. Reidenbach, V.G.; Harris, P.C.; Lee, Y.N.; Lord, D.L. Theological study of foam fracturing fluids using nitrogen and carbon dioxide. *SPE Prod. Eng.* **1986**, *1*, 34–41.
3. Craft, J.R.; Waddell, S.P.; McFatridge, D.G. CO_2-foam fracturing with methanol successfully stimulates canyon gas sand. *SPE Prod. Eng.* **1992**, *7*, 219–225. [CrossRef]
4. Ahmed, S.; Hanamertani, A.S.; Hashmet, M.R. CO_2 foam as an improved fracturing fluid system for the unconventional reservoir. In *Exploitation of Unconventional Oil and Gas Resources-Hydraulic Fracturing and Other Recovery and Assessment Techniques*; IntechOpen Limited: London, UK, 2019.
5. Ahmed, S.; Alameri, W.; Ahmed, W.W.; Khan, S.A. Rheological behavior of $scCO_2$-Foam for improved hydrocarbon recovery: Experimental and deep learning approach. *J. Petrol. Sci. Eng.* **2021**, *203*, 108646. [CrossRef]
6. Zhu, D.Y.; Bai, B.J.; Hou, J.R. Polymer gel systems for water management in high-temperature petroleum reservoirs: A chemical review. *Energy Fuels* **2017**, *31*, 13063–13087. [CrossRef]
7. Yang, J.B.; Bai, Y.R.; Sun, J.S.; Lv, K.H.; Han, J.L.; Dai, L.Y. Experimental study on physicochemical properties of a shear thixotropic polymer gel for lost circulation control. *Gels* **2022**, *8*, 229. [CrossRef]
8. Wang, L.-L.; Wang, T.-F.; Wang, J.-X.; Tian, H.-T.; Chen, Y.; Song, W. Enhanced oil recovery mechanism and technical boundary of gel foam profile control system for heterogeneous reservoirs in Changqing. *Gels* **2022**, *8*, 371. [CrossRef]
9. Sun, X.; Liang, X.B.; Wang, S.Z.; Lu, Y. Experimental study on the rheology of CO_2 viscoelastic surfactant foam fracturing fluid. *J. Petrol. Sci. Eng.* **2014**, *119*, 104–111. [CrossRef]
10. Li, Y.; DiCarlo, D.; Li, X.; Barati, R.; Liang, J.T. An experimental study on application of nanoparticles in unconventional gas reservoir CO_2 fracturing. *J. Petrol. Sci. Eng.* **2015**, *133*, 238–244. [CrossRef]
11. Ahmed, S.; Elraies, K.A.; Hanamertani, A.S.; Hashmet, M.R.; Shafian, S.R.; Hsia, I.C. Investigation of carbon dioxide foam performance utilizing different additives for fracturing unconventional shales. In Proceedings of the Abu Dhabi International Petroleum Exhibition & Conference, Abu Dhabi, United Arab Emirate, 11–14 November 2019.
12. Barati, R.; Liang, J.T. A review of fracturing fluid systems used for hydraulic fracturing of oil and gas wells. *J. Appl. Polym. Sci.* **2014**, *131*, 2765. [CrossRef]
13. Wanniarachchi, W.; Ranjith, P.; Perera, M. Shale gas fracturing using foam-based fracturing fluid: A review. *Environ. Earth Sci.* **2017**, *76*, 91. [CrossRef]
14. Verma, A.; Chauhan, G.; Baruah, P.P.; Ojha, K. Morphology, rheology and kinetics of nano-silica stabilized gelled foam fluid for hydraulic fracturing application. *Ind. Eng. Chem. Res.* **2018**, *57*, 13449–13462. [CrossRef]
15. Arezoo, S.E.; Ibrahim, A.F.; Hisham, A.; Nasr, E.D. Mobility control using nanoparticle-stabilized CO_2 foam as a hydraulic fracturing fluid. In Proceedings of the SPE Europec Featured at 79th EAGE Conference and Exhibition, Paris, France, 12–15 June 2017. SPE-185863-MS.
16. Wang, S.Z.; Luo, X.R.; Jing, Z.F. Rheological properties of BCG-CO_2 fracturing fluid for shale gas. *Earth Environ. Sci.* **2018**, *186*, 12026. [CrossRef]
17. Jing, Z.F.; Feng, C.C.; Wang, S.Z.; Xu, D.H.; Xu, G.X. Effect of foam quality on flow behavior of liquid CO_2-based foam fracturing fluid stabilized by hydrofluoroether. *J. Petrol. Sci. Eng.* **2017**, *159*, 710–716. [CrossRef]

18. Ahmed, S.; Elraies, K.A.; Hashmet, M.R.; Hanamertani, A.S. Viscosity models for polymer-fre CO_2 foam fracturing fluid with the effect of surfactant concentration, salinity and shear rate. *Energies* **2017**, *10*, 265. [CrossRef]
19. Ahmed, S.; Elraies, K.A.; Tan, I.M.; Hashmet, M.R. Experimental investigation of associative polymer performance for CO_2 foam enhanced oil recovery. *J. Petrol. Sci. Eng.* **2017**, *157*, 971–979. [CrossRef]
20. Amit, V.; Geetanjali, C.; Keka, O. Synergistic effects of polymer and bentonite clay on rheology and thermal stability of foam fluid developed for hydraulic fracturing. *Asia-Pac. J. Chem. Eng.* **2017**, *12*, 872–883.
21. Li, C.; Huang, Y.Q.; Sun, X.; Gao, R.M.; Zeng, F.H.; Tontiwachwuthikul, P.; Liang, Z.W. Rheological properties study of foam fracturing fluid using CO_2 and surfactant. *Chem. Eng. Sci.* **2017**, *170*, 720–730. [CrossRef]
22. Edrisi, A.; Kam, S.I. A new foam model in pipes for drilling and fracturing application. *SPE J.* **2014**, *19*, 576–585. [CrossRef]
23. Fei, Y.; Pokalai, K.; Johnson, R.; Gonzalez, M.; Haghighi, M. Experimental and simulation study of foam stability and the effects on hydraulic fracture proppant placement. *J. Nat. Gas Sci. Eng.* **2017**, *46*, 544–554. [CrossRef]
24. Gu, M.; Mohanty, K. Rheology of polymer-free foam fracturing fluids. *J. Petrol. Sci. Eng.* **2015**, *134*, 87–96. [CrossRef]
25. Luo, X.; Wang, S.; Wang, Z.; Jing, Z.; Lv, M. Experimental research on rheological properties and proppant transport performance of GRF–CO_2 fracturing fluid. *J. Petrol. Sci. Eng.* **2014**, *120*, 154–162. [CrossRef]
26. Cong, Z.Y.; Li, Y.W.; Pan, Y.S.; Liu, B.; Shi, Y.; Wei, J.G.; Li, W. Study on CO_2 foam fracturing model and fracture propagation simulation. *Energy* **2022**, *238*, 121778. [CrossRef]
27. Luca, G.; Ulrik, V. *An Overview of Hydraulic Fracturing and Other Formation Stimulation Technologies for Shale Gas Production—Update 2015*; Publications Office of the European Union: Luxembourg, 2015.
28. Fu, C.; Liu, N. Rheology and stability of nanoparticle-stabilized CO_2 foam under reservoir conditions. *J. Pet. Sci. Eng.* **2021**, *196*, 107671. [CrossRef]
29. Li, Q.; Wang, F.; Forson, K.; Guo, Y.; Ma, L.; Wang, Q.H. Affecting analysis of the rheological characteristic and reservoir damage of CO_2 fracturing fluid in low permeability shale reservoir. *Environ. Sci. Pollut. Res.* **2022**, *29*, 37815–37826. [CrossRef] [PubMed]
30. Kadafur, I.; BinGhanim, A.; Aljawad, M.S.; Kamal, M.S.; AlYousef, Z.; Mahmoud, M. Rheological study of CO_2 foamed chelating stimulation fluids under harsh reservoir conditions. *J. Pet. Sci. Eng.* **2022**, *208 Pt B*, 109201. [CrossRef]
31. Tariq, Z.; BinGhanim, A.; Aljawad, M.S.; Kamal, M.S.; Mahmoud, M.; AlYousef, Z. AI-driven foam rheological model based on HPHT foam rheometer experiments. *J. Pet. Sci. Eng.* **2022**, *213*, 110439. [CrossRef]
32. Li, X.; Pu, C.S.; Chen, X. A novel foam system stabilized by hydroxylated multiwalled carbon nanotubes for enhanced oil recovery: Preparation, characterization, and evaluation. *Colloids Surf. A Phys. Eng. Asp.* **2022**, *632*, 127804. [CrossRef]
33. Sheng, Y.J.; Peng, Y.C.; Zhang, S.W.; Guo, Y.; Ma, L.; Wang, Q.H.; Zhang, H.L. Study on thermal stability of gel foam co-stabilized by hydrophilic silica nanoparticles and surfactants. *Gels* **2022**, *8*, 123. [CrossRef]

gels

MDPI

Article

Simulated Investigation in Wormhole Expansion Law of Gelling Acid Etching and Its Influencing Factors in Deep Carbonate Reservoirs

Mingwei Wang [1,2], Wen Zhou [1], Song Li [3,*] and Wen Wu [4]

1 State Key Laboratory of Oil and Gas Reservoir Geology and Exploitation, Chengdu University of Technology, Chengdu 610059, China; 201999010137@swpu.edu.cn (M.W.); zhouwen62@cdut.edu.cn (W.Z.)
2 School of Oil & Natural Gas Engineering, Southwest Petroleum University, Chengdu 610500, China
3 Engineering Research Institute of PetroChina Southwest Oil, Gasfield Company, Chengdu 610017, China
4 Southwest Oil and Gas Field Company of CNPC, Chengdu 610041, China; wuwen2016@petrochina.com.cn
* Correspondence: li_song@petrochina.com.cn

Abstract: Acidizing with gelling acid is the key technology in developing a carbonate reservoir successfully. It is difficult for the laboratory to carry out the radial displacement experiment with a large rock core. It is necessary to establish the gelling acid wormhole expansion model under the radial conditions, simulate the gelling acid wormhole expansion law under the radial conditions, optimize the construction parameters, and provide the basis for the optimal design of carbonate reservoir matrix acidizing. The research objective is to simulate the gelling acid etching wormhole expansion in a deep carbonate reservoir and make clear its influencing factors, which are helpful for reservoir stimulation. The mathematical model of gelling acid wormhole expansion was established, considering the influence of pore microscopic characteristics on acid flow and acid rock reaction. The simulation results indicated that viscosity, surface reaction rate, and hydrogen ion diffusion coefficient have different effects on gelling acid etching wormhole. The spatial distribution of pores determines the trend of gelling acid solution and thus the shape of the armhole. Perforation completion has a significant impact on the expansion of gelling acid etching wormhole. The wormhole extends forward along the perforation hole, and perforation increases the length of the wormhole. This wormhole expansion law is very suitable in situations where a deep carbonate reservoir is needed for gelling acid fracturing.

Keywords: reservoir stimulation; gelled acidizing; wormhole propagation; carbonate rocks; simulation

Citation: Wang, M.; Zhou, W.; Li, S.; Wu, W. Simulated Investigation in Wormhole Expansion Law of Gelling Acid Etching and Its Influencing Factors in Deep Carbonate Reservoirs. *Gels* **2022**, *8*, 470. https://doi.org/10.3390/gels8080470

Academic Editor: Mario Grassi

Received: 25 June 2022
Accepted: 21 July 2022
Published: 27 July 2022

1. Introduction

Gelling acidizing is carried out under the radial conditions of the wellbore. It is difficult for the laboratory to carry out the radial displacement experiment with a large rock core. It is necessary to establish the gelling acid wormhole expansion model under the radial conditions, simulate the gelling acid wormhole expansion law under the radial conditions, optimize the construction parameters, and provide the basis for the optimal design of matrix gelling acidizing [1].

At present, there are a capillary model, network model, single pore model, two-scale model, and lattice Boltzmann model [2–5], they also investigate the dynamics of model polymer networks formed by the condensation of linear poly precursor and PDCA ligand and the subsequent complexation with different metal ions at various pH values and oxidation states [6]. Huang et al. (2000) [7] proposed a single wormhole model to represent the wormhole via a cylindrical tube and considered the effects of fluid leakage and reaction kinetics on wormhole growth. Panga et al. (2004, 2005) [8,9] established the two-scale continuous model to explore the wormhole's initiation and propagation in acidizing carbonate reservoirs. The model considered the effect of reaction rate regime, wormhole density, and dimension on wormhole expansion. Cohen et al. (2007) [10] extended this model to radial flow to study the effect of flow geometry on the PVBT curve. They showed that the optimum injection rate in radial flow is higher than that of linear flow. Akanni and Nasr-El-Din (2015, 2016) [11–13] confirmed that gelling acidizing can reduce the reservoir skin factor by injecting gelling acid, which can consequently be created in the wormholes. The wormhole models were developed to simulate the wormhole patterns, which were observed in the experiment and account for physical phenomena of acidizing such as diffusion, convection, pore dissolution, and rock heterogeneities very well. They were also extended to radial coordinate or spatially correlated pore distributions (Liu et al., 2012; Zhang et al., 2014; Mou et al., 2015) [14–16]. Qiu et al. (2018) [17] completed a series of radial acid injection and linear acid flow experiments, which are conducted at various injection rates while maintaining the same operating conditions. It is found that the pore volume to breakthrough values of the linear acid flow case is lower for the radial acidizing case. QI Dan et al. (2019) [18] established a calculation model of wormhole to analyze wormholes quantitatively, optimize assumed gelling acid volume, and pumping rate and obtain the best production rate, which can be used to describe the method of the fractal. sensitivity analysis of influencing parameters and field acidification calculation was carried out. L Wang et al. (2019) [19] conducted research on wormhole propagation under radial flowing conditions in naturally fractured carbonates, natural fracture models were established by using the Monte Carlo method, and extensive numerical simulation was conducted to study wormhole characteristics considering the effects of natural fractures. N.Z. Liu et al. (2016) [20] developed a novel model to simulate wormhole propagation through VES acidizing. The simulation results showed that the injection velocity of VES acid was a significant factor for wormhole propagation. They also considered that the diversion effectiveness of wormhole was determined by the effect of both viscosifying and wormhole. W Wei et al. (2020) [21] established a two-scale continuum model in a 3D compositional reservoir simulator considering fractures by EDFM (Embedded Discrete Fracture Model), which describes convection and dispersion while IPhreeqc is responsible for the dissolution calculation both in the matrix and on the fracture surface.

N Qi et al. (2019) [22] studied the wormholes of the two-scale continuum model and pseudo-fracture model considering different fractures, and it is found that the gelling acid can be concentrated and accelerated penetrating the formation if fractures parallel to the injection direction. The effect of straight fractures and arc fractures on wormholes was defined. Mustafa et al. (2022) [23] established acid efficiency curves which included different rock lithologies (chalk, limestone, and dolomite) that react with gelling acid at different injection rates. The rock surface hardness at ambient pressure was measured by the impulse hammer technique, and the dynamic Young's modulus and Poisson ratio were measured through acoustic measurements at high confining pressures. Ghommem et al. (2015) [24] developed and validated a predictive model for carbonate acidizing, benchmarked by linear core flooding experiments, which used the two-scale continuum approach to simulate gelling acid flowing and wormhole propagation in carbonate rocks. H Yoo et al. (2021) [25] developed the wormhole propagation behavior to optimize an efficient acid treatment in carbonate acidizing, which used micro X-ray computerized tomography (CT) to observe wormhole visualization clearly. It is found that the wormhole diameters were gradually reduced along with the wormhole propagation axis in all cases, and the wormhole diameter tended to increase according to the higher acid concentration and permeability. A continuum two-scale model for linear and radial flow geometry was developed to simulate dissolution patterns. The simulation results showed that the in-situ generated hydrochloric at low injection rates could create longer wormholes with less pore volume to breakthrough, due to low face dissolution [26].

Research on gelling acid for reservoir stimulation has focused on middle and deep carbonate reservoirs and other areas [27–30]. Yuman Wu et al. (2022) [31] proposed a betaine-based gel used for hydraulic fracturing, it was proved that KCl improves the temperature resistance and increases the viscosity of the optimized fracturing fluid, and the viscoelastic surfactant gel had high shear resistance and high sand-carrying performance. Fei Ding et al. (2022) [32] developed a PHRO gel that was composed of gelatinizing agents, crosslinking agents, and crosslinking promoting agents (oxalic acid). The performance evaluation showed that it has good salt-resistance properties and is suitable for conformance control of low-temperature and high-salinity reservoirs.

At present, more two-scale models are used. This model can take into account the physical phenomena such as gelling acid flow, gelling acid rock reaction, and pore structure change, and can better simulate the gelling acid wormhole morphology obtained in the laboratory. Double scale refers to the Darcy scale and pore scale, that is, the flow is calculated on the Darcy scale; gelling acid rock reaction considers the characteristics of the pore scale, and the reaction speed of gelling acid rock is controlled by specific surface and mass transfer speed, which is the affected by pore size. Therefore, it is necessary to establish a model on a pore scale and consider the influence of micro characteristics on flow and reaction.

2. Results and Discussion

2.1. Effect of Injection Rate on the Morphology of Vermicular Foramen

Displacement (injection rate) is an important parameter in gelling acidizing construction. It is necessary to study the impact of injection rate on wormhole expansion. Figures 1–8 show the changes in wormhole morphology and displacement differential pressure under different gelling acid injection rates. Dimensionless differential pressure is defined as the ratio between the differential pressure at both ends of the core and the initial value during gelling acid injection, which reflects the flow of gelling acid in the rock core. Zero dimensionless differential pressure means that the gelling acid breaks through the core.

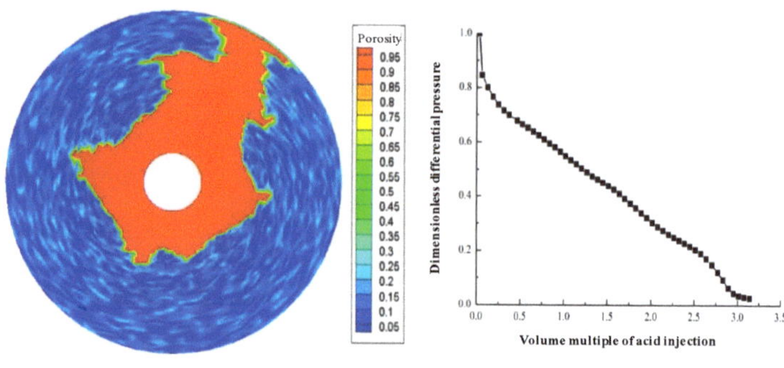

Figure 1. Change of dissolution form and differential pressure with gelling acid injection volume (gelling acid injection speed 0.0065 cm/min).

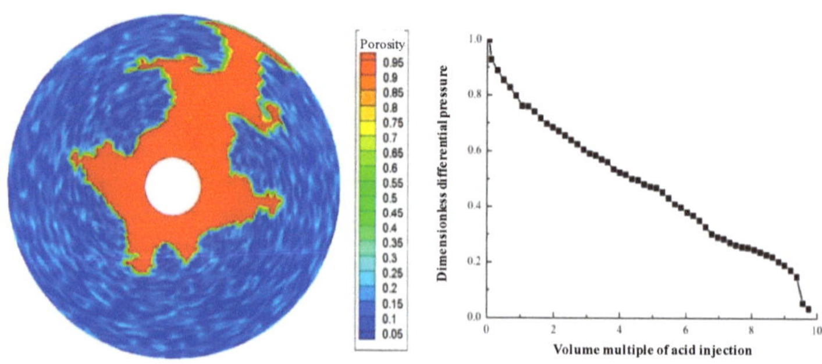

Figure 2. Change of dissolution form and differential pressure with gelling acid injection volume (gelling acid injection speed 0.0135 cm/min).

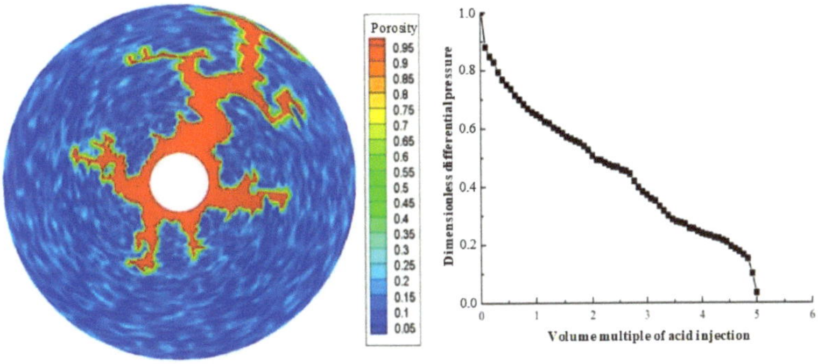

Figure 3. Change of dissolution form and differential pressure with gelling acid injection (gelling acid injection speed 0.035 cm/min).

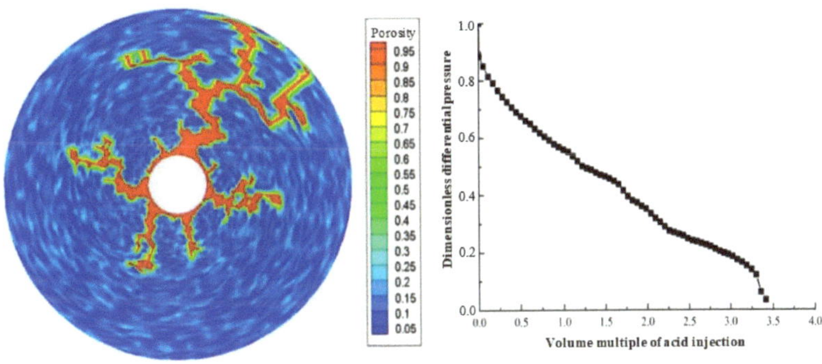

Figure 4. Change of dissolution form and differential pressure with gelling acid injection (gelling acid injection speed 0.065 cm/min).

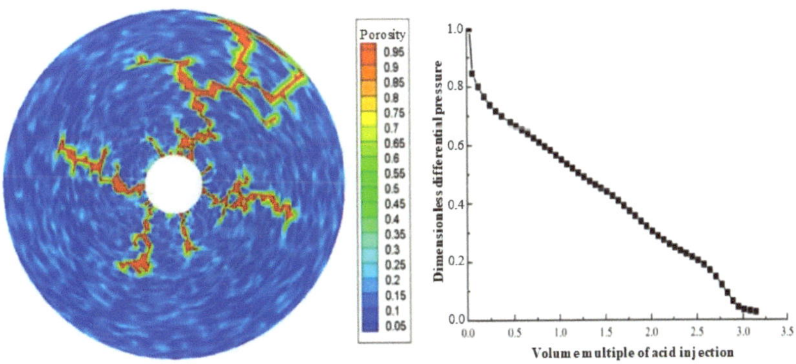

Figure 5. Change of dissolution form and differential pressure with gelling acid injection (gelling acid injection speed 0.65 cm/min).

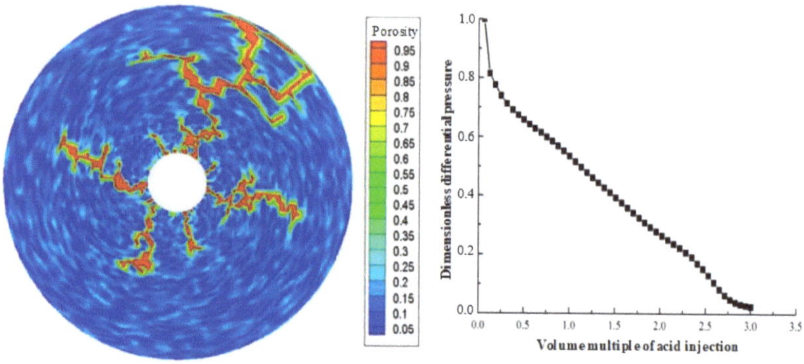

Figure 6. Change of dissolution form and differential pressure with gelling acid injection (gelling acid injection speed 1.35 cm/min).

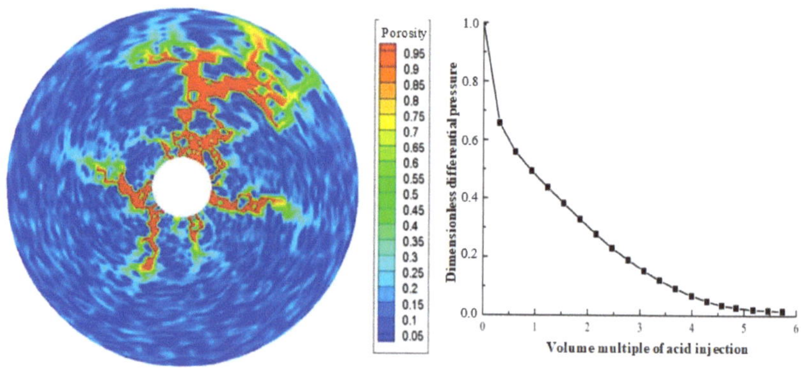

Figure 7. Change of dissolution form and differential pressure with gelling acid injection (gelling acid injection speed 6.5 cm/min).

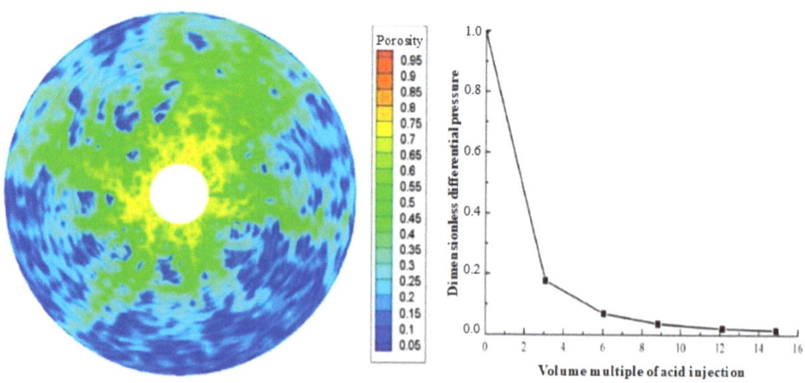

Figure 8. Change of dissolution form and differential pressure with gelling acid injection (gelling acid injection speed 65 cm/min).

The gelling acid injection rate increased from 0.0065 cm/min to 65 cm/min, and the model simulation results are shown in Figures 1–8. The gelling acid corrosion morphology showed the phenomena observed in the experiment: surface corrosion, main wormhole, branch wormhole, and uniform corrosion. When the gelling acid injection speed is very small because the diffusion effect of hydrogen ions in the gelling acid solution is greater than the convection effect, the hydrogen ions are more likely to diffuse upward (i.e., the hole wall), which prevents the gelling acid solution from dissolving the wormhole, and there is less gelling acid reaching the front end of the wormhole. Under this injection speed, increasing the wormhole size is not conducive to increasing the wormhole length, and the excessive dissolution of rocks on the well wall may easily cause the collapse of the well wall. With the increase in gelling acid injection speed, as are shown in Figures 1–8, the convection effect of hydrogen ions gradually increases, but it is still less than the diffusion effect. More gelling acid is consumed on the wall of the earthworm hole. However, due to the enhancement of convection, the dissolution form gradually transits to the earthworm hole. As shown in Figure 5, when the gelling acid injection speed is relatively moderate, the diffusion and convection of hydrogen ions are equivalent. More gelling acid reaches the front end of the wormhole, and the gelling acid reaching the front end of the wormhole not only completely reacts with the rock to promote the growth of the wormhole, but also forms the main wormhole. With the further increase in gelling acid injection speed, the convection effect of hydrogen ions is gradually greater than the diffusion effect. The gelling acid liquid flows forward before it has time to completely react with the encountered rock. The wormholes formed due to the influence of heterogeneity are branched. When the gelling acid injection speed is high (Figure 8), the convection effect of hydrogen ions is much greater than the diffusion effect. The gelling acid liquid flows into the pores before consumption, so that live gelling acid can be obtained everywhere, forming a more uniform dissolution form, and the efficiency of removing pollution is low.

The effect of injection rate on breakthrough volume multiple is shown in Figure 9. The trend of this curve is similar to that obtained in the laboratory. When the injection speed gradually increases from the lowest, the breakthrough pore volume multiple first decreases and then reaches a minimum value. The injection speed corresponding to this point is called the optimal injection speed and then continues to increase the injection speed, resulting in the slow increase in breakthrough pore volume multiple. On the left side of the optimal injection speed, the breakthrough pore volume multiple increases sharply with the decrease in the injection speed, while on the right side of the optimal injection speed, the breakthrough pore volume increases slowly with the increase in the injection speed. In actual construction, it is difficult to obtain the optimal injection speed. For the sake of safety, try to keep the injection speed on the right side of the optimal injection speed. If it is on the left, the injection speed fluctuates, which has a great impact on the breakthrough pore volume. On the right side of the optimal injection velocity, the fluctuation of injection velocity has little effect on the breakthrough pore volume. The simulation results show that for this simulation condition, the optimal injection rate is about 1 cm/min, and the corresponding breakthrough pore volume multiple is about 1.4. The simulation of gelling acid wormhole expansion under radial conditions shows that it is necessary to select an appropriate injection rate. The injection rate is too low, the gelling acid solution dissolves near the wellbore, the action distance is very short, and the plugging removal effect cannot be achieved. The gelling acid injection speed is too fast, forming uniform corrosion, unable to form wormhole, and the transformation effect is not good. At the optimal injection rate, the main wormhole is formed, and the seepage resistance in the wormhole can be ignored. The wormhole passes through the pollution zone to remove the pollution, which is equivalent to expanding the wellbore radius and increasing the production capacity.

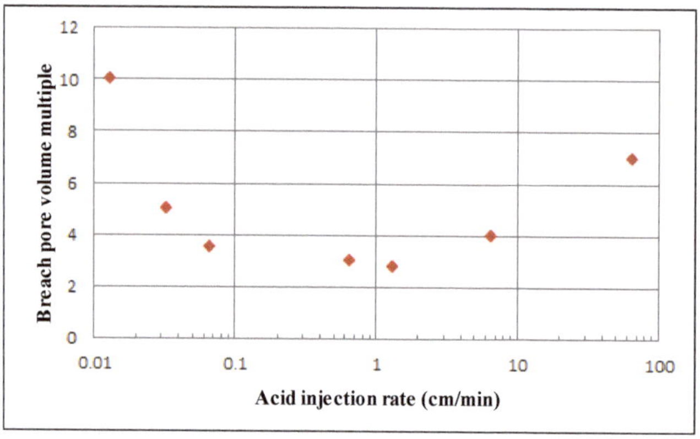

Figure 9. Effect of injection rate on breakthrough volume multiple.

2.2. *Influence of Gelling Acid Properties*

The properties of gelling acid solution mainly include these aspects: viscosity, surface reaction rate, and hydrogen ion diffusion coefficient. In matrix gelling acidification, to facilitate injection, low viscosity gelling acid solution is used, and viscosity is not the main factor. The reaction between carbonate rock and gelling acid is mainly affected by the mass transfer rate, and the hydrogen ion diffusion coefficient is the controlling factor of the mass transfer coefficient. Therefore, only the influence of hydrogen ion diffusion coefficient is analyzed. Figure 10 shows the effect of the hydrogen ion diffusion coefficient on wormhole morphology, and other conditions are the same. When the diffusion coefficient is low, the wormhole is fine. With the increase in the diffusion coefficient, the wormhole near the well becomes thicker, and more gelling acid is consumed near the well; when the diffusion coefficient is high, there is more dissolution in the wellbore wall and the wormhole is thicker. When the diffusion coefficient is high, to form the main wormhole, the injection speed needs to be increased to match the injection speed with the reaction.

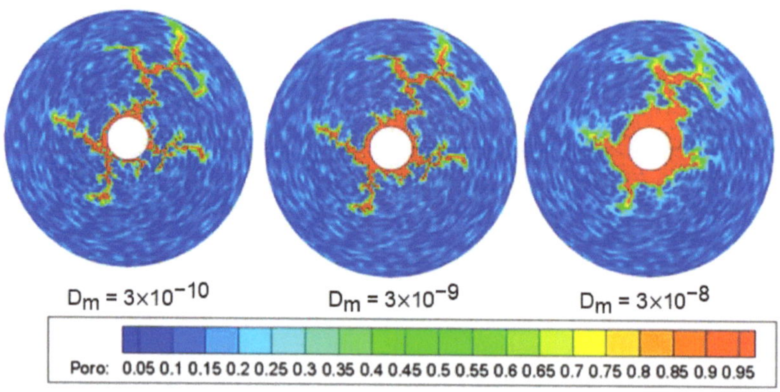

Figure 10. Effect of hydrogen ion diffusion coefficient on wormhole morphology.

2.3. *Effect of Pore Space Distribution on Wormhole Morphology*

The spatial distribution of pores determines the trend of gelling acid solution and thus the shape of the wormhole. C_v is the coefficient of variation of porosity. The smaller the C_v is, the weaker the heterogeneity is, on the contrary, it is about strong. When C_v is 1, more high porosity places can be seen on the pore distribution map. Figure 11 shows the wormhole morphology when the heterogeneity gradually becomes stronger from left to right. For the mean value case, there are more wormholes, and each wormhole is similar; when the heterogeneity is strong, the non-uniform competition of gelling acid flow and reaction is stronger, and the numerical simulation of the main wormhole is less. When $C_v = 0.7$ and 1, the simulated wormhole is more like the wormhole observed in the experiment.

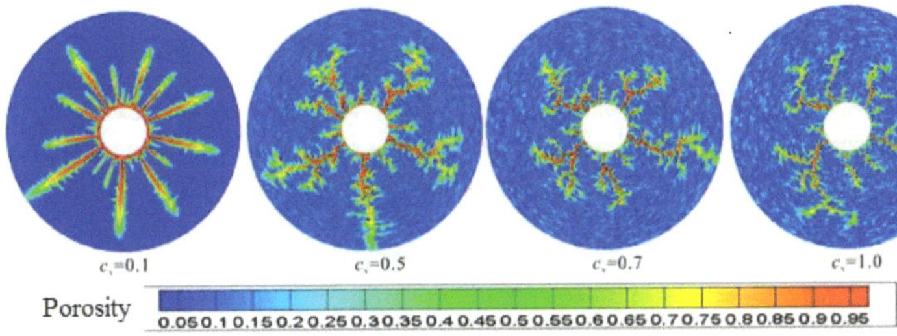

Figure 11. Effect of pore space distribution on wormhole morphology.

Figure 12 shows the wormhole shape when adding holes. When the holes are dispersed, the influence of the convection field is small, so it has little influence on the direction of wormhole. If the holes are continuously distributed, the gelling acid easily flows along hole distribution area, thus affecting the wormhole shape.

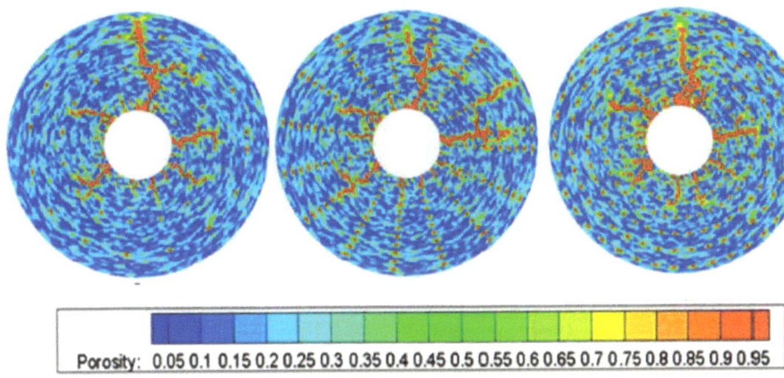

Figure 12. Effect of hole distribution on wormhole expansion.

Figure 13 left shows the real pore spatial distribution of the core. The figure shows that the core porosity distribution is not random, but spatial correlation. There is a high seepage band from the core inlet to the outlet. The figure on the right shows the simulated gelling acid etching wormhole diagram. In the simulation, the wormhole is formed from the upper left and gradually expands to the lower right corner along the high seepage zone. The wormhole path is the distribution area of the high seepage zone, indicating that the pore space distribution determines the direction of the wormhole.

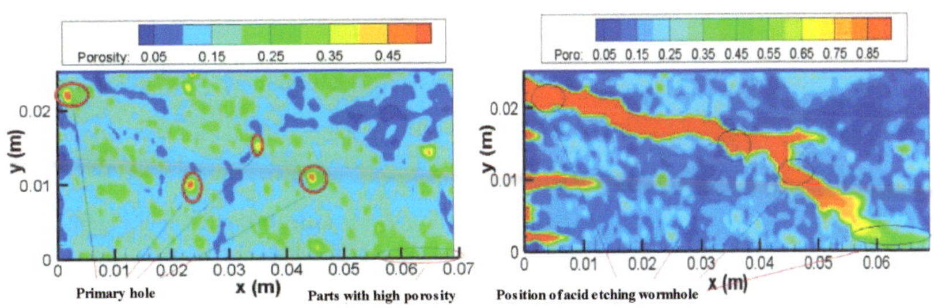

Figure 13. Effect of pore space distribution on the expansion of gelling acid wormhole.

2.4. Effect of Perforation on Wormhole Shape

The perforation hole is a macro channel. For the perforation completion wellbore, the seepage resistance in the hole can be ignored. The injected gelling acid flows to the formation through the perforation hole and bypasses the perforation coverage area. Therefore, the perforation completion will have a significant impact on the expansion of the gelling acid wormhole. Figures 14 and 15 simulate the shape of gelling acid-etched wormholes under perforation conditions. Wormholes are formed and expanded at the front end of perforations. When perforating with a 60° phase angle, 6 wormholes expand outward, and when perforating with a 90° phase angle, 4 wormholes expand outward. As the outward radius increases, the overflow cross-sectional area increases, the gelling acid in the wormhole is filtered around, and the wormhole branches. The more outward, the more branches; the larger the perforation phase angle is, the less the initial expanded wormhole digital modulus is, and the wormhole branch increases because the gelling acid fluid loses more to the surrounding. Through comparative analysis, it is found that the 6 wormholes with a 60° phase angle perforation expand outward shorter than the 4 wormholes with a 90° phase angle perforation. The perforation hole is equivalent to a very large wormhole. The perforation completion is when the wormhole extends forward along the perforation hole, and the perforation increases the length of the wormhole.

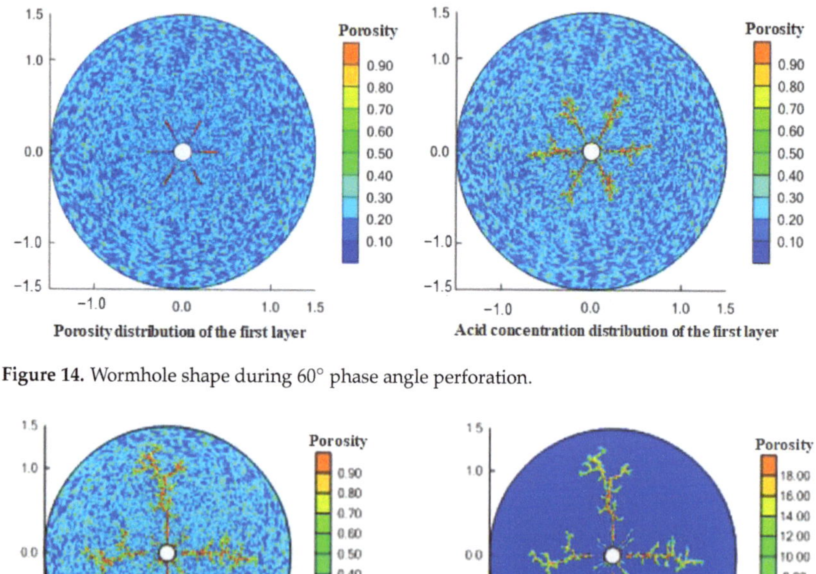

Figure 14. Wormhole shape during 60° phase angle perforation.

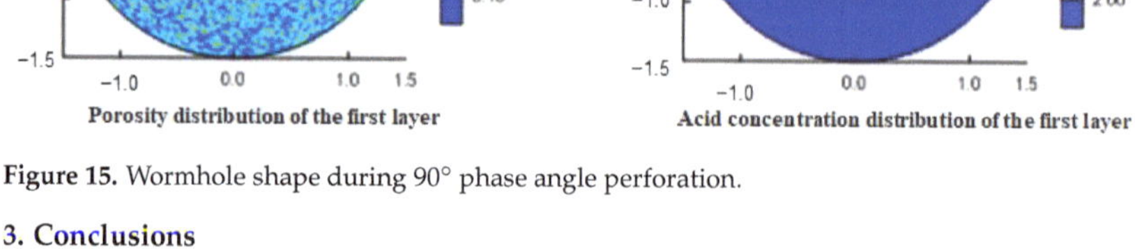

Figure 15. Wormhole shape during 90° phase angle perforation.

3. Conclusions

(1) The mathematical model of gelling acid etching wormhole is established, and the expansion experiment of wormhole in the radial core is carried out with a real core to verify the correctness of the model.

(2) The gelling acid injection rate increased from 0.0065 to 65 cm/min, corresponding to the gelling acid corrosion morphology of wormhole: surface corrosion, main wormhole, branch wormhole, and uniform corrosion.

(3) Viscosity, surface reaction rate, and hydrogen ion diffusion coefficient have different effects on gelling acid etching wormhole. Viscosity and surface reaction rate are not the main factors. The reaction between carbonate rock and gelling acid is mainly affected by mass transfer rate. The hydrogen ion diffusion coefficient determines the mass transfer coefficient. When the diffusion coefficient is low, the wormhole is fine; with the increase in the diffusion coefficient, the wormhole near the well becomes thicker, and more gelling acid is consumed near the well. When the diffusion coefficient is high, there is more dissolution in the wellbore wall and the wormhole is thicker. When the diffusion coefficient is high, to form the main wormhole, the injection speed needs to be increased to match the injection speed with the reaction.

(4) The spatial distribution of pores determines the trend of gelling acid solution and thus the shape of the armhole. The smaller the coefficient of variation of porosity is, the weaker the heterogeneity is, and the more the numerical models of wormholes are, and the wormholes are similar. When the heterogeneity is strong, the non-uniform competition of gelling acid flow and reaction is stronger, and the numerical simulation of the main wormhole is less.

(5) Perforation completion has a significant impact on the expansion of gelling acid etching wormhole. The wormhole extends forward along the perforation hole, and perforation increases the length of the wormhole. When 60° phase angle perforation, 6 wormholes expand outward, and when 90° phase angle perforation, 4 wormholes expand outward. As the outward radius increases, the overflow cross-sectional area increases, the gelling acid in the wormhole is filtered around, and the wormhole branches. The more outward, the more branches. The larger the perforation phase angle is the less the initial expanded wormhole digital modulus is, and the wormhole branch increases.

4. Materials and Methods

Two-scale continuum model was considered to simulate acid flow, acid-rock reaction, porosity change, and wormhole expansion during acidizing of carbonate reservoirs. The detail of the model has been discussed by Panga et al. (2005). The model of Gelling Acid Etching Wormhole that we established is considering the influence of pore microscopic characteristics on acid flow and acid rock reaction. It is considered a natural transition between porous media flow and Stokes free flow in wormholes; fractures are influenced by pressure and velocity fields.

4.1. Continuity Equation

According to the fluid mass balance, inflow term-outflow term + source-sink term = cumulative term. As shown in Figure 16, the micro hexahedron unit is taken in the stratum, and the components of mass velocity of point P in unit on the three coordinates are ρu_x, ρu_y and ρu_z, then the mass flow of point P' in the x-direction is:

$$\rho u_x - \frac{\partial(\rho u_x)}{\partial x}\frac{dx}{2}$$

The mass flow flowing in through the face of P' in dt time is:

$$[\rho u_x - \frac{\partial(\rho u_x)}{\partial x}\frac{dx}{2}]dydzdt$$

Similarly, the mass flow of the plane where P'' is located in the x-direction is

$$[\rho u_x + \frac{\partial(\rho u_x)}{\partial x}\frac{dx}{2}]dydzdt$$

Then the mass flow difference of hexahedron flowing in and out from x-direction in dt time is

$$-\frac{\partial(\rho u_x)}{\partial x}dxdydzdt$$

Similarly, the mass flow difference of inflow and outflow from y and z directions in dt time is

$$-\frac{\partial(\rho u_y)}{\partial y}dxdydzdt-\frac{\partial(\rho u_z)}{\partial z}dxdydzdt$$

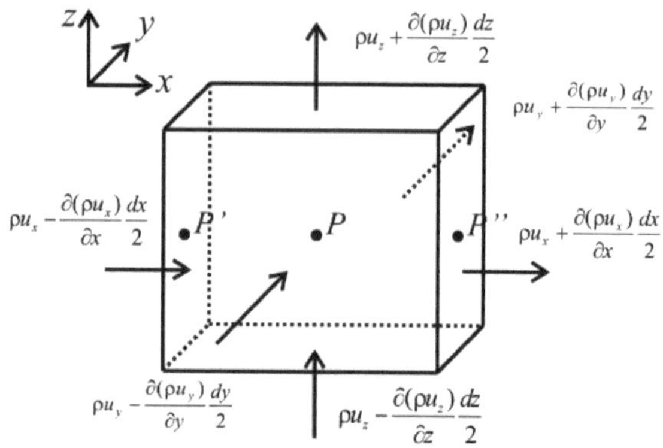

Figure 16. Schematic diagram of micro hexahedron unit.

Then the total mass flow difference of inflow and outflow in the hexahedron in dt time is

$$-[\frac{\partial(\rho u_x)}{\partial x}+\frac{\partial(\rho u_y)}{\partial y}+\frac{\partial(\rho u_z)}{\partial z}]dxdydzdt$$

The total change of fluid mass in dt time is

$$\frac{\partial(\rho\phi)}{\partial t}dxdydzdt$$

where, ρ is the fluid density, kg/m^3; ϕ is formation porosity, %.

Then the total mass change of hexahedron in dt time equals to the mass difference between inflow and outflow of hexahedron in dt time:

$$\frac{\partial u_x}{\partial x}+\frac{\partial u_y}{\partial y}+\frac{\partial u_z}{\partial z}+\frac{\partial\phi}{\partial t}=0 \quad (1)$$

The above formula is the continuity equation of incompressible fluid seepage in porous media. If the z-direction is ignored and the above formula is converted to the polar coordinate system, the two-dimensional continuity equation can be obtained as:

$$\frac{1}{r}\frac{\partial(ru_r)}{\partial r}+\frac{1}{r}\frac{\partial u_\theta}{\partial\theta}+\frac{\partial\phi}{\partial t}=0 \quad (2)$$

where, u_r and $u_{\theta R}$ are speed in direction r and θ direction, m/s.

4.2. Equation of Motion

The motion equation is the Darcy seepage equation, and the motion equation in the polar coordinate system is as follows:

$$(u_r,u_\theta)=-\frac{K}{\mu_a}(\frac{\partial p}{\partial r},\frac{1}{r}\frac{\partial p}{\partial\theta}) \quad (3)$$

where, K is the rock permeability, μm^2; μ_a is gelling acid viscosity, mPa·s; p is reservoir pressure, MPa.

4.3. Gelling Acid Distribution Equation

The flow of gelling acid in porous media is affected by both convection and diffusion. Similar to the derivation of the continuity equation, suppose a hexahedral element, the mass flow caused by diffusion at point P is u_i, and then the x-direction passes through P'

$$u_i - \frac{\partial u_i}{\partial x}\frac{dx}{2}$$

The mass flow of the outflow unit through the face of P'' is:

$$u_i + \frac{\partial u_i}{\partial x}\frac{dx}{2}$$

Then the mass flow difference of inflow and outflow units in the x direction is:

$$-\frac{\partial u_i}{\partial x}dx$$

Similarly, the mass flow difference of inflow and outflow units in y and z directions is:

$$-\frac{\partial u_j}{\partial y}dy, -\frac{\partial u_k}{\partial z}dz$$

The mass flow of point P caused by convection in the x-direction is $u_x C_f$, and the mass flow difference in the X direction through the P' and P'' planes is:

$$-\frac{\partial(u_x C_f)}{\partial x}dx$$

where, C_f is gelling acid concentration, mol/L.

Similarly, the mass flow difference in the y, z direction of inflow and outflow unit is:

$$-\frac{\partial(u_y C_f)}{\partial y}dy, -\frac{\partial(u_z C_f)}{\partial z}dz$$

It can be seen from the above that the mass change caused by convection and diffusion in dt time is:

$$-(\frac{\partial u_i}{\partial x} + \frac{\partial u_j}{\partial y} + \frac{\partial u_k}{\partial z})dxdydzdt - [\frac{\partial(u_x C_f)}{\partial x} + \frac{\partial(u_y C_f)}{\partial y} + \frac{\partial(u_z C_f)}{\partial z}]dxdydzdt$$

The mass change caused by convection-diffusion in dt time must be equal to the total mass change in dt time:

$$-(\frac{\partial u_i}{\partial x} + \frac{\partial u_j}{\partial y} + \frac{\partial u_k}{\partial z}) - [\frac{\partial(u_x C_f)}{\partial x} + \frac{\partial(u_y C_f)}{\partial y} + \frac{\partial(u_z C_f)}{\partial z}] = \frac{\partial(\phi C_f)}{\partial t}$$

According to Fick's law and considering the effective diffusion in porous media, the above formula can be written as:

$$\nabla(\phi D_e \cdot \nabla C_f) - \nabla(UC_f) = \frac{\partial(\phi C_f)}{\partial t} \quad (4)$$

where, D_e is the effective diffusion coefficient of the gelling acid solution, m^2/s.

Due to the gelling acid rock reaction, the quality change caused by the chemical reaction must be considered. Here, assuming that the gelling acid flow to the wall of porous media completely reacts with the rock, the gelling acid concentration on the wall of the pore can be regarded as 0, and the gelling acid concentration in the pore is the original gelling acid concentration. In this way, there is a concentration gradient from the center of the pore to the wall of the pore. The size of the concentration gradient depends on the mass transfer rate from the fluid mass transfer to the fluid-solid surface and the reaction rate on the pore surface. If the reaction speed is less than the mass transfer speed, the concentration gradient can be ignored. At this time, the speed of the whole gelling acid rock reaction is controlled by the reaction speed of the liquid-solid interface. When the reaction rate is greater than the mass transfer rate, a large concentration gradient appears in the pores, which can be described by a simple concentration difference, as follows:

$$k_c(C_f - C_s) = k_s C_s = R(C_s) \tag{5}$$

where, k_c is the mass transfer coefficient of the gelling acid solution, m/s; C_s is gelling acid concentration on liquid-solid surface, mol/L; k_s is the gelling acid rock reaction rate constant, m/s.

Therefore, the mass change caused by a reaction can be expressed as:

$$k_c a_v (C_f - C_s)$$

where, a_v is specific surface area, m^{-1}.

Bring the above formula into Equation (4):

$$\nabla(\phi D_e \cdot \nabla C_f) - \nabla(U C_f) - k_c a_v (C_f - C_s) = \frac{\partial(\phi C_f)}{\partial t} \tag{6}$$

The above formula is the gelling acid mass balance equation considering convection, diffusion, and reaction, which is converted to the polar coordinate system:

$$\begin{aligned} &\frac{\partial(\phi C_f)}{\partial t} + \frac{1}{r}\frac{\partial}{\partial r}(r u_r C_f) + \frac{1}{r}\frac{\partial}{\partial \theta}(u_\theta C_f) \\ &= \frac{1}{r}\frac{\partial}{\partial r}(r\phi D_{er}\frac{\partial C_f}{\partial r}) + \frac{1}{r}\frac{\partial}{\partial \theta}(\frac{\phi D_{e\theta}}{r}\frac{\partial C_f}{\partial \theta}) - k_c a_v (C_f - C_s) \end{aligned} \tag{7}$$

where, D_{er} and $D_{e\theta}$ are, respectively, effective diffusion coefficients in the r and θ direction, m^2/s.

Equation (5) can be converted into the following formula:

$$C_s = \frac{C_f}{1 + k_s/k_c} \tag{8}$$

$k_s << k_c$, when the reaction rate constant is much less than the mass transfer coefficient, C_s approximately equal to C_f; $k_s >> k_c$, when the reaction rate constant is much greater than the mass transfer coefficient, C_s approximately equal to 0. Since the reaction rate constant is approximately determined for a specific gelling acid solution, the control conditions of the whole gelling acid rock reaction are determined by the mass transfer coefficient. For porous media, due to the existence of heterogeneity and gelling acid rock reaction, the mass transfer coefficient also changes with time and space.

4.4. Model Validation

To verify the correctness of the model, it is necessary to make some experiments to simulate, and the simulation results are compared with the experimental results to verify the model. Wormhole expansion experiments are conducted in radial cores using real formation cores obtained from carbonate rock in Longwangmiao zone Sichuan of China, and the experiments' results are shown in Figure 17. The data required in the simulation are taken from the experimental data of tardy et al. [33], as is shown in Figure 18, for example, the outer diameter and inner diameter of the core are 7.04 cm and 0.32 cm, respectively.

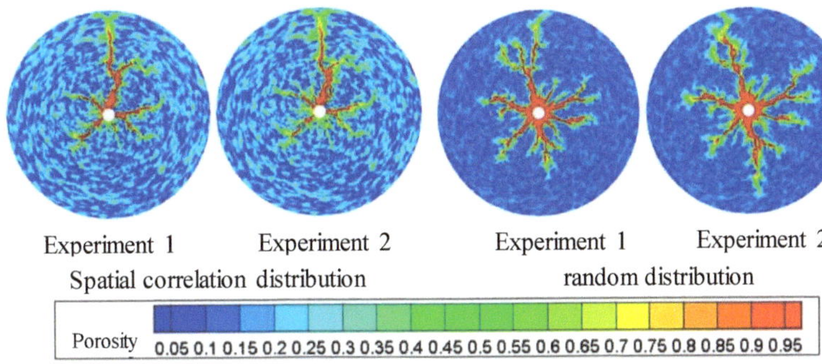

Figure 17. Simulated etching wormhole morphology obtained by CT scanning the rock end face gelled acid treated.

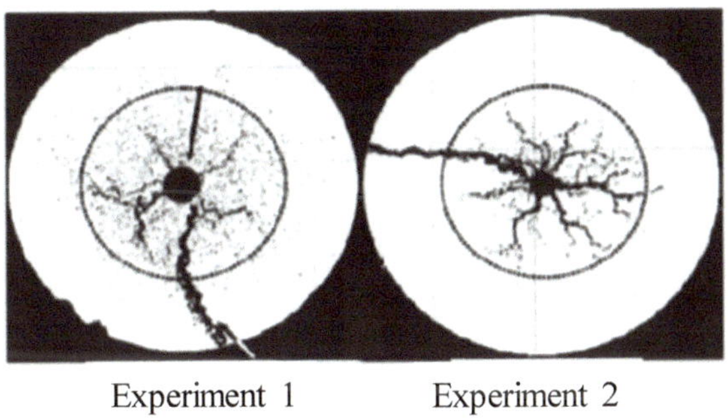

Figure 18. Radial wormhole morphology (adapted with permission from Ref. [27]. 2007, copyright Tardy et al.).

Different porosity distributions will produce different wormhole expansion patterns. Although random porosity distribution is mostly used to simulate in the literature, a large number of experimental results show that the porosity value of carbonate rocks conforms to the normal distribution law of spatial correlation. To study the influence of different porosity distribution laws on wormhole expansion, two porosity distributions are used in the simulation: random distribution porosity and spatially related porosity distribution. Figure 3 shows the wormhole morphology under the spatial correlation distribution and random distribution of porosity. It can be seen that although only one main wormhole breaks through the core, the dissolution forms of different porosity distributions are quite different. For the spatial correlation porosity distribution, there is only one relatively developed wormhole in the rock core, while other wormholes are relatively short and relatively undeveloped. For random porosity distribution, there are several relatively developed wormholes in the rock core at the same time. This is because the heterogeneity of spatially correlated porosity distribution is stronger than that of random porosity distribution. When the gelling acid is injected into the highly heterogeneous rock core, the gelling acid will produce uneven competition at the inlet and tend to enter the hole with the lowest resistance to produce wormholes. Once the wormhole is generated, most of the gelling acid will enter the wormhole, and then the main wormhole will be generated, resulting in the lack of development in other later-formed wormholes. For randomly distributed porosity, gelling acid liquid produces uniform competition at the inlet, resulting in multiple wormholes of the same size. Compared to the experimental results of Tardy et al., it is obvious that the wormhole morphology under the spatially correlated porosity distribution is similar to the experimental results, which verifies the correctness of the model.

Author Contributions: M.W.: Conceptualization, Funding acquisition, Project administration, Resources, and Software. W.Z.: Conceptualization, Funding acquisition, Project administration, Methodology. S.L.: Data curation, Formal analysis, Methodology, Writing—original draft and Writing—review and editing. W.W.: Investigation, Methodology, Software and Visualization. All authors have read and agreed to the published version of the manuscript.

Funding: This work was supported by the support of the National Natural Science Foundation of China: Study on dynamic characteristics of methane/carbon dioxide in shale heterogeneous reservoir under multi-field coupling (Program No. 41772150).

Institutional Review Board Statement: Not applicable.

Informed Consent Statement: Not applicable.

Data Availability Statement: Data are contained within the article.

Acknowledgments: The authors are grateful to Xuangang Meng (Yanchang Oilfield Co., Ltd.) for his modifications, and as well as reviewers and editors for careful review of this manuscript.

Conflicts of Interest: The authors declare no conflict of interest.

References

1. Hung, K.M.; Hill, A.D.; Sepehrnoorl, K. A mechanistic model of wormhole growth in carbonate matrix acidizing and acid fracturing. *J. Pet. Technol.* **1989**, *41*, 59–66. [CrossRef]
2. Gdanski, R. A fundamentally new model of acid wormhole in carbonate. In Proceedings of the SPE European Formation Damage Conference, The Hague, The Netherlands, 31 May–1 June 1999. Paper SPE-54719-MS.
3. Fredd, C.N.; Fogler, H.S. Influence of transport and reaction on wormhole formation in carbonate porous media. *AIChE J.* **1998**, *44*, 1933–1949. [CrossRef]
4. Kang, Q.J.; Zhang, D.X.; Chen, S.Y.; He, X. Lattice Boltzmann simulation of chemical dissolution in porous media. *Phys. Rev. E* **2002**, *65*, 036318. [CrossRef]
5. Golfier, F.; Zarcone, C.; Bazin, B.; Lenormand, R.; Lasseux, D.; Quintard, M. On the ability of a Darcy scale model to capture wormhole formation during the dissolution of a porous medium. *J. Fluid Mech.* **2002**, *457*, 213–254. [CrossRef]
6. Ahmadi, M.; Panahi, F.; Bahri-Laleh, N.; Sabzi, M.; Pareras, G.; Falcone, B.N.; Poater, A. pH-responsive gelation in metallo-supramolecular polymers based on the protic pyridinedicarboxamide ligand. *Chem. Mater.* **2022**, *34*, 6155–6169. [CrossRef]
7. Huang, T.; Hill, A.D.; Schechter, R. Reaction rate and fluid loss: The keys to wormhole initiation and propagation in carbonate acidizing. *SPE J.* **2000**, *5*, 287–292. [CrossRef]
8. Panga, M.K.; Ziauddin, M.; Balakotaiah, V. A new model for predicting wormhole structure and formation in acid stimulation of carbonates. In Proceedings of the SPE International Symposium and Exhibition on Formation Damage Control, Lafayette, LA, USA, 18–20 February 2004. Paper SPE 86517.
9. Panga, M.K.; Ziauddin, M.; Balakotaiah, V. Two-scale continuum model for simulation of wormholes in carbonate acidization. *AIChE J.* **2005**, *51*, 3231–3248. [CrossRef]
10. Cohen, C.E.; Ding, D.; Quintard, M.; Bazin, B. From pore scale to wellbore scale: Impact of geometry on wormhole growth in carbonate acidization. *Chem. Eng. Sci.* **2007**, *63*, 3088–3099. [CrossRef]
11. Akanni, O.O. *Mathematical Modeling of Wormhole Propagation during Matrix Acidizing of Carbonate Reservoir*; Texas A&M University: College Station, TX, USA, 2015.
12. Akanni, O.O.; Nasr-El-Din, H.A. The accuracy of carbonate matrix-acidizing models in predicting optimum injection and wormhole propagation rates. In Proceedings of the SPE Middle East Oil & Gas Show and Conference, Manama, Bahrain, 8–11 March 2015. Paper SPE 172575.
13. Olatokunbo, O.A.; Nasr-El-Din, H.A. Modeling of wormhole propagation during matrix acidizing of carbonate reservoirs by organic acids and chelating agents. In Proceedings of the SPE Annual Technical Conference and Exhibition, Dubai, United Arab Emirates, 26–28 September 2016. Paper SPE-181348-MS.
14. Liu, M.; Zhang, S.; Mou, J.Y. Effect of normally distributed porosity sities on dissolution pattern in carbonate acidizing. *Pet. Sci. Eng.* **2012**, *94*, 28–39. [CrossRef]
15. Zhang, Y.; Yang, S.; Zhang, S.; Mou, J.Y. Wormhole propagation behavior and its effect on acid leakoff under in situ conditions in acid fracturing. *Transp. Porous Media* **2014**, *101*, 99–114. [CrossRef]
16. Mou, J.Y.; Liu, M.; Zhang, S. Diversion conditions for viscoelastic surfactant-based self-diversion acid in carbonate acidizing. *SPE Prod. Oper.* **2015**, *30*, 191–199. [CrossRef]
17. Qiu, X.; Aidagulov, G.; Ghommem, M.; Edelman, E.; Brady, D.; Abbad, M. Towards a better understanding of wormhole propagation in carbonate rocks: Linear vs. radial acid injection. *J. Pet. Sci. Eng.* **2018**, *171*, 570–583. [CrossRef]
18. Qi, D.; Zou, H.L.; Ding, Y.H.; Zhang, H.W. Wormhole calculation model of carbonate acidizing based on Hausdorff fractal dimension. *Sci. Technol. Eng.* **2019**, *19*, 105–110.

19. Wang, L.; Mou, J.Y.; Mo, S.Q.; Zhao, B.; Liu, Z.Y.; Tian, X.X. Modeling matrix acidizing in naturally fractured carbonate reservoirs. *J. Pet. Sci. Eng.* **2020**, *186*, 106685. [CrossRef]
20. Liu, N.Z.; Liu, M. Simulation and analysis of wormhole propagation by VES acid in carbonate acidizing. *J. Pet. Sci. Eng.* **2016**, *138*, 57–65. [CrossRef]
21. Wei, W.; Yu, W.; Chen, Y.G.; Sepehrnoori, K. Geochemical modeling of wormhole propagation during carbonate acidizing with consideration of fractures. In Proceedings of the SPE Improved Oil Recovery Conference, Virtual, 31 August–September 2020. Paper SPE-200319-MS.
22. Qi, N.; Chen, G.B.; Liang, C.; Guo, T.K.; Liu, G.L.; Zhang, K. Numerical simulation and analysis of the influence of fracture geometry on wormhole propagation in carbonate reservoirs. *Chem. Eng. Sci.* **2019**, *198*, 124–143. [CrossRef]
23. Mustafa, A.; Alzaki, T.; Aljawad, M.S.; Solling, T.; Dvorkin, J. Impact of acid wormhole on the mechanical properties of chalk, limestone, and dolomite: Experimental and modeling studies. *Energy Rep.* **2022**, *8*, 605–616. [CrossRef]
24. Ghommem, M.; Zhao, W.S.; Dyer, S.; Qiu, X.D.; Brady, D. Carbonate acidizing: Modeling, analysis, and characterization of wormhole formation and propagation. *J. Pet. Sci. Eng.* **2015**, *131*, 18–33. [CrossRef]
25. Yoo, H.; Kim, Y.; Jang, H.; Lee, J. Propagation characteristics of optimum wormhole in carbonate matrix acidizing using micro X-ray CT imaging. *J. Pet. Sci. Eng.* **2021**, *196*, 108010. [CrossRef]
26. Kiani, S.; Jafari, S.; Apourvari, S.N.; Mehrjoo, H. Simulation study of wormhole formation and propagation during matrix acidizing of carbonate reservoirs using a novel in-situ generated hydrochloric acid. *Adv. Geo-Energy Res.* **2021**, *5*, 64–74. [CrossRef]
27. Elsafih, M.; Fahes, M. Quantifying the effect of multiphase flow on matrix acidizing in oil-Bearing carbonate formations. *SPE Prod. Oper.* **2021**, *36*, 795–806. [CrossRef]
28. Nicolella, P.; Lauxen, D.; Ahmadi, M.; Seiffert, S. Reversible hydrogels with switchable diffusive permeability. *Macromol. Chem. Phys.* **2021**, *222*, 2100076. [CrossRef]
29. Makeiff, D.A.; Cho, J.-Y.; Smith, B.; Carlini, R.; Godbert, N. Self-assembly of alkylamido isophthalic acids toward the design of a supergelator: Phase-selective gelation and dye adsorption. *Gels* **2022**, *8*, 285. [CrossRef]
30. Wang, L.-L.; Wang, T.-F.; Wang, J.-X.; Tian, H.-T.; Chen, Y.; Song, W. Enhanced oil recovery mechanism and technical boundary of gel foam profile control system for heterogeneous reservoirs in Changqing. *Gels* **2022**, *8*, 371. [CrossRef] [PubMed]
31. Wu, Y.; Zhang, J.; Dong, S.; Li, Y.; Slaný, M.; Chen, G. Use of Betaine-based gel and its potential application in enhanced oil recovery. *Gels* **2022**, *8*, 351. [CrossRef] [PubMed]
32. Ding, F.; Dai, C.; Sun, Y.; Zhao, G.; You, Q.; Liu, Y. Gelling behavior of PAM/phenolic crosslinked gel and its profile control in a low-temperature and high-salinity reservoir. *Gels* **2022**, *8*, 433. [CrossRef] [PubMed]
33. Philippe, M.J.; Tardy, B.L.; Christanti, Y. An experimentally validated wormhole model for self-diverting and conventional acids in carbonate rocks under radial flow conditions. In Proceedings of the European Formation Damage Conference, Scheveningen, The Netherlands, 30 May–1 June 2007. Paper SPE-107854.

MDPI AG
Grosspeteranlage 5
4052 Basel
Switzerland
Tel.: +41 61 683 77 34

Gels Editorial Office
E-mail: gels@mdpi.com
www.mdpi.com/journal/gels

www.ingramcontent.com/pod-product-compliance
Lightning Source LLC
Chambersburg PA
CBHW041354210926
51850CB00013B/3079
* 9 7 8 3 7 2 5 8 2 8 8 4 5 *